T0175123

Microbial Ecology

EDITORS

Allen I. Laskin, Ph.D
Esso Research and Engineering Company
Linden, New Jersey

Hubert Lechevalier, Ph.D
Institute of Microbiology
Rutgers University
New Brunswick, New Jersey

CRC Press
Taylor & Francis Group
Boca Raton London New York

CRC Press is an imprint of the
Taylor & Francis Group, an **informa** business

CRC Press
Taylor & Francis Group
6000 Broken Sound Parkway NW, Suite 300
Boca Raton, FL 33487-2742

First issued in paperback 2020

© 1974 by Taylor & Francis Group, LLC
CRC Press is an imprint of Taylor & Francis Group, an Informa business

No claim to original U.S. Government works

ISBN-13: 978-1-315-89543-7 (hbk)
ISBN-13: 978-0-367-65748-2 (pbk)

A Library of Congress record exists under LC control number: 73094086

Publisher's Note
The publisher has gone to great lengths to ensure the quality of this reprint but points out that some imperfections in the original copies may be apparent.

Disclaimer
The publisher has made every effort to trace copyright holders and welcomes correspondence from those they have been unable to contact.

Visit the Taylor & Francis Web site at
http://www.taylorandfrancis.com

and the CRC Press Web site at
http://www.crcpress.com

AUTHORS AND CONTRIBUTORS

David Pramer is Professor of Microbiology and Director of Biological Sciences at Rutgers University. He received his B.Sc. and Ph.D. (Microbiology) degrees from Rutgers.

Robert M. Pfister is Associate Professor, Department of Microbiology, Ohio State University. He received his A.B., M.S. (Microbiology), and Ph.D. (Microbiology) degrees from Syracuse University.

Jean-Marc Bollag is Associate Professor of Soil Microbiology, Department of Agronomy, The Pennsylvania State University. He received his Ph.D. (Plant Physiology) degree from the University of Basel, Switzerland.

Guenther Stotzky is Professor, Department of Biology, and Head, All University Department of Biology, New York University. He received his B.S. (Soil Science) degree from California State Polytechnic College, and his M.S. (Agronomy-Microbiology) and Ph.D. (Agronomy-Biochemistry) degrees from Ohio State University.

J. L. Meers is Project Leader, Biomass Production from Methanol, Imperial Chemical Industries Ltd., Agricultural Division, Billingham Teesside, England. He received his B.Sc. degree from the University of London and his Ph.D. degree from London University.

THE EDITORS

Allen I. Laskin is Head of Microbiological Research at the Esso Research and Engineering Co., Linden, New Jersey. He received his B.S. degree from the City College of New York and his M.S. and Ph.D. degrees from Texas University.

Hubert Lechevalier is Professor of Microbiology at the Institute of Microbiology, Rutgers University, New Brunswick, New Jersey. He received his M.S. degree from Laval (Canada) and his Ph.D. degree from Rutgers University.

This book originally appeared as articles in Volume 2 Issue 1 and Volume 2 Issue 2 of *CRC Critical Reviews in Microbiology,* a quarterly journal published by CRC Press, Inc. We would like to acknowledge the editorial assistance received by the journal's editors, Dr. Allen I. Laskin, Esso Research and Engineering Co., and Prof. Hubert Lechevalier, Institute of Microbiology, Rutgers University. The referees for these articles were Fumio Matsumura, University of Wisconsin, Madison; H. Geissbühler, CIBA-GEIGY Ltd., Basle, Switzerland; D. Pramer, Rutgers University, New Brunswick, N.J., and H. W. Jannasch, Woods Hole Oceanographic Institute, Woods Hole, Mass,

INTRODUCTION

No individual exists as an isolated entity. All life is dependent on its surroundings, both animate and inanimate, and the biological specialty that is concerned with the interactions of organisms and environments is called ecology. Interest in ecology has intensified in recent years, and society increasingly demands that ecologists contribute to decisions regarding resource management and environmental quality.

Because of their small size, microorganisms are more intimately related to their environment than most other forms of life. Microbial ecosystems, therefore, have many attractive features and provide exceptional possibilities for extending our understanding of structure and function in nature. Microbial ecology is exceedingly complex, however, since it includes not only the individual, but also the community and the environment as they vary and interact in place and time. Ideally, microbial ecology is the integration of all knowledge concerning the microbe and its surroundings.

Ecology is among the oldest of the microbiological arts and the youngest of the sciences. In the past, food microbiologists, soil microbiologists, marine microbiologists and their colleagues in other fields were concerned with the characterization of microorganisms and their activities in natural habitats. Today, however, it is fashionable to deemphasize the diversity of habitats, to generalize rather than specify, and to focus on basic phenomena that are common to most if not all ecological systems.

Only special topics in microbial ecology are considered here. The book is not intended to be definitive, but the contributions of the four authors are somewhat related, and they mirror various levels of sophistication encountered in contemporary microbial ecology. Dr. Meers' paper defines and describes associative and antagonistic interrelationships. It is initially theoretical but ultimately concerned with microbial interactions in various natural environments, including sludge, wine, beer, soil, water, and the rumen. Dr. Stotzky focuses on soil. He discusses the influence of physicochemical and biological factors on the growth and activity of microorganisms, and describes selected methods for studying microorganisms in soil. All this is then used as background for an analysis of clay minerals as ecological determinants, an area of investigation which Dr. Stotzky pioneered. The two additional authors deal with the subject of pesticides. Dr. Pfister summarizes present knowledge of the effects of halogenated compounds on the growth and survival of microorganisms in the laboratory as well as in aquatic and terrestrial environments. He also considers the ability of microorganisms to absorb, accumulate, and metabolize DDT and related substances. Dr. Bollag categorizes pesticide detoxification reactions mediated by soil fungi. His contribution is biochemical in emphasis. It warns that concern for environmental pollution and public health must extend beyond the effects of the pesticides themselves and include studies of biodegradation products and their toxicology.

The essays that comprise this anthology of the best in ecology from *Critical Reviews in Microbiology* describe principles and practices in considerable detail. There is no attempt, however, at a balanced presentation of the different groups of microorganisms or their activities. Likewise, some areas of current concern are considered cursorily and others not at all. Nevertheless, the book is an interesting and informative introduction to a growing endeavor. The combined experience and insight of the contributing authors will surely aid the reader to develop an ecological attitude, and to better appreciate microorganisms as determinants of environmental quality.

David Pramer
Rutgers University
New Brunswick, N.J.
October 1973

TABLE OF CONTENTS

GROWTH OF BACTERIA IN MIXED CULTURES
J. L. Meers

INTERACTIONS OF HALOGENATED PESTICIDES AND MICROORGANISMS: A REVIEW

Author: **Robert M. Pfister**
Department of Microbiology
Ohio State University
Columbus, Ohio

THE EFFECTS OF HALOGENATED COMPOUNDS ON MICROORGANISMS

The effects of DDT on the soluble salts in the soil, the accumulation of nitrate, and the production of ammonia from organic materials on a population of organisms in soil were tested by Wilson and Choudhri (1946). In addition, the presence of nodulation on leguminous plants and the effect of DDT on the growth of a variety of pure cultures of legume bacteria were also examined. A number of molds including *Aspergillus*, *Penicillium*, and the green alga *Chlorella* were examined. High concentrations of DDT were used (up to ½%) and the results indicated that with these quantities there was no evidence that the pesticide was injurious to any of the microorganisms tested, to ammonification, or to the accumulation of nitrate or the normal concentration of salts in the soil. The authors concluded that, from their work, any further examination of the DDT seemed useless.

Samples of two types of soil were air dried and sieved (Jones, 1952). Each was seeded with moistened DDT in concentrations of 0.001 up to 0.1%. Sulfur oxidation, nitrate production, ammonification, plate counts, and nitrogen fixation determinations were made. There was no injury to nitrifiers, ammonifiers, and sulfur oxidizing microorganisms from concentrations of DDT ordinarily added to soil. In all cases, toxicity began to be noticed in concentrations of about 0.1%. Nitrogen fixing bacteria in soil containing DDT as high as 1% appeared to be unaffected and in fact may have been stimulated, as determined by plate counts. DDT added to the soil was remarkably stable during the first year of storage, but after two and three years, a decrease was noted.

Samples for plate counts were taken from a series of experimental plots laid out on several different soil types to examine the effects of insecticides on soil microorganisms in the field (Bollen et al., 1954).

Different insecticides were applied at different rates of application. The results indicated that different isomers of BHC when added to clay soil at 1000 ppm in the laboratory caused different numbers of bacteria and molds to develop after incubation. The isomers of BHC also had a different influence on ammonification and nitrification. The gamma isomer increased the bacterial population except for *Streptomyces*. Where mold development was favored after the addition of dextrose, α and β forms of the compound caused a depression in numbers. In the absence of dextrose, where molds were fewer in number, all isomers except alpha were inhibiting. The delta and gamma isomers of BHC increased ammonification in peptone, while other isomers had no effect. Nitrification appeared to be stimulated by beta and gamma isomers of BHC except in the field where heavier concentrations were used. The evidence indicated that these compounds were insufficiently intensive to materially influence fertility of the soil.

Soil fertility and microorganisms were studied further by Fletcher and Bollen (1954). Ten Oregon soils and six soil classes were examined in the laboratory with aldrin applied at 200 and 1000 ppm. Aldrin appeared to stimulate the bacterial populations with the exception of *Streptomyces* and other molds. Aldrin did not affect the development of *Azotobacter*. The effects of aldrin on ammonification and nitrification of peptone were minor and irregular. Any losses from the laboratory treated soils were largely a result of factors other than microbial. These authors concluded that for insect control aldrin would not significantly affect the soil microorganisms and adversely affect soil fertility.

Gray and Rogers (1954) reported on a study to determine if the addition of BHC to the agar medium used to estimate bacterial numbers in the soil would interfere with the formation of colonies or might enhance the development of certain groups of bacteria. BHC greatly reduced the number of microorganisms in soil, but there appeared to be no outstanding differences in numbers of BHC resistant bacteria in different soils. This suggested to the authors that the indigenous microfloras of these soils were similar with regard to their degree of resistance. In addition, BHC prevented the growth of colonies of most groups of bacteria except for Gram negative short rods. There did not appear to be any outstanding differences in the physiological characters that were examined with respect to BHC resistance in comparison to the controls.

The toxicity of DDT, chlordane, BHC, dieldrin, aldrin, endrin, and methoxychlor to soil microorganisms was examined (Jones, 1956). None of these compounds was excessively inhibitory to the ammonifying organisms or to those responsible for the decomposition of organic matter with the liberation of ammonia. Toxicity appeared to be manifested at a concentration of about 0.1% which was equivalent to a ton of insecticide per acre thoroughly distributed throughout the first 6 in. of soil. Jones concluded that there was little likelihood of these compounds reaching concentrations toxic to ammonifying organisms. A soil fumigant, ethylene dibromide, stimulated the ammonifying organisms over a wide range of concentrations. The author concluded that the herbicide 2,4-D would not be toxic to ammonifying organisms in the concentrations used. Dieldrin, aldrin, and chlordane were the most toxic and inhibited nitrifying microorganisms at concentrations of 0.1%. Soil organisms capable of oxidizing sulfur compounds were less resistant than microbes decomposing organic matter and liberating ammonia. The author concluded that if these compounds were used in the recommended amounts, they can be used safely without serious injury to sulfur oxidizing microorganisms. There was no correlation between the numbers of organisms in the treated soil and the nitrifying, ammonifying, and sulfur-oxidizing activities of soil microbes. The addition of DDT, chlordane, and BHC in sufficient concentration to inhibit these transformations caused significant increase in the number of microorganisms. Ordinarily no harm would result to soil nitrifiers from the use of 2,4-D and the presence of organic matter in the soil tended to protect the organisms there against the toxic action of pesticides used.

In a study by Eno and Everett (1958), two gallon glazed pots were filled with sand at pH 6.65. Ten insecticides, heptachlor, chlordane, methoxychlor, lindane, aldrin, toxophene, dieldrin, TDE, DDT, and BHC were added at concentrations of 12.5, 50, and 100 ppm. Each was seeded with black valentine beans two weeks after the soil treatment and examination of the germination of the seeds was made at that time. Aliquots were taken to determine the numbers of fungi, bacteria, amount of carbon dioxide produced, nitrification, and the pH. A second crop of beans was planted 11 months after applying the insecticides and the experiment was repeated. In the first sample the numbers of fungi were not significantly changed from those of the control for any insecticide except dieldrin which appeared stimulatory. In addition, the release of carbon as

CO_2 was increased by toxophene, dieldrin, TDE, and DDT. No significant difference in bacterial numbers was detected. The nitrate data indicated a decrease in nitrification rate for heptachlor, lindane, and BHC and an increase for soils treated with toxophene, TDE, and DDT. Sixteen months after the application of the insecticides, another evaluation of microbial response was made, and there appeared to be no difference in the numbers of fungi, bacteria, or amount of carbon dioxide produced at that time. Nitrate production, however, was significantly decreased by DDT and BHC. The authors concluded that applications of as much as 200 lb per acre of most of these insecticides caused little or no damage to the microbial population as indicated by measurements of CO_2 evolution and total numbers of fungi and bacteria. They suggest that microorganisms may be stimulated by the addition of these compounds.

In an experiment where five annual applications of insecticides to two southern California soils and the cumulative effect of these applications on the soil microbe population were examined (Martin et al., 1959), dosages were chosen to approximate the maximum use in normal field practice. Insecticides used and their approximate annual application rates in pounds per acre were aldrin, 5; chlordane, 10; DDT, 20; dieldrin, 5; endrin, 5; heptachlor, 5; lindane, 1; and toxaphene, 20. These chemicals were applied to the surface of the soil in the fall of the year as a spray and then the ground was turned over to a depth of 6 in. The soil plot was sampled to a depth of 6 in. with a 1 in. coring tube and plating to estimate the number and kinds of microorganisms was done immediately. The soil was stored for further use. The authors concluded from their experiments that the microbial populations of the plots were remarkably uniform. There was no indication that pesticide applications influenced numbers or kinds of soil fungi or numbers of bacteria. In addition, the decomposition of organic materials as measured by the amount of CO_2 evolved from enriched soil, suggested that the decomposition rate was the same in all treated soils as in the control samples. The same results applied to the oxidation of ammonium to nitrate. In the opinion of the authors, the chemicals aldrin, chlordane, DDT, dieldrin, endrin, heptachlor, lindane, and toxaphene exerted no measurable effect on numbers of soil bacteria and fungi, on kinds of soil

fungi developing, or on the ability of the soil population to perform the normal functions of organic matter decomposition and ammonia oxidation.

Some conditions which accelerate the conversion of DDT to DDD in soil and some inhibitory effects of DDT and DDD on soil microorganisms were reported on by Ko and Lockwood (1968). The effect of DDT and DDD on soil bacteria, actinomycetes, and fungi was studied in three ways. Diluted cell or spore suspensions were transferred onto nutrient agar containing 1, 10, or 100 ppm of DDT or DDD. Visual estimations of growth after three to five days incubation were made by comparison of a medium containing the same amounts of the solvent acetone. Secondly, soil suspensions were plated on selected media containing different concentrations of the compounds and after three to ten days incubation, numbers were counted, and thirdly, DDT and DDD were added to small quantities of soil which was incubated and examined for total numbers after 8 days incubation. Soils were prepared by either moistening 5 g of dried soil with 0.5 ml water or submersion in 3 ml distilled water which resulted in a waterlogged condition. The authors concluded that anaerobic conditions appeared to enhance the decomposition of some chlorinated hydrocarbon pesticides and that this rate was enhanced by the addition of organic materials such as alfalfa. The supplement of alfalfa did not appear to affect the rate of conversion, under aerobic conditions. The authors concluded that the conversion of DDT to DDD in soils under anaerobic conditions may explain the detection of DDD in soils where only DDT was originally applied. This is dangerous in that DDD in the soil is more stable than DDT and this may contribute to the persistance of residues of this type in soil and water. These workers also showed that at concentrations as low as 1 to 10 ppm DDT was highly toxic to soil bacteria and *Actinomycetes* in culture. DDD, which is also used as an insecticide, had a broader antimicrobial spectrum and was more toxic to microorganisms in vitro than was DDT. Perhaps, due to the stability and high potency of these compounds against microorganisms, such habitats as the rumen in cattle and the rhizosphere where microbial activity is concentrated may be important.

A study has been made of the effects of DDT,

dieldrin, and heptachlor on selected bacteria (Collins and Langlois, 1968). *E. coli, Pseudomonas fluorescens*, and *Staphylococcus aureus* in either autoclaved trypticase soy broth (TSB) or dehydrated skim milk with pesticides were plated immediately after inoculation and 1, 2, 4, and 7 days of incubation. Results of their work indicate that in concentrations of 50 and 100 ppm neither DDT, dieldrin, or heptachlor affected the normal growth of *E. coli* in either TSB or skim milk. *Pseudomonas fluorescens* was inhibited by both 50 and 100 ppm DDT in TSB after 2 days. Growth in the skim milk was not affected by any of the three pesticides. The growth of *S. aureus* was not affected by dieldrin in either TSB or in skim milk containing any of the pesticides, but was sharply inhibited by heptachlor.

The factors affecting the inhibition of growth of *Staphylococuss aureus* by heptachlor were examined further (Langlois and Collins, 1970). The pesticides used were 4 different concentrations of heptachlor, from 72 to 99% and heptachlor epoxide at a concentration of 99%. Fifteen to eighteen-hour TSB cultures of *Staphylococcus aureus* were incubated in the presence of the chemicals. Growth of the organism in the medium containing the pesticide and the appropriate control was determined by plating each sample on plate count agar. Direct microscopic counts were also made to check results of plating and in all cases the results obtained by both methods were in agreement. The final concentrations of pesticides used were between 1 and 100 ppm. The researchers summarized their results by stating that growth of the organisms was inhibited for 1 day by less than 3 ppm technical heptachlor, and almost completely stopped for 7 days by 10 ppm or more. Heptachlor epoxide appeared to have no effect, while inhibition of growth was more pronounced for technical heptachlor than similar levels of the pure compound. Growth in skim milk was unaffected by up to 10 ppm of heptachlor epoxide or up to 100 ppm heptachlor or technical heptachlor. The addition of 3% casein to the broth medium eliminated heptachlor inhibition, but did not eliminate it completely when trypticase was omitted.

Using the standard TSB growth medium and static cultures incubated at 25 or 37°C, Langlois and Collins (1970) concluded that the growth of *Pseudomonas fluorescens* and *Staphylococcus*

aureus in the medium was affected by DDT exceeding 50 μg/ml. The organisms growing in the skim milk appeared to be protected in concentrations up to 100 μg/ml of DDT. The concentration of 3 μg/ml of heptachlor increased the lag phase of growth of *S. aureus* up to 18 hr, while 15 to 20 μg/ml prevented growth up to 7 days. Inhibition of *S. aureus* was not observed in skim milk or in basic medium plus 3% whole casein or a mixture of α and β casein fractions containing up to 100 μg/ml of heptachlor. Heptachlor epoxide appeared to be noneffective, while 72% heptachlor was more inhibitory than the 99%. The authors concluded from their experiments that the levels of organo-chlorine pesticides, which may be in milk or milk products, will not affect the growth of the resident microflora.

In the work of Hurlburt et al. (1970), the compound Dursban was added 4 times at 2 week intervals to small ponds in California at a concentration of 0, 0.01, 0.05, 0.1, and 1 lb per acre. The authors concluded from their experiments, which were done mainly to study duck mortality, that the Dursban treatment caused a decline in the total amount of zooplankton and that the population required at least two weeks and probably a month to recover completely even at the lowest dosages.

Whether microorganisms which had been isolated from different marine environments would vary in response to three chlorinated insecticides was determined by Menzel et al. (1970). Four species in culture were examined for their response to DDT, endrin, and dieldrin. According to Weaver et al. (1965), these compounds are the most widely distributed chlorinated hydrocarbons in major U.S. river basins. The species of organisms examined included *Skeletonema costatum*, a costal centric diatom isolated from Long Island Sound, *Dunaliella tertiolecta*, a green flagellate typical of tidal pools and estuaries, a Cocolithaforid, *Coccolithus huxleyi*, and the diatom *Cyclotella nana*, both from the Sargasso Sea. In these experiments, carbon 14 was added to cultures which were illuminated by fluorescent lights. The cell carbon concentration of the medium was adjusted to between 100, 250, and 500 μg of carbon/l. and was considered to be within the range of naturally occurring carbon concentrations in surface oceanic water. After 24 hr of exposure to the light, the cells were filtered and counted in a Geiger Muller end window counter. Longer-term effects of the DDT and endrin on cell division were studied by counting cells for seven days. Cultures were grown in 125 ml flasks, containing 50 ml portions of medium and 100 ppb of the insecticide was added daily to each flask. The authors showed that none of the insecticides that were tested up to 1000 ppb affected the cultures of *Dunaliella* and no change in the rate of cell division could be detected over the 7 day period. The rate of C^{14} uptake in *Skeletonema* and *Coccolithus* was significantly altered at concentrations above 10 ppb in the case of all three of the insecticides. At 100 ppb DDT had some effect on cell division in *Skeletonema*, but no effect on *Coccolithus*. Endrin appeared to have little effect on the final concentration of *Skeletonema* cells, but the rate of growth over the first five days was slower than the controls. Reduced growth rates were also noticed with the *Coccolithus*. In contrast to these other species, *Cyclotella* was inhibited by all three insecticides and the slopes of the dose response curves for dieldrin and endrin suggest that quantities as low as 0.01 ppb may have been effective. Cell division was completely inhibited by dieldrin and endrin, while cells exposed to DDT divided more slowly than the controls. Because of the solubility difficulties with these compounds, the authors could not conclude much about the incorporation of these substances into the cell material, but felt because some species responded to concentrations above solubility limits that this indicated they were capable of incorporating these compounds as small particulates or that the saturation was quickly maintained while they concentrated the pesticide from solution. Although chlorinated hydrocarbons may not be universally toxic to all species, they may exert an influence on the dominance or succession of the individual forms.

Cowley and Lichtenstein (1970) used 17 species of fungi isolated from Wisconsin prairie soils. These organisms were grown on Czapek nutrient agar that had been treated with the chlorinated insecticides aldrin and lindane. Aldrin at a concentration of 20 ppm inhibited growth of *Fusarium oxisporum* by 37 to 44%. The addition of yeast extract, asparagine, ammonium sulphate, ammonium nitrate, or ammonium sulphamate to the medium resulted in a suppression of the growth inhibitory effect. Substitution of a vitamin mixture in place of the yeast extract had no effect.

The interaction of DDT and river fungi and the effects of the insecticide on the growth of four aquatic hyphomycetes have been studied (Dalton et al., 1970). *Heliscus submersus, Tetracladium setigerum, Varicosporium elodeae*, and *Clavariopsis aquatica* were obtained from river water as single spore isolates. DDT was added to a basal medium in concentrations that ranged from 0.1 to 60 μg of DDT/ml. The chlorinated hydrocarbon content of the medium before inoculation was determined by gas liquid chromatography. The extent of culture growth was determined at three day intervals by filtering through previously dried and weighed filter paper. The results indicate that the growth of *T. setigerum, V. elodeae*, and *C. aquatica* were unaffected by DDT concentrations below 2 μg/ml. At higher concentrations the growth rates were enhanced. *H. submersus* exhibited an enhanced rate of growth at each DDT concentration tested. Speculation about the results obtained (Dalton et al., 1970) suggests that the increase in growth may be the result of the use of DDT as a carbon source; that DDT may affect the permeability of the fungal cells to other nutrients; that DDT may increase the metabolic rate of the fungi in the capacity of a cofactor; and that this compound could have ecological significance since these fungi decompose organic debris and would be involved in the recycling of nutrients in the fresh water environment.

Dougherty et al. (1971) questioned what the sensitivity of *Bacillus thurengiensis* was to a number of pesticides as compared to other substances already known, such as antibiotics and sulfonomldes. The method used for testing was to inoculate the surface of a petri plate containing nutrient agar with bacillus spores at a concentration of 7000 per plate. Sterile paper discs containing the appropriate insecticide or herbicide were placed aseptically on the surface of the medium. The following compounds did not inhibit the bacillus-DDT; methoxychlor; lindane; dieldrin; 2, 4-D and 2, 4, 5-T Significant inhibition was obtained with chloropropam.

METABOLISM OF HALOGENATED PESTICIDES

In experiments by Okey and Bogen (1965), metabolism was followed with a Warburg respirometer and by direct observation of substrate removal. Culture material or cellular material was derived from soil and sewage inoculum and maintained on a single carbon source which was the unsubstituted homolog of the chlorinated test substrate. Their purpose was to develop a biological system capable of assimilating a specific carbon frame so that steric and substituent effects of a molecule could be properly evaluated. As the result of examination of a variety of chlorinated parafins, chlorinated napthalene, benzene, chlorinated benzoic acids, catechol, chlorinated catechol, chlorinated benzaldehyde, benzoic acid, and chlorinated phenoxyacetic acids, it was concluded that the presence of chlorine on a microbial substrate does not necessarily impede microbial metabolism. The effect of the chlorine is apparently regulated by such features as molecular size, the nature, and the number and position of other substituents. It appeared that there was no necessary development of any special enzyme system for the removal of chlorine and that the length of the chlorinated molecule had a significant bearing on its assimilability. No metabolism of the four carbon chlorinated parafins was detected while longer substances were completely utilized. The controlling feature was the number of carbon atoms between the site of chlorine substitution and the terminal carbon. In aromatic compounds, the chlorine substitution is altered by the oxidative state of the molecule. One chlorine atom affected the fate of chlorobenzene, chlorobenzaldehyde, and napthalene, but if the molecule was presented in a more oxidized form, monochlorination appeared to become less significant, until as in the case of chlorocatachol, little or no effect was observed at all. The location of the chlorine also appeared to affect the rate of metabolism. Metachlorobenzoate was metabolized completely, while ortha and para analogs were attacked partially. Because chlorine is such a strongly electronegative substituent, it is known to have a profound effect on chemical reactivity. Chlorine is known to alter the resonant properties of aromatic substances and in turn alter the electron densities of specific sites. This alteration of electron densities may increase or reduce the activation energy barrier through which a molecule passes during a reaction. For further clarification of these theories, the reader is referred to the work of Nakamura.[1] After searching the literature on detoxification, Okey and Bogen agree with the concept that hydrolysis is not the basic mechanism for the effect. Available reports indicate that chlorine is removed from alkyl substances in later catabolic steps (chloroacetic acid) and from aromatic substances during one of the electron rearrangements between muconic and adipic acid. The authors concluded that the metabolism of monochlorinated parafins was impeded when six carbons or less are between the site of substitution and the terminal carbon, or when the presence of three or more chlorine atoms are an aromatic ring. They believe that the inhibition noted during microbial metabolism of these substrates can be explained by the electronic effects of the chlorine on the substrate. The initial attack on an intact aromatic ring appears to be electrophilic in nature. In general, an increase in oxidative state of chlorinated molecule appears to render it more available for attack. The completeness of metabolism of refractory substances may be regulated in part by the quantity of cellular material present and by the availability of the easily assimilable substrates present. The resistance to attack previously reported for lindane and dieldrin type compounds and 2, 4, 5,-T might be explained by the electron poor nature of the unsubstituted carbon atoms.

Stenersen (1965) performed experiments to investigate whether absorption, metabolism, or excretion of DDT was the cause of resistance in the stable fly *Stomoxys calcitrans*. These particular flies develop very high resistance to DDT, methoxychlor, and DDD when kept under selection pressure in the laboratory. *Serratia marcesens, Alcaligenes faecalis,* and one unidentified bacterial strain were isolated from excreta of these flies kept in sterile cages. These isolated strains and other laboratory strains of *E. coli, Bacillus brevis,* and *Aerobacter aerogenes* were cultured under anaerobic conditions in meat extract under nitrogen atmosphere, in unshaken cultures (oxygen deficiency), and in fully aerated cultures. The medium contained C^{14} labeled DDT and after incubation the mixture was extracted with hexane and analyzed for radioactivity. When growing anaerobically or in oxygen deficiency, the facultative anerobes *S. marcesens* and *E. coli* and their unidentified strain converted DDT almost completely to DDD (about 90%) and DDE (5%). In the aerated cultures, neither the facultative anaerobes nor the obligate aerobes had any effect.

An attempt was made (Matsumura and Boush,

1967) to discover microorganisms that were capable of degrading dieldrin, known to be extremely stable in the environment. Microorganisms isolated from various soils were exposed to 0.01 µM of C^{14} labeled dieldrin. These mixtures were incubated for 30 days without shaking and the reactions stopped by adding trichloroacetic acid to each tube. The tubes were extracted with chloroform and distribution of radioactivity between the aqueous and solvent phases was reported. Eight cultures which appeared to be promising in their ability to degrade dieldrin were extracted with chloroform and analyzed with thin layer chromatography. The results indicate the definite existence of a variety of breakdown products of dieldrin and suggest that in soil there are microbes capable of degrading dieldrin.

In a search for metabolites which might be produced by microorganisms detoxifying dieldrin, Matsumura et al., (1968) isolated and chemically identified the chief products of degradation of dieldrin involved with one microorganism. A *Pseudomonas* species, originally isolated from a soil sample near a dieldrin factory belonging to the Shell Chemical Co. near Denver, Colorado, was shaken in liquid culture with dieldrin in acetone for seven days. The experimental culture and the soil known to contain 100 ppm of dieldrin were extracted and the metabolites purified using a column of Florisil and thin layer chromatographic separation with silica gel HF. The metabolites that were separated from others were purified on a gas chromatographic system. In all cases the identity of the metabolites that were produced in the *Pseudomonas* culture were compared to those from soil by matching the R_f values and the retention times on various gas chromatographic or thin layer systems. Infrared analysis of the metabolites and mass spectroscopic analysis were carried out with the result that the following (Figure 1) proposed degradation for dieldrin was established. With the exception of aldrin, the metabolites that were discovered do not exist in nature and the authors felt that it was extremely important to study their toxological effects on various ecosystems in our environment.

In 1968 the question of whether the conversion of DDT to DDD involved a step in reductive dechlorination or was a two step reaction with DDE as an intermediate product was examined (Plimmer et al., 1968). *Aerobacter aerogenes* was the organism of choice and was grown at 37°C in shake flasks containing 3% trypticase soy broth medium. Logarithmically growing cells were exposed to C^{14} and deuterium labeled DDT in separate experiments. Incubation of these cultures was continued in still culture for 86 hr. Extracts from the medium were examined gas chromatographically and with silica gel G chromatographic plates. In addition, for mass spectroscopic studies, thin layer chromatographic plates were developed and bands scraped from the plate, extracted with ether and portions of these extracts injected into the gas chromatograph. At the same time, a portion of the sample was placed in a glass capillary tube in the probe of the mass spectrometer. The use of the deuterium labeled DDT enabled the investigators to demonstrate that with *Aerobacter aerogenes*, conversion of DDD occurred with the replacement of chlorine by hydrogen, indicating that there were no unsaturated intermediates involved. If an alternate

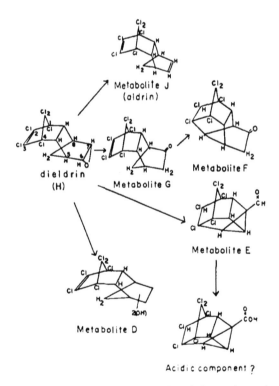

FIGURE 1. Proposed dieldrin degradation pathways by the microorganism in soil. The structures of metabolites G and F were established, F being one of the stable "terminal residues." The structures shown for E and the resulting acid metabolite are the most likely ones. The structural arrangement of those two hydroxyl groups in metabolite D is still unknown except that one of them is at the 6 carbon position. (From French, A. L. and Hoopingarner, R. A., *J. Econ. Entomol.*, 63, 756, 1970. With permission.)

route for the breakdown had occurred (the path through DDE to DDD), then the deuterium which was attached to the carbon atom would have been released and detected in their experiments. It was apparent that *Aerobacter aerogenes* affected the conversion of DDT to DDD by a process of reductive dechlorination. The thin layer chromatography indicated that the ether layer extracted in the experiment contained DDT and DDD as the only radioactive materials. Continuous ether extraction of the aqueous layer and the cellular debris revealed also only DDT and DDD.

Miles et al. (1969) attempted to isolate and identify the microorganisms which convert heptachlor to its epoxide in soil, and also to determine if there were other pathways of chemical and microbial degradation of heptachlor. The study was performed with 92 species of microorganisms originally isolated from soil. One ppm of the insecticide and/or its metabolites were incubated at $28°C$ in an aqueous basal medium for a period up to 6 weeks. Additions of insecticide were made at intervals along that incubation period. Products were extracted from the medium with hexane and analyzed using GLC and electron capture detectors. Analyses of the products produced from heptachlor indicated that there was a formation of 1-hydroxychlordene, heptachlorepoxide, chlordene, chlordeneepoxide, 1-hydroxy 2, 3-epoxychlordene and one unknown. The chemical and soil microbial degradation of heptachlor proceeded by at least three pathways — epoxidation, hydrolysis, and reduction. Thirty-five out of forty-seven fungi and twenty-six of forty-five bacteria and *Actinomycetes* produced the epoxide from heptachlor. Heptachlor hydrolyzed chemically to 1-hydroxychlordene in aqueous medium. Soil microorganisms were capable of producing 1-hydroxy-2, 3-epoxychlordene in this way. Heptachlor was dechlorinated by bacteria to chlordene which was epoxidyzed to chlordene epoxide and the insecticide and its by-products were concentrated in the fungal mycelium. The presence of cyclodiene insecticides in the media appeared to influence some microbial metabolic processes. Previous to this work, it had been assumed heptachlor could be degraded in two ways, either by volitilization or by conversion to heptachlor epoxide. This work suggests that there are two additional pathways of degradation: chemical hydrolysis to 1-hydroxychlordene followed by microbial epoxidation to

1-hydroxy-2, 3-epoxychlordene and conversion to an unknown product and bacterial dechlorination to chlordene and then oxidation to chlordene epoxide. The authors felt that the pathway leading to the unknown product through hydrolysis and microbial decomposition was the major degradation route (Figure 2).

Twenty microbial isolates previously found capable of degrading dieldrin were tested on their ability to degrade endrin, aldrin, DDT, Baygon, and gamma BHC (Patil et al., 1970). The isolation of the microorganisms had been described by Matsumura and Boush (1967). Each isolate was cultured in a 10 ml solution of yeast extract and mannitol and maintained at $30°C$ for 57 hr. These suspensions were inoculated with labeled insecticide and incubated for 30 days without shaking. In the final analysis, the distribution of radioactivity between aqueous and solvent phases was recorded and suggested that all the isolates were capable of degrading DDT and endrin, while 13 degraded aldrin. None were able to degrade Baygon or Gamma BHC. Since a number of Gamma BHC degrading microorganisms did exist, the authors concluded that it was likely that the enzyme systems metabolizing dieldrin, endrin, and DDT were different from those for Gamma BHC. It is not entirely surprising that dieldrin degrading microorganisms could degrade endrin, because endrin is an endo isomer of dieldrin.

Metabolism of the insecticide lindane by *Chlorella vulgaris* and *Chlamydamonas reinhardtii* in axenic culture was ascertained by Sweeny (1968). The organisms were grown for 14 days in bristol solution containing lindane. After incubation the medium was extracted with hexane and the extract examined using gas and thin layer chromatographic analysis. A substance was identified as 1, 3, 4, 5, 6-pentachlorocyclohex-1-ene a known nontoxic lindane metabolite which suggested that these organisms under certain conditions could utilize lindane.

The comparative metabolism of methoxychlor, methiochlor, and DDT was studied in mice, insects, and in a model ecosystem (Kapoor et al., 1970). A model ecosystem constructed for evaluating pesticide biodegradability consisted of a 10 x 12 x 18 in. glass aquarium containing a shelf of 15 kg of washed white quartz sand which was molded into a sloping soil air water interface. The lower portion was covered with 12 l. of standard reference water which had satisfactory nutrient

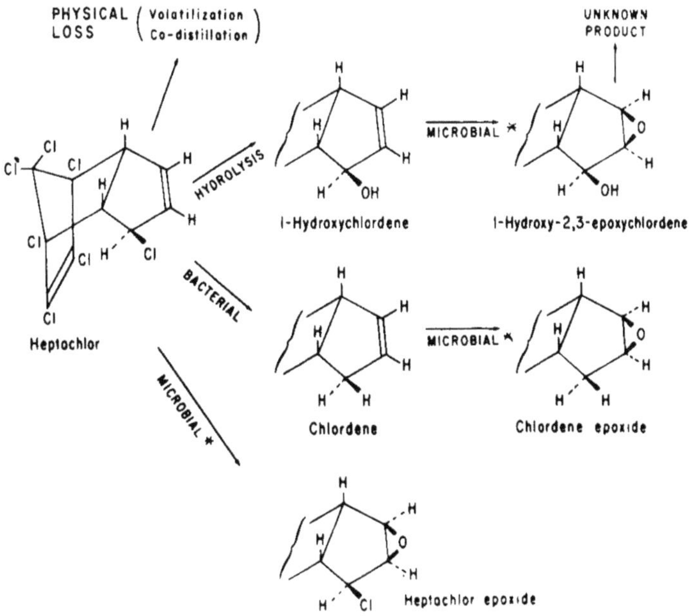

FIGURE 2. Scheme for chemical and microbial degradation of heptachlor. (From Matsumura, F., Boush, G. M., and Tai, A., *Nature*, 219, 965, 1968. With permission.)

material for the growth of *Sorghum halpense*, on the aerial portion and the alga *Oedogonium cardiacum* in the aquatic portion. The latter was seeded with a complement of plankton and also contained *Daphnia magna* and snails. The aquarium was provided with aeration and kept in an environmental growth chamber at 80°F with 12-hour exposure to 5000 foot candles of light. This model system was not fully evaluated but the workers were able to provide evidence of the relative biodegradability of the various pesticides examined.

About 150 isolates from various soil samples were screened to investigate the role of soil microorganisms in degrading endrin (Matsumura et al., 1971). Using techniques that they had described prior to this (Matsumura et al., 1968) they found that of the total cultures tested 25 were active in degrading endrin. At least seven metabolites of endrin were further isolated from the mass culture of a *Pseudomonas* species. In addition, all dieldrin degrading microbes which had earlier been reported to be highly active in degrading dieldrin were found to be also active in degrading endrin (Patil et al., 1970). On the basis of spectroscopic analyses, the conclusion was that the majority of the microbial metabolites of endrin were ketones and aldehydes with five to six chlorine atoms. Only one of the metabolites was

identified positively, and that was their metabolite IV (ketoendrin).

The herbicide 2, 3, 6-trichlorobenzoate (TBA) has been shown to be a potent growth regulator producing drastic formative effects similar to those of 2, 4-D, but TBA appears to be more resistant to bacterial degradation. The recalcitrant nature of this compound appears to be a result of the number of chlorines attached to the aromatic nucleus and the location of these substituents on the benzene ring. Monochlorobenzoates have been shown to undergo rapid degradation by McCrea and Alexander (1965). This same soil suspension would not attack 2, 4; 3, 4; and 2, 5-dichlorobenzoates; 2, 3, 4; 2, 3, 5; 2, 3, 6; and 2, 4, 5-trichlorobenzoates or 2, 3, 4, 5-tetrachlorobenzoate. Moreover, the chlorine in the 3, or 5 position in a ring appears to impart resistance to bacterial degradation of the molecule. Because the isolation of any organism capable of using TBA as a sole source of carbon and energy had not been reported, Horvath (1971) thought that the process of cometabolism may be an important phenomenon in this case. His investigation attempted to demonstrate cometabolic degradation of TBA, its mechanism, and the pathway of its degradation with an organism identified as *Brevebacterium* sp. cultured on benzoate salts agar. Standard manometric techniques were employed to measure oxygen uptake and CO_2 release. The microorganism had been previously described as being able to oxidize TBA with the uptake of one μM of oxygen per μM of TBA (Horvath and Alexander, 1970). The same organism was not able to use this herbicide as a sole source of carbon and energy for growth and it was considered that this represented a cometabolic mechanism. Measurement of inorganic chloride released during the oxidation of TBA indicated that one μM of chloride was released per μM of substrate oxidized. This discovery, coupled with manometric data indicated that a dichlorocatachole could be a product of TBA oxidation. Thin layer chromatography and infrared spectrophotometry provided evidence for the formation of 3, 5-dichlorocatachol. It appears that the oxidative pathway proceeded through 2, 3, 6-trichloro-4-hydroxybenzoate and 2, 3, 5-trichlorophenol to the end product 3, 5-dichlorocatachole which accumulated in the medium. 3, 5-dichlorocatachol was toxic to whole cells, but did not inhibit 2, 3, 6-trichlorobenzoate oxidizing enzymes or the pyro catechase enzyme. Horvath concluded that the specificity of pyrocatechase enzyme for an unsubstituted catechol was the cause of the phenomenon of cometabolism which this isolate exhibited.

The early experiments of Audus (1955) using *Bacterium globiforme* isolated from garden soil indicated that there were at least two intermediates produced in 2, 4-D breakdown which demonstrate phytotoxic properties.

While studying the metabolism of 2, 4-D by *Aspergillus niger*, the main product of the 2, 4-D metabolism was 2, 4-dichloro-5-hydroxyphenoxyacetic acid (Faulkner and Woodcock, 1964). Through the use of infrared spectrophotometry and mixed melting points, a second metabolite was identified as 2, 5-dichloro-4-hydroxyphenoxyacetic acid.

The products formed while the biological degradation of chlorinated catechols occurred and the establishment of the pathway by which the chlorophenoxyacetic acids and intermediates are metabolized was of interest to Tiedje et al., (1969). An enzyme preparation was prepared from 40 l. of actively growing *Arthrobacter* sp. grown in the presence of 2, 4-D. The enzymatic preparation catalyzed the conversion of 4-chloro and 3, 5-dichlorocatechols to *cis*-3-chloro and *cis, cis*-2, 4-dichloromuconic acids, respectively, by an ortho

fission mechanism. Catechol, 3- and 4-methylcatechols, were also converted to the corresponding muconic acids. Upon acidification the muconic acids formed butenolides which along with the corresponding chlorinated *cis, cis* muconic acids, and chlorocatechols were converted enzymatically to identical products identified tentatively as maleylacetic and chloromaleylacetic acids, respectively. Ring labeled 2, 4-D was metabolized to succinic acid.

The herbicide 2, 2-dichloropropionate has been shown to be an effective inhibitor of panothenic acid production and pyruvate utilization in microorganisms. Monochlorinated and dichloronated aliphatic acids have also been shown to be metabolized by microorganisms. Kearney (1964) was concerned with a cell free preparation of an *Arthrobacter* species which was able to dehalogenate 2,2-dichloroproprionate; and a number of other pure cultures which were capable of decomposing this compound. These are listed as representatives of the following genera: *Pseudomonas, Agrobacterium, Bacillus, Alcaligenes, Arthrobacter,* and *Nocardia*. The enzyme was partially purified by ammonium sulfate precipitation and showed a substrate specificity with nine chlorinated aliphatic acids. The enzyme (the substrate specificity was tested with nine) had its greatest activity on 2, 2-dichloroproprinate with less activity on 2-chloroproprionate, dichloroacetate, and 2, 2-dichlorobutyrate. There was no activity on any beta-chloro substituted aliphatic acid. The basis of activity was made upon the number of μg of chloride ion liberated per mg of protein per 10 min. The biotransformation of 2, 4-dichlorophenoxy alkanoic acids and related compounds by soil microflora is perhaps the most extensively studied. A fine reference source for the metabolism of these compounds and pesticides in general has been compiled by Menzil (1969).

The soil bacterium *Arthrobacter* sp. can cleave the 2, 4-D molecule enzymatically to the corresponding phenol. In this case, the 2, 4-D and related compounds such as 4-chlorophenoxyacetate and 4-chloro-2-methyl phenoxyacetate (MCPA) are converted to 2, 4-dichlorophenol, 4-chlorophenol and 4-chloro-2-methylphenol. In other work (Bollag et al., 1968a), it was shown that these particular phenols can be converted to catechols by a mixed function oxidase. In the case of 4-chlorophenol, 4-chlorocatechol resulted and in the case of 2, 4-dichlorophenol, 3, 5-dichlorocatechol was generated. In order to examine the microbiological fate of these catechols a cell extract of *Arthrobacter* sp. was prepared by culturing the microorganism with aeration at 25°C in 18 l. of MCPA medium. Enzyme activity was demonstrated in the extract by shaking a reaction mixture containing phosphate and substrate at neutral pH. The reaction was terminated by the addition of sulfuric acid and sodium Na_2WO_4. Where anaerobic incubation was desired, thunberg tubes were evacuated and flushed with nitrogen gas. In order to detect the products of catechol metabolism, the acidified protein free reaction mixture (proteins were removed by centrifugation) were extracted with diethyl ether and dried with anhydrous Na_2SO_4. Formation of catechol degradation products in the reaction mixtures was then measured at 260 mμ in a spectrophotometer. The ether extract was also examined with a gas chromatograph and the products were converted to their tri methyl silyl derivatives by treatment of the ether extract with bis (trimethylsilyl) acetamide at room temperature. Mass and infrared spectra were also utilized during the course of the investigations. The bacterial cell extract did metabolize 3, 5-dichlorocatechol and 4-chloro-catechol as shown by the disappearance of catechol, chloride formation, or changes in absorbancy in the residue from the ether extract. The formation of an ultraviolet absorbing compound and chloride were noted during the metabolism of the 2 catechols. At low enzyme levels the chlorocatechols were metabolized completely and small amounts of chloride were formed. Oxygen was required but iron and pyridine nucleotides did not appear to be stimulatory. The data suggest that there were 2 compounds in the reaction mixture after incubation of the enzyme with 3, 5-dichloro-catechol. One of these compounds appeared to be alpha chloro-gamma-carboxymethylene Δ alpha-butenolide. The product generated from 4-chloro-catechol appeared to be gamma-carboxymethelene Δ alpha-butenolide. The authors proposed a scheme shown in Figure 3 as a degradation pathway for chlorocatechols by *Arthrobacter*. The fate of carboxymethylene butenolides was not known. The formation of a chlorosubstituted beta-keto-adipate which could be cleaved to give acetate and chloromaleate resulting in acetate and chlorosuccinate was expected. Toxicity of these products to biological systems is not known. If

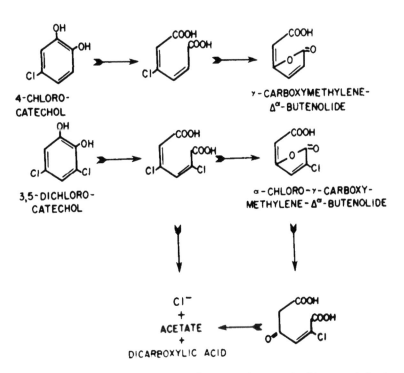

FIGURE 3. Proposed pathway for degradation of chlorocatechols by *Arthrobacter* sp. (Reprinted from *J. Agr. Food Chem.*, 16, Sept./Oct., 1968, 833. Copyright 1968 by the American Chemical Society. Reprinted by permission of the copyright owner.)

they are found in cultures of a microorganism normally found in soil, they could accumulate during the decomposition of these types of herbicides and be significant with respect to physiological activity.

Bollag et al. (1968) were able to show that a chlorophenoxyacetate degrading *Arthrobacter* sp. contained an enzyme which was able to convert 2,4-dichlorophenol and other chlorophenols to catechols. Oxygen and NADH were required in the reaction. The compounds that were formed from 2,4-dichlorophenol and 4-chlorophenol were identified as 3,5-dichlorocatechol and 4-chloro-catechol, respectively. Actually, NADH or NADPH were used with the latter, having a greater effect on the reaction. The addition of FAD or ferrous ions did not enhance the activity and quinacrine was not inhibitory. Up to 90% of the phenol metabolized was recovered as a catechol. The reaction was dependent on the presence of oxygen and the requirements of both oxygen and NADPH that suggest a mixed function oxidase system was required.

Previous work characterized the pathway for the conversion of 2,4-D to chloromaleylacetic acid and for the conversion of 4-chlorophenoxy-acetic acid to maleylacetic acid. Duxbury et al. (1970) were concerned with the enzymatic breakdown of the maleylacetic acids that are formed during biological decomposition of 2, 4-D. Tiedje et al. (1969) did report the enzymatic formation of succinic acid from 2, 4-D and suggested that

chloromaleylacetic acid was cleaved to a 2 and 4 carbon fragment. Enzymes which were obtained from *Arthrobacter* grown in the presence of 2, 4-D converted chloromaleylacetic acid and maleyl-acetic acid to succinic acid. This conversion was stimulated when NADH or NADPH was added to the substrate. The authors proposed the pathway shown in Figure 4 for the conversion of maleyl-acetic acids to succinic acid in the terminal phase of phenoxyacetate herbicide detoxication.

BIODEGRADATION OF PESTICIDES

For a number of years many microbiologists had been certain that any organic compound could be degraded if the compound were exposed to microorganisms in the environment long enough. In reality, many forms of both fungi, bacteria and protozoa are able to resist breakdown in the soil; for example, sclerotia, chlamydospores, and *Bacillus* endospores persist for many years with very little loss in viability. Human hair, tree stumps, rope, etc. have been found to remain largely intact for thousands of years. These prolonged periods of stability suggest that structural configurations of the molecule or environmental conditions associated with the particular storage are probably associated with molecular recalcitrance. It had been felt for many years that given enough time, the microbial inhabitants of the terrestrial or aquatic habitats

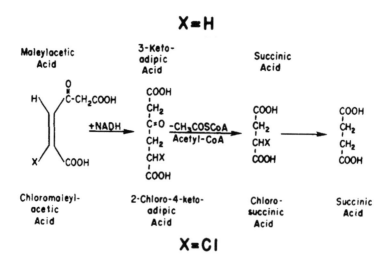

FIGURE 4.   Proposed pathway for enzymatic degradation of maleylacetic acids. (Reprinted from *J. Agr. Food Chem.*, 18, March/April, 1970, 201. Copyright 1970 by the American Chemical Society. Reprinted by permission of the copyright owner.)

would be able to adapt or mutate to modify the appearance of any new carbonaceous nutrients. Because the number of compounds known to be degraded by microorganisms is enormous, most scientists had felt that given the right conditions certain microbiological phenomena would persist and degrade whatever compounds were available. In addition, environmental factors such as anaerobiasis, adsorption, complexing, encrustation, substrate movement, and ecological determinants such as lack of water, osmotic (salt) effects, an unattractive pH, or extremes in temperature, etc. are important. Alexander (1965) concludes that there are factors which probably affect the biodegradability or decomposition rate of organic materials and that these are "inaccessability of the substrate, absence of some factor essential for growth, toxicity of the environment, inactivation of the requisite enzymes, a structural characteristic of the molecule which prevents the enzyme from acting, and the inability of the community of microorganisms to metabolize the compound because of some physiological inadequacy."

MacRae and Alexander (1965) were concerned with the microbial degradation of herbicides by utilizing C^{14} labeled CO_2 release. Radioactive herbicides were applied to the soil. A spectrophotometric measurement of the cleavage of the aromatic ring and plant bioassays were also used. The results of their studies with the benzoates, phenols, and phenoxy compounds demonstrated that the number of chlorines on the aromatic ring determine their susceptibility to microbial degradation in the case of the benzoates. In contrast with the phenol and phenoxy compounds, the position rather than the number of halogens appears to influence the susceptibility or resistance to decomposition. It was concluded that susceptibility of the chlorobenzoates to decomposition was related to the number of chlorines on the aromatic ring; benzoate metabolizing microorganisms are inactive on monochlorobenzoates and the resistance of dichlorophenols is associated with the presence of a chlorine in the position meta to the phenolic hydroxyl group.

The results of microbial degradation experiments on two herbicides, CIPC, and 2, 4-D, in aqueous solutions are discussed by Schwarz (1967), who concluded that breakdown of the CIPC was definitely of biological origin. The addition of carbon (nutrient broth) to the microbial system increased the rate at which the isopropyl carbon atoms were metabolized, but it did not appear to have any effect on the degradation of the phenol group. This suggests a possible metabolic pathway for this degradation involving the following steps: "initially the molecule is hydrolyzed to 3-chloroanaline and isopropyl alcohol. The latter is completely degraded. The 3-chloroanaline is modified to 4-chlorocatechol which is degraded via the muconic acid keto adipic acid pathway typical for aromatic compounds." Results concerning 2, 4-D suggest that the molecule was partially degraded by the bacteria in a dilute salts medium. The author used the number 2 carbon atom of the acetic acid moiety as the labeled compound in this study and suggests that the chlorinated ring structure would be more resistant to biological attack than the acetic acid residue. This is in apparent agreement with the literature which suggests that no destruction of the aromatic segment of 2, 4-D occurs in lake water, while a limited degradation of the acetic acid group does occur in a similar environment. The author concluded that only a small amount of 2, 4-D was degraded in a mixed microbial population under the conditions of the experiment, and that the addition of large amounts of nutrient broth as an additional carbon source had no appreciable effect on the rate of decomposition.

When grown in trypticase soy broth containing 20 ppm of dieldrin and incubated aerobically for five days, *Aerobacter aerogenes* could bring about the degradation of dieldrin to aldrin diol (6, 7-transdihydroxydihydroaldrin) (Wedemeyer, 1968). The author utilized electron capture gas chromatography and thin layer chromatography to establish the presence of this compound.

The interaction of clay, microorganisms, and diquat was studied by Weber and Coble (1968). A mixture of polypeptone and beef extract nutrient solution was treated with labeled diquat. Soil organisms were inoculated and the release of labeled CO_2 was measured. The addition of clay (montmorillonite) adsorbed the diquat and inhibited the release of CO_2, while the use of kaolinite clay appeared to have no significant effect on the total diquat decomposed in the system.

Ninety-two pure cultures of soil microorganisms were screened for aldrin degrading activity (Tu et al., 1968). The majority analyzed had some ability to convert aldrin to dieldrin. Species of *Fusarium, Trichoderma, Penicillium, Nocardia, Streptomyces,* and *Thermoactinomyces* all converted aldrin to dieldrin during a six week incubation period. It is a reasonable conclusion that no single species or genus was solely responsible for this conversion.

In order to examine the interaction of bacteria and chlorinated pesticides in sterile milk, one ppm of aldrin, DDT, or lindane was inoculated with strains of *Streptococcus lactus, Streptococcus cremoris, Streptococcus diacetilactis,* and *Lactobacillus casei,* then incubated for 14 days at 32°C. After testing the milk using gas chromatography, the gram positive bacteria used were unable to bring about any measurable biodegradation in the chemical configuration of any of the three compounds.

Lindane (gamma isomer of benzene hexachloride) persists in nonflooded soils and is considered to be susceptible to biodegradation in submerged soil. The degradation of lindane by a bacterium isolated from a Phillipine rice soil and the characterization of the unknown compound which was formed during this degradation was examined by Sethunathan et al. (1969). A species of *Clostridium* which had been previously shown to degrade the insecticide under anaerobic conditions was exposed to gamma BHC and gamma pentachlorocyclohexene considered to be a degradation of lindane. During the incubation periods, samples were withdrawn periodically, extracted with hexane, and analyzed for gamma BHC content by gas liquid chromatography and thin layer chromatography. The "degradation product of α lindane and alpha lindane differed from gamma pentachlorocyclohexene a direct product of dehydrochlorination of αBHC." It was assumed that reductive dechlorination as in the case of DDT to DDD was the mechanism by which α BHC was degraded to pentachlorocyclohexane.

CONVERSION OF DDT TO DDD

In an experiment designed by Kallman and Andrews (1963), commercial yeast was cultured in a defined medium which contained approximately 8 μg of DDT labeled with C^{14} in the phenol group. After incubation at 25°C the yeast was harvested, washed, ground, and extracted with acetone. The extract was dissolved in *n*-hexane and analyzed using paper chromatography. Radioactivity was determined in the strips of paper from the chromatograms. Analysis of their data show that the yeast had the ability to convert DDD from DDT by reductive dechlorination and the reaction did not require the intermediate formation of DDE.

In 1965, Barker et al. tested the hypothesis that the degradation of DDT to DDD in animals is due at least in part to the microflora in the gut. The microorganisms from a DDT resistant mouse were isolated and cultured in brain-heart infusion medium. Each of the tubes had been seeded with DDT. After five days at 30°C, the contents of the tube were extracted with methanol and chloroform. An analysis of the products using paper chromatography was carried out with the result that an isolate identified as *Proteus vulgaris* was shown to dechlorinate DDT to DDD. The experiment was repeated with a known culture of *P. vulgaris,* with the results that extracts of the medium after incubation showed that DDD was the main metabolite present. The authors suggested that the DDD that was produced is further metabolized to other products. Culturing these organisms on DDE did not result in any yield of DDD being produced and it can be concluded that as in DDT detoxication with yeast, DDE does not appear to be an intermediate in the production of DDD from DDT.

Twenty adult male and female rats were raised to maturity on a diet low in chlorinated pesticide residues (Mendel and Walton, 1966). These animals were matched according to sex and weight and were administered DDT dissolved in corn oil at a concentration of 2 mg/ml. The pesticide was administered to the test animals at the rate of 4 mg/kg body weight either by a stomach tube or intraperitoneal injection. Feces were collected for 48 hr, and then the animals killed and livers removed. Feces and livers were extracted with ethyl ether, cleaned up, and chromatographed on "florisil." The purified extracts were analyzed by gas-liquid chromatography. Animals which had been fed DDT via the stomach tube had a major residue of DDD with at most a trace of DDT in them. No chlorinated pesticides of any consequence were found in the feces of the rats injected intraperitoneally. The livers of the animals to which DDT had been administered by stomach tube indicated the presence of both DDT and DDD, while the rats having received the compound intraperitoneally had principally DDT in their livers. The authors concluded that gram negative bacteria in the gastrointestinal tract could be responsible for the conversion of DDT to DDD and that the conversion of DDD or the conversion of DDT to DDD did take place in intact rats. The production of DDD as a post mortem artifact caused by tissue decomposition does not appear to be accurate.

Chacko et al. (1966) initiated experiments to discover microbes which might degrade dieldrin, pentachloronitrobenzene (PCNB), or DDT and to identify degradation products. The research was carried out by culturing the microorganisms involved, particularly soil microorganisms, in a nutrient medium in which 5 to 10 μg of the desired pesticide was added. At the conclusion of incubation, the cultures were extracted with hexane and isopropyl alcohol and the extract examined by gas chromatography. Nine Actinomycetes and eight fungi were tested for their ability to degrade the compounds. All the test organisms except *Streptomyces albus* degraded PCNB to an unknown metabolic product. None of the organisms degraded detectable amounts of dieldrin. The DDT was dechlorinated to DDD by six Actinomycetes but not by any of the fungi. The most effective dechlorinating Actinomycetes appeared to be *Nocardia* and several *Streptomyces* species. Most of the test organisms appeared to be unaffected by the concentrations of the three pesticides that were used and it was concluded that the microbial degradation product of PCNB might be penta-chloro analine. The authors claimed that theirs was the first report of any soil microorganisms degrading DDT or of any organism at all degrading PCNB and that, since microorganisms in the soil tend to be largely inactive because of a deficiency of available carbon, these compounds may persist in the soil in spite of the presence of organisms that could partially degrade them.

A wide spectrum of pathogenic and saprophytic bacteria are normally associated with plants. The ability of these bacteria to degrade DDT was studied by Johnson et al. (1967). The organisms were grown either aerobically in brain heart infusion medium, or anaerobically in thioglycolate medium. Pure DDT at a concentration of 100 μg was added to 10 ml of bacterial culture containing

10^6 bacterial cells. At the conclusion of incubation extractions from the cells were examined using the gas chromatograph for additional identification. For verification of DDT and its metabolites, thin layer chromatography was used. In evaluating 27 different bacterial species for their ability to degrade DDT the occurrence of any degradation aerobically was essentially nil. In general, the dechlorination of DDT to DDD was widespread among bacteria which were grown under anaerobic conditions and appeared to occur most actively during longer incubation periods up to 14 days. In addition, the experiments demonstrated that strict anaerobes could also be variable with regard to their ability to degrade DDT. The authors suggested that there were metabolites that were still unidentified and that a broad range of plant pathogens and saprophytes have the ability to convert DDT to DDD under anaerobic conditions in vitro.

Braunberg and Beck (1968) maintained rats on diets containing 100 ppm of DDT. This did not appear to alter the microflora of the gastrointestinal tract. The bacterial population of the intestinal tract was determined immediately after death and the numbers and different types of bacteria were estimated. Representative colonies of the types found were selected and then cultured in trypticase soy broth containing 2.5 μg of DDT. Incubation was carried out anaerobically for 20 hr and at the conclusion the entire contents of the tubes were extracted with diethyl ether. Gas chromatographic analysis was used to determine DDT conversion. The results suggest that most of the genera studied degraded the pesticide to DDD with the exception of some gram positive cocci and that DDD in the feces of rats which had been fed DDT was probably due to microbial rather than mammalian metabolism.

The degradation of DDT in ensiled pasture herbage is of extreme importance and has been examined by Henzel and Lancaster (1969). The objectives of their experiments were to measure the extent of DDT degradation and to characterize any derivitives which might appear in the silage. These workers had the idea to assess the effect of the type of silage fermentation on the degradation of DDT. Two main types of herbage were selected: one basically rye grass and the other a coxfoot clover rye grass mixture. Commercial DDT had been applied to the areas where the herbage was collected while unsprayed areas were used for control material. After suitable mixing, samples were placed in laboratory vacuum silos and fermentations maintained for 90 days. At the conclusion, DDT was extensively decomposed in the process with the accumulation of degradation products, such as DDD. The concentration of DDT was reduced by about 50% and suggested that this ensilage could be a useful management procedure in dealing with DDT contaminated pastures.

The possibility of the degradation products of certain pesticides interacting when placed in perhaps unexpected combinations can be of very practical importance. Bartha (1969) experimented with the herbicides Propanyl and Solan and discovered that when incubated in soil an unexpected residue which was 3,3', 4-trichloro-4'-methylazo benzene was formed and it was determined that each of the herbicides contribute ½ of the asymetric azobenzene molecule. This presents an interesting question with respect to the variety of soil treatments being carried out and the fact that such unexpected hybrid products may be formed in soil; and certainly increases the complexity of pesticide residue problems. Whether or not such interactions can be mediated by microorganisms will need to be studied further. Ledford and Chen (1969) exposed representative microorganisms of the flora associated with a number of varieties of surface ripened cheese to lindane, DDE, and DDT. The organisms were grown in liquid culture with the pesticides for a period of 10 days, after which the concentrations of residual pesticides were determined using electron capture gas chromatography. The *Streptococci* and *Micrococci* did not appear to cause any change in pesticide level, and they themselves were apparently unaffected by the pesticides. However, other isolates did dechlorinate DDT. DDT and DDE might be degraded if they were present in certain types of surface ripened cheeses.

An interesting phenomenon pertinent to microbiology is the report of Singh and Malaiyandi (1969) who reported that DDT was reduced to yield DDD as one of the products in an aqueous medium under pressure. DDT was treated in an aqueous medium with granular metallic tin and ammonium chloride in a sealed borosilicate tube, and heated to $115°C$ for a period of 8 hours. All the DDT was degraded to DDD, DDE, and probably DDA along with at least three other major unknown components.

DDT and DDE have also been shown to be photooxidized in methanol (Plimmer et al., 1970). Photolytic generation of free radicals may react with oxygen or abstract hydrogen from either the solvent or from substrate and result in further decomposition of other intermediates. Oxidation products included benzoic acids, aromatic ketones, and chlorinated phenols.

The question of whether the intestinal microflora of the northern anchovy is capable of dechlorinating DDT and the importance of bacteria and fungi in this particular process was examined by Malone (1970). It had been shown in previous work that reductive dechlorination of DDT to DDE and DDD in three species of salmonid fishes, Atlantic salmon, cutthroat trout, and rainbow trout could occur. The intestinal contents of 25 adult anchovies were placed into sterile test tubes to which nutrient broth was added. Cultures were divided into 4 different groups; one, unaltered; the second, inoculated with 100 units of penicillin and streptomycin to suppress bacterial activity; the third, inoculated with 100 units of mycostatin to suppress fungal activity; and the fourth, 100 units of penicillin, streptomycin and mycostatin. Each culture was inoculated with 0.05 ml of 111 ppm of C^{14} labeled DDT and incubated anaerobically under nitrogen. DDT residues were extracted with hexane concentrated by descending paper chromatography. The chromatograms were cut into strips and counted with a scintillation counter. The culture filtrate was also extracted and separated into a particulate fraction and polar fraction or filtrate with a millipore filter and also counted with a scintillation counter. Most of the activity was found in the hexane fraction. No C^{14} labeled DDD or DDE appeared in the controls and activity found in the polar and the particulate fractions was less than the test cultures. All the test cultures did yield high concentrations of labeled DDD, but little or no DDE. The highest level of labeled DDD occured in the first group in which neither the bacteria nor the fungi were suppressed. The lowest yield of DDD occurred in the fourth group in which both types of organisms were suppressed. An analysis of the data showed that there was some differential activity of the intestinal fungi and bacteria. In group two, where bacterial activity was suppressed, the DDD levels were lower and the polar phase activity higher than in the third group where the fungal activity was suppressed. Thus, while both bacteria and fungi dechlorinated the DDT to DDD, the fungi may have been primarily responsible for further degradation of DDD to water soluble products in anaerobic conditions.

E. coli was grown in broth culture and skim milk to which DDT had been added to study the reductive dechlorination process (Langlois, 1967). After 7 days of incubation, in the 99.3% DDT test in broth, *E. coli* could dechlorinate DDT to DDD. In the case where skim milk was used, very small amounts of DDT were detectable.

Bacillus cereus, Bacillus coagulans, Bacillus subtilis, Escherichia coli, and *Enterobacter aerogenes* grown in trypticase soy broth were capable of degrading DDT into a variety of metabolites (Langlois, 1970). Seven metabolites were identified from aerobic growth of *Bacillus* and from anaerobic growth of *E. coli* and *E. aerogenes. Psuedomonas fluorescens* and *Staphylococcus aureus* were not able to degrade DDT under aerobic conditions, and none of the species degraded DDT in skim milk or in broth containing casein. None of the species tested degraded dieldrin or heptachlor. The fact that the authors were unable to detect any degradation of dieldrin may be due to the fact that other workers had examined intracellular products of isolates which had been taken from dieldrin treated soils. Degradation of DDT in milk would not be expected since it is becoming clear that casein may complex DDT and prevent degradation.

Focht and Alexander (1970) reported on the microbiologically induced cleavage of certain ring structures of metabolites known to be generated from DDT, as well as certain structural analogs, and again demonstrated the influence of chemical substituents. An enrichment culture technique using diphenol methane as carbon source was used to isolate a *Hydrogenomonas* species from sewage. It was found that *Hydrogenomonas* could cleave one of the rings of *p, p'*-dichloro diphenyl methane, a product of DDT metabolism and *p*-chlorophenyl acetate was identified as a product which was metabolized further. Substituents on the methylene carbon and para-chloro substitution were critical factors governing the resistance to DDT and related compounds to aerobic metabolism and decomposition by the bacterium. With the understanding that sometimes the benzene rings are cleaved and that microbial modification of these chemicals may occur, the ultimate fate of DDT in the environment can be more favorably understood. Microorganisms cultured under anaerobic conditions are known to be able to convert DDT to compounds of the type able to be utilized by *Hydrogenomonas* and the authors suggested that a biological model now exists for the tracing of the pathway of DDT decomposition in nature.

The cellular components of the bacterium *E. coli* have been used to examine the metabolic processes involved in reductive dechlorination of DDT to DDD (French and Hoopingarner, 1970). Spheroplasts of *E. coli* were harvested and subjected to osmotic shock. The particulate components were separated from the cytoplasm by centrifugation. The metabolic studies were carried out in a Warburg respirometer using glass beads which had been coated with C^{14} uniformly labeled DDT. Incubation components were held at 37°C under nitrogen. After 4 hr, the particulate and supernatant fractions were separated by centrifugation at 20,000 x g and DDT and its metabolites were identified by thin layer and gas liquid chromatography. The experiments were designed to determine the site of DDT metabolism in *E. coli*. Table 1 shows the results of these experiments. The membrane system alone and the cytoplasm alone were not effective in the dechlorination process. The same was true for boiled membrane and cytoplasm. The membrane systems plus cytoplasm, however, were fairly active in producing substantial amounts of DDD. This suggested that the membranes in *E. coli* were the site of reductive dechlorination and that the cytoplasm contained essential factors. When NAD, NADP, FAD, ADP, inorganic phosphate, malate, and pyruvate were added, the level of DDD was not significantly different than that of membranes alone. When NAD, NADP, or the malate and pyruvate were omitted from the incubation mixtures, increases in DDD production did occur. Omission of ADP and inorganic phosphate or FAD did not help and significantly substantial increases of DDD were noted when preparations containing FAD, ADP, and inorganic phosphate were used. Additional experimentation suggested that isolated membranes required the addition of exogenous FAD to obtain any significant DDD production. The authors conclude from their experiments that DDD production by microorganisms is an enzymatic process and not passive in nature. Additionally, the observation that the addition of FAD to membrane fractions enhanced DDD

TABLE 1

Effect of Membrane and Cytoplasm of *E. coli* on Conversion of DDT

	% ^{14}C found as DDT metabolites[a]				
Components	DDE	o,p'DDT	TDE	p,p'DDT	Unknown
Membrane only[b]	0.4	1.8	4.6	90.5	2.7
Membrane and cytoplasm[c]	0.3	1.3	29.8	61.9	6.7
Cytoplasm only	0.1	2.0	2.4	92.5	3.0
Boiled membrane and cytoplasm	0.3	1.5	3.8	90.0	4.4
DDT-^{14}C and buffer	0.4	1.9	0.6	93.9	3.2

[a]Means of 2 experiments, 3 replicates each.
[b]The membrane fractions (3 ml aliquots) consisted of Tris buffer, at pH 8.0, containing membranes at 25.0 mg/ml original dry weight of cells.
[c]The membrane fractions were resuspended in the cytoplasmic fraction at 25.0 mg/ml original dry weight of cells.

production suggested that the capacity to metabolize DDT to DDD residues is in the membrane portion of the bacterial cell and is not cytoplasmic. The cell walls were removed or depolymerized through the action of lysosyme and did not appear to play a direct role in the reductive dechlorination of DDT. The presence of Kreb cycle activity appeared to inhibit DDD production during the study, and FAD enhancement of the DDD production was dependent on anaerobic conditions which suggested that normally operating oxidative pathways preclude the reductive dechlorination of DDT. This work appears to refute any indication that the cytochrome oxidase system is involved in the conversion of DDT to DDD because of the fact that addition of malate pyruvate and cofactors of the Kreb cycle to isolated membrane fractions did not enhance DDD production. Membranes are known to be capable of metabolizing such intermediates and their results would not have been expected if reduced cytochrome oxidase had been responsible. Since a variety of increments of exogenous FAD added to the membrane fractions did not increase DDD production, there may be another factor or factors limiting this rate of DDD production. Thus, reductive dechlorination appears to occur in the membrane portion of the cell; is stimulated by component or components in the cytoplasm; does not use electrons produced by Kreb cycle intermediates; is dependent upon reduction of FAD; and occurs only under anaerobic conditions.

MICROBIAL ACCUMULATION OF PESTICIDES

Chacko and Lockwood (1967) tried to determine whether soil microorganisms could accumulate DDT or dieldrin from a culture solution. Three fungal cultures *Mucor ramannianus*, *Glomerella cingulata*, *Trichoderma viride*; three *Streptomyces* – *S. lavendulae*, *S. griseus*, and *S. venezuelae* and three bacteria – *Bacillus subtilis*, *Serratia marcescens*, *Agrobacterium tumefaciens* were tested for their ability to accumulate DDT and dieldrin from distilled water containing either DDT or dieldrin. The test organisms were grown for six days in shake culture in a nutrient medium and then harvested by centrifugation. Pesticide 0.1 to 1 mg/ml was added to distilled water along with the cells or mycelia and the mixture shaken for 24 hr. The cells were filtered, washed, and extracted with hexane and isopropyl alcohol and the supernatant liquid which originally had pesticide in it was also extracted. Dieldrin was accumulated up to 76% in the case of the fungi; 83% in the case of the *Streptomycetes* and in the case of the *Agrobacterium tumefaciens*, 90%. *Bacillus* and *Serratia* cultures only accumulated 16% of the dieldrin. The same picture was reflected in the uptake of DDT with a maximum of 100% in the *Agrobacterium* culture. The authors concluded that the process of accumulation did not involve metabolism because heat killed cells and mycelia responded similarly to their living counterparts and suggested that the ability to accumulate these compounds may be wide spread among organisms. Additionally, the pH range over which this accumulation occurred appeared to be quite wide and they postulated that accumulation by microbes may be a factor in the retention of these compounds in soil.

The accumulation and concentration of dieldrin, DDT, and the fungicide pentachloronitrobenzene (PCNB) by *Actinomycetes* and fungi in the soil was reported by Ko and Lockwood (1968). Fungal and *Actinomycete* mycelia were added to soil containing dieldrin, DDT, and PCNB and were found to accumulate these compounds to concentrations above those in the soil. The amount of dieldrin accumulated in a species of *Rhizoctonia* increased with increasing amounts of mycelium concentration of dieldrin and incubation time of mycelium in the soil. It also appeared that the presence of this organism in soil restored the ability of the soil to retain the pesticides against aqueous leaching.

Chlorella sp. was cultured under batch and continuous culture conditions and the accumulation and mechanism of accumulation of C^{14} labeled DDT was studied by Södergren (1968). The rate of uptake of labeled DDT was compared in both living and dead cells. The uptake of DDT by *Chlorella* was rapid and in fact completed in less than 15 sec. This uptake appeared to be mainly due to physical absorption dependent upon diffusion in the water since the rate of absorption was the same as that of diffusion. The DDT was taken up according to a theoretical rate based on calculations. In addition, growing *Chlorella* in the DDT solution for a period of three days in the continuous flow situation created a morphological change in the algae followed by an agglomeration of the cells and a change in the rate of uptake of the pesticide. When the supply of DDT was discontinued, and when the chemical had been removed from the culture by dilution (washout), the cell concentration did return to its original level and the agglomeration disappeared with a gradual conversion to the normal unicellular condition.

In a study designed to investigate the accumulation and effects upon the metabolism of several commonly used chlorinated pesticides on selected algal populations four unialgal cultures, *Microcystis aeruginosa*, *Anabaena cylindrica*, *Scenedesmus quadricauda*, and *Oedogonium* sp. were grown in the presence of four chlorinated pesticides; aldrin, dieldrin, endrin, and DDT (Vance and Drummond, 1969). The pesticides were used in concentrations of 5, 10, and 20 μg/ml. At the conclusion of the experiment, cells were removed, washed, and extracted with hexane. Pesticides were cleaned using a "florisil" column and analyzed in a dual column gas chromatograph equipped with an electron capture detector. Results indicate that the types of algae studied were resistant to the pesticides tested. The algae did appear to accumulate pesticides in detectable quantities within 30 min after exposure and in the case of DDT the breakdown products one would expect such as DDD, DDE did occur. Exposure of up to one ppm of these compounds to cells in a Warburg respirometer indicated that these compounds had no significant effect on the respiration of the organisms that were studied. The researchers concluded that while algae may concentrate pesticides many times, they may be more resistant to these compounds than are higher members in the food chain.

An autoradiographic technique was developed to detect the accumulation of DDT in microorganisms cultured on a peptone agar plate. Carbon 14 labeled DDT was added to an inoculated medium and incubated at 25°C. When colonies had developed, an x-ray film was placed over the agar. The film was exposed at 3°C for 5 days and then developed (Kokke, 1970). Colonies which accumulated DDT or products appeared as dark spots after development. The background was slightly grey due to the distribution of DDT throughout the medium and colonies which had not accumulated DDT were white in appearance. The tap water that Kokke worked with contained about 1% of the microorganisms capable of accumulating DDT. When polluted surface water was used, as many as 7% of the microorganisms were capable of accumulation and with garden soil samples up to 70% were capable. In the case of nursery soil, which had been recently treated with pesticide, as many as 95% accumulated DDT. A portion of DDT-resistant bacteria in a population was measured using the pour plate technique. A peptone agar with various concentrations of the pesticide was mixed with a water or soil suspension sample. Daily counts indicated differences in retardation and inhibition between 2 samples which had been shown by autoradiography to contain 3% and 93% DDT accumulating bacteria. DDT sensitive bacteria were detected by the use of peptone grown colonies retransferred on fresh plates with or without DDT. It was found that 5 ppm of DDT had a definite inhibition and that the larger proportion of DDT-resistant bacteria was found in soil samples, as compared to those in aquatic samples. This may be due in part to the insolubility of the compound in water.

Experiments have been conducted to determine the partition coefficients of DDT at concentrations similar to those in the natural environment because high concentrations may affect the apparent partition coefficient of an organism for DDT residues in water. C^{14} labeled DDT uptake was measured in three species of marine phytoplankton in pure culture. Algal suspensions (100 ml) were added to flasks with stock solutions of the isotope to give desired concentrations. The mixtures were stoppered, agitated, and allowed to equilibrate for a few minutes after which the contents of the flask were filtered onto a glass fiber filter or a membrane filter and dried for 24 hr. The filters were removed from the dessicator and examined in a scintillation counter. The results indicate that between 16 and 54% of the initially added labeled DDT was removed from the water by the algal mass. Partition coefficients were calculated on the basis of final equilibrium concentration of DDT in the medium. Using an estimate of 1.9×10^5 for a relative partition coefficient and the estimate of DDT residue in sea water at 15 ppt, the authors obtained an expected value of about 30 ppm (oxidizable carbon) DDT residues.

Lindane and dieldrin were used to examine the degree of adsorption and desorption of these types of compounds to the yeast *Saccharomyces cerviciae*. Bakers' yeast was mixed with a known concentration of insecticide for periods up to 18 hr. After equilibration, the yeast were removed by centrifugation and the concentrations of insecticides remaining in the supernatant were determined. Killed yeast were utilized by warming known amounts in a volumetric flask at 96°C for 3 min. After cooling, an insecticide solution was added and treated in the same way. The quantity of insecticide adsorbed per gram of yeast was calculated from the decrease in the concentration in the water. The desorbtion of the compounds was estimated by shaking both the living and dead yeast with a solution of lindane or dieldrin for two hours. The yeast were centrifuged and the water removed. Organic material was extracted with

fresh water by shaking two hours at each period. The water extracts were analyzed for concentrations of desorbed lindane or dieldrin. Voerman and Tammes concluded from their experiments that the pesticide adsorption is a fast process being completed in less than 30 min; adsorption could be described by an equation; that lindane was adsorbed less than dieldrin and that this was probably due to the solubility; that killing the yeast aided in the adsorption capacity, especially for dieldrin; and that these insecticides could be removed by washing with fresh water.

The diatom *Cylindrotheca closterium* was capable of storing or adsorbing DDT from a liquid culture on the average of about 265 times (Keil and Priester, 1969). In general, the conclusions of these workers were that this diatom was able to adsorb and concentrate DDT above the level in sea water and that the main metabolite (even though small concentrations were detected) was DDE.

In experiments designed to study the absorption of dieldrin by *Chlorella pyrenoidosa*. Wheeler (1970) concluded that radioactive dieldrin was taken up by the *Chlorella* cells in increasing amounts for varying periods of time. The label became more difficult to extract with time, suggesting that there may have been a distribution of label through subcellular organelles. Comparative results with those of Södegren (1968) suggest that the rate of dieldrin absorption is slower than that of DDT with *Chlorella*. The solubility of the two compounds is substantially different, with dieldrin being 100 times more soluble in water than DDT which may have a greater affinity for a cell-water interface with subsequent cellular absorption more rapidly. The dieldrin did penetrate the algal cells rapidly to a maximal level at from 6 to 24 hr after the introduction of the insecticide and there were no metabolites detected.

Shin et al. (1970) carried out an empirical experiment to seek out actively absorbing fractions which may be ecologically significant in soil. Three different soil types were extracted with solvents of increasing polarity. Residual soil fractions were used as adsorbants in batch equilibrium experiments to derive adsorbtion isotherms for DDT. Adsorption isotherms were also derived for tissues of *Rhizoctonia solani*, alfalfa, and unextracted soil and soil residues after H_2O_2 digestion. The partitioning of DDT and other hydrocarbons from soil into microbial tissues had been shown prior to this work (Ko and Lockwood, 1968), and represents a concentrating mechanism for trophic transfer of pesticides in the environment. With these concepts in mind, the authors compared the distribution coefficients for fungal mycelia and alfalfa tissues under the same conditions of equilibrium partitioning as were used for soils and soil fractions. The distribution coefficients were similar to those in other soils and the partitioning and concentration of DDT reported by Ko and Lockwood could have been predicted from these data. The authors data "illustrate three probable sources of anomaly in attempts to relate adsorption of non-ionic pesticides to soil organic matter content; structural and water repellant effects of lipoidal deposits on accessibility of sorptive surfaces; differences in nature of complex minerals and the extent to which they substitute for structural carbon or induced differences in surface properties of organic materials intimately associated with them; and nature and proportion of non-humified to humified organic matter. Stepwise elimination of non-humified materials holds promise for identifying and characterizing actively partitioning surfaces."

In an examination of aerobic heterotrophic bacteria isolated from Lake Erie, 33 of the isolates tested in 6 different growth media revealed that 19 cultures were capable of forming flocs in at least one medium. Ten of the cultures formed flocs in two or more of the media. Leshniowsky et al. (1970) were concerned with the ability of two of these isolates to concentrate and accumulate the pesticide aldrin from solution. The bacteria used were identified tentatively as a *Flavobacterium* or a *Protaminobacter* and a Gram positive bacillus. Bacterial flocs were placed on a rotary shaker with the pesticide to give a final concentration of 1 ppm. At the conclusion of a desired time period, the flocs were removed, separated by centrifugation, washed and supernatant fractions extracted separately with a mixture of heptane and acetone. Samples were analyzed in an electron capture gas chromatograph. Almost all of the aldrin adsorbed to the floc was adsorbed within the first 20 min of contact. All of the aldrin added to the Gram positive bacteria was recovered from the floc with none appearing in the supernatant. Essentially the same results were obtained with the gram negative organism. Calculations indicated that the concentrating effect of these bacteria is quite high.

For example, the Gram positive floc adsorbed pesticide from water with a concentration factor about 625 to 1 within the 20 min period. Analogous findings have been reported for algae and discussed previously (Kiel and Priester, 1969). In another experiment, samples of natural sediments that were in the process of settling and accumulating in Lake Erie were collected in a specially designed sediment collecting device, placed on a submerged reef. The sediment was analyzed in a manner similar to that of the bacterial floc and the presence of both aldrin and dieldrin was detected in these samples, using both gas chromatography and microcoulometry. When additional aldrin was added experimentally, it was quickly adsorbed (about 10 min). The authors concluded that floc forming microorganisms act as adsorbants for other suspended microparticles, including chlorinated hydrocarbons and that this adsorption is a natural process in the removal of microparticles from water. Once these particles are on the bottom of the lake, their fate is in question, but it is likely that they may be degraded under anaerobic conditions. The effect of the concentration of these compounds on the flora and fauna on the bottom of the lake is significant.

The change in content of DDT residues in phytoplankton samples over a collection period of 1955 through 1969 has been analyzed by Cox (1970). All samples had been maintained in a 3% solution of formalin in sea water and estimated DDT residues for the samples were based on the carbon content as determined by wet combustion. The question of whether or not DDT residues are increasing in coastal waters or marine environments was examined by the analysis of phytoplankton samples because they are particularly suited in that they represent the first link in pelagic food chains, and that trends in their concentrations of DDT residues would appear to be relevant to higher order consumers in the food chain and apparently DDT taken up by the phytoplankton is done so, rapidly and essentially irreversibly. After preparation, each sample was extracted with *n*-hexane and examined in the gas chromatograph. Results suggest that there were higher concentrations of DDT residues in more recent samples, but because of the sample storage may have had an effect on this result, an experiment was performed using ring labeled C^{14} DDT to test the effect of decomposition on the relative proportions of 3 DDT constituents found in the samples. The pesticide was placed in a sealed ampule along the portions of phytoplankton sample preserved in formalin and stored in the dark, since the originals had been stored that way. Elevated temperatures were used for short periods of time to duplicate any such fluctuations which may have occurred during storage periods. At the conclusion of a six day period, samples were removed from the ampules and analyzed. The results indicated that no breakdown of the C^{14} labeled DDT had occurred in the samples that were heated at $30°C$. In samples heated at $60°$, 28% of the DDT did break down. In the sample heated at $75°$, 38% of the C^{14} labeled DDT broke down to polar compounds, but DDT comprised only 15% of the nonpolar material, while DDD and DDE made up 83%. They concluded that a change in the relative proportions of the DDT residues in the sample would be expected if any net decomposition to nonpolar products had occurred during storage. It was found that the relative proportions of the DDT constituents found in the samples were constant in that their actual trend toward an increase in DDT residues was real, rather than a loss of analyzable residues during sample storage.

TOXICITY

Ragland et al. (1971) found that the exposure of mallard drake ducks to DDT increased the hepatic microsomal mixed function oxidase activity. When duck hepatitis virus (DHV) was injected the activity was decreased. If DHV was injected prior to exposure of DDT, it was found that the mixed function oxidase may actually have been enhanced by the inoculation of the virus. These experiments with ducks suggest that the virus protects them against the lethal effects of DDT by stimulating hepatic microsomal mixed function oxidase activity. In other words, the metabolism of DDT was probably enhanced into the less toxic DDE. The fact that the DHV alone depresses the activity is most difficult to understand and remains unexplained.

Type E botulism has been implicated in the death of water birds on Lake Michigan, and this toxin which is produced by *Clostridium botulinum* has been found in dead fish taken from the lake. To determine whether DDT and its metabolic residues are also present in the birds and the fish, a study was undertaken to test the possibility that there may be an interaction between these materials both of which are neurotoxins (Monheimer, 1969). The experiment was designed as a 4 by 5 factorial design experiment with DDT dissolved in corn oil injected into mice 22 hr before intraperitoneal injection of dilutions of the botulinum toxin. A two way analysis of variance indicated that the DDT treatments alone affected all the mice similarly and at the 97.5% significance level there was an interaction between the DDT and the botulinal toxin. The results suggested that lower levels of toxin were protecting mice from DDT. The author concluded that his study did not appear to have any ecological significance for the Lake Michigan water bird mortalities. "Low levels of botulinal toxin may protect some individual birds from DDT; but large numbers of birds are unlikely to obtain the precise quantities of the two materials that this experiment indicates are necessary for an antagonism."

Bounds et al. (1969) experimented with two microorganisms — a *Micrococcus* species which was found to be insensitive to the herbicide "dalapon" and an organism identified as an *Agarbacterium* species which was sensitive and inhibited by the compound. When the *Micrococcus* was placed in the environment containing "dalapon," which would normally be toxic to the *Agarbacterium*, the inhibitory effect was lessened and the *Agarbacterium* capable of growth. Neither of the organisms was able to degrade the herbicide as determined by chloride release or disappearance from the medium. The addition of calcium pantothenate in the concentration of 0.05 µg per ml allowed the same amount of growth of the *Agarbacterium* in the "dalapon" medium, as did a filtrate from the *Micrococcus*. A bioassay of the *Micrococcus* filtrate revealed the presence of beta-alanine, valine, pantothenic acid, and alpha-keto-isovalerate. This finding suggested that an "associative action" of microorganisms may play an important role in microbial resistance to herbicides.

ANAEROBIC ACTIVITY

Hill and McCarty (1967) made use of an active anaerobic condition under an idealized situation to examine whether a group of chlorinated hydrocarbon pesticides were capable of being degraded. The pesticides were mixed continuously at 35°C with biologically active anaerobic digested waste water sludge. Samples for analysis were extracted by organic solvent extraction process and were analyzed for pesticides in both the supernatant liquor and the suspended solid fractions. The results of their study suggest that many chlorinated hydrocarbon pesticides were degraded under suitably active anaerobic conditions. "Degradation of most of the chlorinated hydrocarbon pesticides studied was more rapid under anaerobic than under corresponding aerobic conditions; exceptions were heptachlor epoxide and probably dieldrin, which were very persistent in both environments." The authors ranked the compounds in an order of increasing persistence under anaerobic conditions. It was "lindane, heptachlor, endrin, DDT, DDD, aldrin, heptachlor epoxide and dieldrin." The increase of temperature from 20 to 35° did not appear to increase the rate of pesticide degradation except in the case for anaerobic degradation of lindane and DDT. The authors commented that the adsorption of the pesticides was "inversely related to their solubilities", a conclusion which has also been reached by a number of other investigators.

In order to determine if DDT could be degraded anaerobically by soil microflora and what the decomposition products were, Guenzi and Beard (1967) added C^{14} labeled DDT to soils which were wetted at a 28% level of water. These were incubated anaerobically in an atmosphere of 20% CO_2 and 80% nitrogen at 30°C. The DDT was dechlorinated to DDD by soil microorganisms under anaerobic conditions, and to determine whether or not this conversion was microbial, a complete set of soil samples was sterilized by autoclaving and dually incubated. DDT was not degraded in these soils, suggesting that the conversion was due to the action of microorganisms. The DDT appeared to be converted directly to DDD with no other appreciable build-up of by-products detected. There was some evidence that there were further degradation products which were water soluble.

An anaerobic bacterium was isolated in pure culture from samples of flooded soils in which lindane was the only added carbon source in the enrichment medium (MacRae et al., 1969). The bacterium was a Gram negative anaerobic spore forming rod-shaped organism which they classify as a *Clostridium* and create a small point of confusion. In order to determine lindane degradation, aliquots of a cell suspension and a solution of lindane in phosphate buffer were placed in separate compartments of a two compartment glass apparatus. After deaeration, the contents were mixed and incubated. Samples for lindane analysis were removed hourly for 6 hr and again after 27 hr. The lindane was extracted with *n*-hexane and determined quantitatively with the use of the gas chromatograph. Inorganic chloride in the supernatant was determined by a microdiffusion method. The cell suspension was active in the degradation of lindane and the ability of the organism to utilize lindane also brought about the release of the covalently linked chlorine of the lindane molecule as chloride ion. Authors have evidence that there was microbial degradation of the insecticide based upon the observation that the bacterium caused a marked loss of lindane from anaerobically incubated reaction mixtures consisting of bacteria, phosphate buffer, and lindane, but not in uninoculated controls. These conclusions are substantiated by the concommitant release of covalently linked chlorine as chloride ion.

A recent effort was designed to test the ability of alfalfa volatiles which are active in increasing soil respiration to accelerate anaerobic decomposition of DDT in soil. Burge (1971) was interested in determining whether or not the breakdown or decomposition of DDT in soil was truly a microbial process and one not just governed by chemical interactions of the complex components in the soil. The anaerobic degradation of DDT in soils was accelerated by glucose and other volatiles found in alfalfa. The order of the effectiveness of the substances was "acetaldehyde = isobuteraldehyde is greater than ethanol is greater than glucose is much greater than methanol." The process itself (DDT degradation) appears sensitive to oxygen and is microbial at least in the conversion to DDD. The authors concluded that it could be that the first step in the conversion of DDT to DDD or DDE involves a complex which reacts with microbial protoplasm or other components of the soil, mineral, or organic matter. It may be useful that additions of compounds like those found in the ground alfalfa to the soil may be useful in decontamination processes for DDT.

PERSISTENCE OF PESTICIDES

Carbon 14 labeled aldrin and dieldrin were both applied to sterile nutrient agar and to empty sterile glass petri plates. Both compounds disappeared rapidly from the agar even under a glass covered petri dish. Much of the evaporating pesticide could be recovered on corn oil soaked strips of filter paper which were placed around and atop the petri dishes in order to collect the vapors. In most of the situations this disappearance of these pesticides was retarded from agar which had been inoculated with either fungal or bacterial cultures. When microorganisms were involved, dieldrin could also be recovered. The rate of volitilization from the agar surface was rapid and occurred at a rate of 50% during the first day of incubation.

Aldrin, dieldrin, DDT, and lindane along with four organophosphorus insecticides were examined for their rate of volitilization from various substrates (Lichtenstein and Schulz, 1970). Substrates were either glass beads 150 microns in diameter, water, or a silt loam soil. The most volatile insecticide was aldrin and the highest volitilization rates were observed from water while relatively small amounts of insecticides volatilized from soil. The addition of soil, algae, or a detergent to the water reduced the volatility of a specific compound in many cases.

THE EFFECTS OF PESTICIDES ON PHOTOSYNTHESIS

The effect of DDT on photosynthesis by marine phytoplankton species was examined by Wurster (1968). A diatom *Skeletonema costatum*, the coccolithophore *Coccolithus huxleyi*, the green alga *Pyranimonas* sp., and the neritic dinoflagellate *Peridinium trochoideum* represented the four main classes of organisms important as food sources in the ocean. The 4 laboratory cultures were grown axenically in an enriched sea water medium in 50 ml portions under fluorescent illumination. Various concentrations of DDT were added to each flask. After 24 hr C^{14} bicarbonate was added and the algae were illuminated for an additional 4 to 5 hr. A few controls were run in darkness. Radioactivity retained by the filtered cells at the conclusion of the experiment is related to the amount of carbon fixed by photosynthesis. The results indicated that even a few ppb of DDT in water was capable of reducing photosynthesis in these four species of coastal and oceanic phyto plankton. Similar results were obtained with a natural phytoplankton community from Woods Hole, Massachusetts. The accurate evaluation of any ecological significance of this is difficult, but it is possible that any imbalance or process that would aggravate the already burdening problem of eutrophication would be of importance to understand. The involvement and intertwining of these effects may be insidious and their causes obscure.

Stadnyk et al. (1971) measured the changes in cell biomass, cell number and C^{14} assimilation in low density populations of the fresh water alga *Scenedesmus quadricaudata*. The pesticide diuron, carbaryl, 2-4D, DDT, dieldrin, toxophene and diazinon were examined under static conditions. Pesticides in concentrations of 0.1 and 1 mg/l. were used. In the case of the diuron, there was a sizable reduction in cell number after the second day of exposure which continued throughout the eight day test period. The decline in growth was reflected with a conspicuous decrease in biomass and the suppression of carbon assimilation. All the other compounds tested had some slight effect on the organism tested, with the exception of diazinon. The results of these workers may suggest that long-term chronic effects of pesticides within the ecosystem should be more properly evaluated. While small additions of organo chlorine insecticides may cause a reduction in population density, and a suppression in carbon fixation, the recovery of cell number does occur. The interference of a lowered production at all trophic levels by repeated stimulation of pesticide must be understood.

EFFECTS ON NITROGEN METABOLISM

Laboratory studies on the influence of insecticides on some of the more important soil microbial functions and results on studies made on bulk soil samples treated with BHC and DDT have been given by Bollen et al. (1954). Their experiments suggest that the gamma isomer of BHC increased the bacterial population although *Streptomyces* decreased. Other isomers of BHC gave smaller increases. Nitrification of ammonium sulfate appeared to increase with the beta and gamma isomers of BHC and there was no difference in the BHC and DDT treated soil as compared to untreated soil with respect to CO_2 evolution and dextrose decomposition. The gamma isomer of BHC was quantitatively and qualitatively more effective than the other isomers in causing a response by soil microorganisms.

The pesticide concentrations damaging to soil nitrification and the relative susceptibilities to pesticide damage of plants were determined by Shaw and Robinson (1960). A soil was treated with aldrin, heptachlor, 2-4D, and chlordane and exposed to weather for about six months. Leachates were analyzed for ammonium, nitrite, and nitrate nitrogen. The results of this experiment indicated there was no effect on nitrification with an 8 lb per acre treatment. Subsequently, pesticide concentrations were increased. Results of the final experiments where rates of 10 to 100 lb per acre were added showed that in the soil that was tested, there was no inhibition of nitrification.

It had been found by a number of investigators that in general, herbicides and insecticides, when used at recommended rates in the fields, appear to have no harmful effects on the total microbial population or on its metabolic activities. *Nitrobacter* had been shown to be inhibited by monuron (a substituted urea herbicide in concentrations similar to those used in soils (Casely and Luckwil, 1965). *Nitrosomonus* was inhibited by *N*-serve (2-chloro-6-(trichloromethyl)-pyridine) (Shattuck and Alexander).[2] With these problems in mind, Garretson and San Clemente (1968) determined the inhibitory concentrations of the insecticides aldrin, lindane and DDD on the oxidation of NH_4-N and NO_2-N by *Nitrosomonas europea* and *Nitrobacter agilis* in pure cultures throughout a two week exposure to the compounds. The insecticides were used at final concentrations ranging from 0.1 to 1000 $\mu g/ml$ and the effect on nitrifying organisms was determined by the degree of oxidation of the nitrogen supplied in the culture medium. Chemolithotrophic nitrifiers were sensitive to low concentrations (less than 10 $\mu g/ml$) to several insecticides. Aldrin, lindane, and DDD in concentrations of 1 $\mu g/ml$ inhibited nitrification by *Nitrobacter agilis*. The authors warned that there is temptation to predict toxicity in field situations

on the basis of laboratory data, but that factors such as adsorption on soil colloids, solubility, and other physical and chemical parameters must be taken into account. This is excellent advice and should be considered by any one interested in this field.

Winely and San Clemente (1970) examined the mode of action of several pesticides on the growth of *Nitrobacter agilis* in aerated cultures and on respiration in cell suspensions. During the study the pesticides that were tested were "aldrin, CIPC, DDD, eptam, heptachlor, lindane, and simazine." The range of pesticide concentrations that were tested was between 1 and 250 $\mu g/ml$. Active cultures (400 ml) were used to inoculate 10 l. fermentors for the production of large volumes of cells. Five of the compounds studied, CIPC, chlordane, DDD, heptachlor, and lindane did interfere with growth at a concentration of 10 $\mu g/ml$. CIPC and eptam appeared to inhibit nitrite oxidation by cell suspensions and the addition of DDD and lindane resulted in a partial inhibition of the oxidation. This was similarly true with heptachlor and chlordane which were more toxic with cell free extract nitrite oxidase. None of the pesticides examined inhibited nitrate reductase activity in the cell free extracts but did cause repression of cytochrome c oxidase activity. The inhibition caused by these pesticides was not characterized, but results indicated that growth studies with low concentrations of pesticides were more suitable than measurements of nitrite oxidation for examining toxicity in *Nitrobacter agilis*. The authors pointed out "with growth studies, inhibition of biosynthetic reactions will be detected whereas only inhibition of energy assimilation is detected by nitrite oxidation studies."

In more recent experiments (Winely and San Clemente, 1971), the effects of CIPC and eptam on oxidative phosphorylation coupled with nitrite oxidation by cell free extracts of *Nitrobacter agilis* were studied. In order to examine whether the electron transport system was affected by these herbicides, $NADH_2$ oxidation was also examined. These experiments demonstrated that CIPC and eptam both had an uncoupling effect on the oxidative phosphorylation linked to nitrite oxidation. Nitrite oxidation was inhibited and even though this was the case, "an intensely severe inhibition of phosphorylation caused a 50% reduction in the T/O ratio." The classical phosphorylation uncoupler 2, 4-dinitrophenol (DNP) also affected the oxidative phosphorylation in *Nitrobacter*. These researchers concluded that the two herbicides and dinitrophenol had an analogous mode of inhibition in this organism, and that the herbicides do outwardly affect the electron transport because $NADH_2$ oxidase activity was not affected by the additions of either of the herbicides.

SUMMARY

In the early experiments (ca. 1950), it was the general conclusion that there were no effects of halogenated compounds on microorganisms. There seemed to be no damage to ammonification, accumulation of nitrate, or other normal soil parameters. In addition, it was noticed that certain compounds, e.g. DDT, were remarkably stable over a period of years.

While some reports claimed that chemicals such as aldrin or isomers of BHC had an effect on soil organisms, it was felt that the presence of such compounds in the quantities applied to the soil were of little consequence as far as soil fertility was concerned. Experiments by a number of investigators did imply that microorganisms may be stimulated by the addition of these compounds to soil.

Later, it was discovered that DDT could be broken down (generally dechlorinated) to DDD and this appears to be most effective under anaerobic conditions, especially when organics such as alfalfa have been added. These additions are not as important in any aerobic degradation process. The presence of DDD in soil may be more significant, since it has a broader effect on microorganisms and is more stable than DDT. When halogenated pesticides are reacted with other compounds, or other pesticides, the results may be surprising (Bartha, 1969). Pesticides may be adsorbed strongly to soil particles, or, as is the case with milk, bound to the colloidal fraction, and not be effective against any microbial group. It does appear that such pesticides in milk may not harm the resident microorganisms, but there does remain much experimentation necessary to elaborate any other more subtle or lasting effects.

In the aquatic environment, Dursban has been shown to have some lasting effect at least over a period of several weeks on the total zooplankton population. Other types of marine microorganisms such as flagellates, diatoms, and cocolithoforids may be capable of storing DDT, endrin, and dieldrin, and be affected by their presence. This could lead to an influence in the dominance or succession of individual forms and have far-reaching consequences for aquatic life and the transfer of energy through trophic levels.

A variety of aquatic fungi have also been shown to be affected by DDT. The presence of this chemical in water could have ecological significance through an effect on saprophytes and parasites of these types. A survey of the work done to date does reflect the fact that there is a great deal of confusion as to just what these halogenated compounds do or do not do in soil and water. Whether they are really important in altering microbial ecology is only hinted at, and much needs to be learned to actually evaluate their presence and their effect on our future.

As far as the metabolism of halogenated compounds is concerned, a beginning has been made in learning what may occur under specific sets of conditions. Initially, microbiologists were confident that the bacteria would be able to degrade man-made pesticides of the chlorinated hydrocarbon type. As Alexander[3] has pointed out so eloquently, microorganisms are fallible, and consequently pesticide residues of DDT, DDD, dieldrin, endrin, etc. have been accumulating in our environment.

It appears that the molecular structure and the distribution of halogen atoms on the molecule are very important in determining whether or not the compound may be susceptible to microbial breakdown. In general, an increase in the oxidative state of the chlorinated pesticide makes it more suitable to microbial attack.

Recently, recalcitrant molecules, such as dieldrin, have been shown capable of degradation and release of metabolites which are not natural products. The toxocological effects of these metabolites need to be evaluated in various ecosystems.

The pesticide DDT has been shown to be broken down to DDD under a variety of conditions, and compounds such as heptachlor have been shown to produce 1-hydroxychlordene, heptachlor epoxide, chlordene, 1-hydroxy 2, 3-epoxychlordene, and at least one other unknown product. A high proportion of soil fungi, *Actinomycetes* and bacteria were able to convert the epoxide from heptachlor. Several microbes (20 cultures) capable of degrading dieldrin were not able to break down DDT, Baygon, or lindane, but could degrade endrin into ketones and aldehydes with 5 to 6 chlorine atoms. One such metabolite has been identified as ketoendrin.

One important phenomenon, which may be very important from an ecological point of view, is that of cometabolism. When the factors leading to an understanding of which compounds may be involved in cometabolic degradation are better understood, then the total picture of coinvolvement in the soil and water ecology will be improved.

Cell-free extracts of the soil bacterium *Arthrobacter* contained an enzyme which showed a substrate specificity with 9 chlorinated aliphatic acids. The biotransformation of 2, 4-D and related molecules has been the most extensively studied. The reference source by Menzie[4] lists about 50 papers concerned with this topic. The degradation of this and other chlorophenols to catechols and the involvement of oxygen and NADPH have been documented. A general conclusion reached is that chlorinated or halogenated molecules in the environment are able to be broken down at least partially and in some cases to small biologically active molecules. It is most important to examine the type of environment that the refractory compounds reside in.

It has been known for some time that higher plants and animals store recalcitrant pesticides for an unknown duration. It is also becoming apparent that microorganisms can store such compounds in one form or another. Whether this storage ability is actually inside the cell or merely the entrapment of halogenated compound in polymer matrices on the external cell surface remains to be seen. In either case, extensive surface area that the microbial system presents for adsorption places them in an important position with respect to a larger ecology.

The area of "associative action" of microorganisms or the interaction of either virus or toxin and pesticides is one of serious interest. These complex but delicate interactions which may affect our biosphere need further clarification.

More recent work on the photosynthetic process and on the ability of soil organisms to inhibit nitrite oxidation only leads one to believe that these interactions are serious since they do touch upon basic energy transfer mechanisms.

The general conclusion which can be reached after reviewing the literature in the field of microbial-halogenated pesticide interactions is that initially, these compounds appeared to be innocuous, with little effect on our health and welfare. As the techniques of biology have become more sophisticated and instrumentation developed to detect more sensitive reactions, there is the realization that these chemicals are not inert to the microbial world and that they may be interacting in an insidiously slow manner changing our already burdened environment. There is sadness in this discovery because the presence and use of these pesticides has done immense good both to crops for foods and to health in the areas of the world where much human suffering exists. There must be programs and interested people to evaluate and conduct further experiments in this interesting and vital area of biological-chemical sciences.

For further detailed information into this interesting subject there are a number of pertinent reviews[3,5-10] which will be most helpful.

LIST OF HALOGENATED PESTICIDES

Acarol[R] – (Isopropyl bromobenzilate) (Isopropyl 2,2-bis(p-bromophenyl glycolate)
Afalon – (Linuron), (N-(2,4-dichlorophenyl-N'-methyl-N'-methoxyurea)
Aldrin – 1,2,3,4,10,10 – Hexachloro-1,4,4a,5,8,8a-hexahydro-1,4-endo, exo 5,8-dimethanonaphthalene
Allisan[R] – (Dichloran, Botran, Ditranil, DCNA) (2,6-Dichloro-4-nitroaniline)
Amiben – (3-Amino-2,5-dichlorobenzoic acid)
Aresin – (Monolinuron), (N(-4-chlorophenyl)-N'-methyl-N-methoxyurea)
Banol[R] – (2-Chloro-4,5-dimethylphenyl N-methylcarbamate)
Banvel D[R] – (Dicamba), (2-Methoxy-3,6-dichlorobenzoic acid)
Barban – (4-Chloro-2-butynyl N-(3-chlorophenyl)carbamate)
Bayer 73[R] – (Ethanolamine salt of N-(2-chloro-4-nitrophenyl)-5-chloro-salicylamide)
Bayer 9015[R] – (3, 3'-Dichloro-5,5'-Dinitro-,'-Biphenol)
Benzene hexachloride – (BHC), (1,2,3,4,5,6-Hexachlorocyclohexane), (Lindane = y-BHC)
Birlane – (Chlorfenvinphos, Supona, SD 7859 GC-4072) (2-Chloro-1-(2, 4-dichlorophenyl)vinyl   diethyl phosphate)
Botran[R] – (Dichloran. Ditranil Allisan, DCNA))
Bromophos – (O-(4-Bromo-2,5-dichlorophenyl), O,O-dimethylphosphorothioate)
Bromoxynil – (3,5-Dibromo-4-hydroxybenzonitrile)
Butonate – (Tribuphon), (O,O-Dimethyl-1-1 butyryloxy-2 2 2 trichloroethylphosphate)
Captan – (N-Trichloromethylthio-4-cyclohexene-1,2-dicarboximide)
Carbophenothion – (Trithion), (O,O-Diethyl S-(p-chlorophenylthio) methyl phosphorodithioate)
Casoron[R] – (Dichlobenil), (2,6-Dichlorobenzonitrile)
CCC[R] – (Cycocel), (2-(Chloroethyl) Trimethylammonium Chloride)
CDAA – (a-chloro-N, N-diallylacetamide)
CDEC – (Vegedex), (2-Chloroallyl-N, N-diethyldithiocarbamate)
CEPA – (Amchem 66-329), (2-Chloroethylphosphonic Acid)
CEPC – (2-Chloroethyl-N-(3-chlorophenyl) carbamate)
Chlorate
Chlordane (Chlordan), (1,2,4,5,6,7,8,8-Octachloro-2,3,3a,4,7,7a-hexahydro-4,7-methanoindene)
Chlordane[R] – (Chlordene, Dihydrochlordene, Dihydroheptachlor, Heptachlor)
Chlordene – (Chlordane, Chlordene, Dihydrochlordene, Dihydroheptachlor, Heptachlor) (4,5,6,7,8,8- (4,5,6,7,8,8-Hexachloro-3a,4,7,7a,tetrahydro-4, 7-methanoindene)
Chlorobenzene
Chlorobiphenyl – (Biphenyl, 4-Chlorobiphenyl, and 2-Hydroxybiphenyl)
2-Chloro-N-Isopropylacetanilide
N-(3-Chloro-4-methylphenyl)-2-methylpentamide
Chloroneb – (1, 4-Dichloro-2, 5-dimethoxybenzene)
p-Chlorophenyl p-chlorobenzenesulfonate
Chlorothion – (O, O-Dimethyl O-(3-chloro-4-nitrophenyl) phosphorothioate)
Chlorothion, isomeric – (Dicapthon. Dicaptan), (O-(2-chloro-4-nitrophenyl) O,O-dimethyl phosphorothioate)
Chlorthiamid Prefix – (2, 6-Dichlorobenzenethioacetamide)
Chloroxuron – (Tenoran), (N-(p-Chlorophenoxyphenyl)-N',N'-dimethylurea)
Ciba C-9491 – (O-(2, 5-Dichloro-4-iodophenyl), O, O-dimethylphosphorothioate)
CIPC – (Chloropropham), (Isopropyl N-(3-chlorophenyl) carbamate)
Citicide – (Polychlorinated turpentine)
Cotoran[R] – (Fluometuron) (3-(m-trifluoromethylphenyl)-1,1-dimethylurea)
Coumaphos – (Coral, Bayer   21/199,   muscatox),   (O,   O-Diethyl-O-(3-chloro-4-methylumbelliferone) phosphorothioate)
2- and 4-CPA – (2- and 4-Chlorophenoxyacetic Acid)
4-(2-CPB) – (4-(2-Chlorophenoxy)butyric Acid) (I)

4-(3-CPB) — (4-(3-Chlorophenoxy)butyric Acid) (II)

4-(4-CPB) — (4-(4-Chlorophenoxy)butyric Acid) (III)

Cyclophosphamide — (N, N-bis (2-chloroethyl)-N',O-propylenediamide)

Cycocel — (CCC) (2-(Chloroethyl) Trimethylammonium Chloride)

2, 4-D — (2,4-Dichlorophenoxyacetic acid)

Daconil® — (TCIN) 2,4,5,6-Tetrachloroisophthalonitrile

Dacthal® — (2,3,5,6-Tetrachlorodimethylphthalate)

Dalapon — (2,2-Dichloropropionic acid)

4-(2, 4-DB) — (4-(2,4-Dichlorophenoxy)butyric acid)

DCPA — (Propanil. Stam-F 34, DPA), (N-(3, 4-Dichlorophenyl)propionamide)

DDC® — (Dimethyldithiocarbamate), Ferric Salt (Ferbam), Sodium Salt (NaDDC), Zinc Salt (Ziram)

DDD — (TDE, Rhothane), (1,1-Dichloro-2,2-bis (chlorophenyl) ethane)

DDP — (di-n-propyl-2, 2-dichlorvinyl phosphate)

DDT — (1,1,1-Trichloro-2, 2-bis (p-chlorophenyl) ethane)

DDVP — (Dichlorvos Vapona), (2,2-dichlorovinyl dimethyl phosphate)

DEP — (Falone), (2,4-Dichlorophenoxyethyl phosphite)

Dibrom® — (1,2-dibromo-2,2-dichloroethyl dimethyl phosphate)

Dichlone — (Phygon), (2,3-Dichloro-1, 4-naphthoquinone)

Dichlorobenzene — (DCB)

3, 4-dichlorobenzyl N-methylcarbamate

1, 3-Dichloropropene, cis- and $trans$-

Dicofol — (1,1-bis (p-chlorophenyl)-2,2,2-trichloroethanol), (Kelthane)

Dicryl — (N-(3, 4-dichlorophenyl)-methacrylamide)

Dieldrin — (1.2.3.4.10.10-hexachloro-6,7-epoxy-1,4,4a,5,6,7,8,8a octa hydro- 1,4-endo, exo 5,8-dime-thanonaphthalene)

Dihydrochlordene — (Chlordane, Chlordene, Dihydrochlordene, Dihydroheptachlor, Heptachlor)(4,5,6,7, 8,8-hexachloro-1,2,3a,4,7,7a-hexahydro-4,7-methanoindene)

Dihydroheptachlor — (Chlordane, Chlordene, Dihydrochlordene, Dihydroheptachlor, Heptachlor, DHC) (3,4,5,6,7,8,8-heptachloro-1,2,3a,4,7,7a-hexachloro-4,7-methanoindene)

Dimefox — (DMF) (bis(dimethylamido)phosphoryl fluoride)

Dinoben — (2, 5-Dichloro-3-nitrobenzoic acid)

Dipterex — (Trichlorfon, Bayer L 13/59) (O, O-Dimethyl 2,2,2-Trichloro-1-Hydroxyethyl Phosphonate)

Diquat — (1, 1'-Ethylene-2, 2'dipyridinium dibromide)

Diuron — (3-(3,4-dichlorophenyl)-1,1-dimethylurea)

Dursban® — (3,5,6-Trichloro-2-pyridyl phosphorothioate)

Dyrene® — (Kemate), (2,4-Dichloro-6-anilino-s-triazine)

Endosulfan — (Thiodan), (6,7,8,9,10,10-Hexachloro-1,5,5a,6,9,9a-hexahydro 6,9-methano-2,4,3-benzo-dioxathiepin-3-oxide)

Endrin — 1,2,3,4,10,10-Hexachloro-6,7,epoxy-1,4,4a,5,6,7,8,8a-octahydro-endo-,1,4-endo, 5,8-dime-thanonaphthalene

Ethylene bromohydrin — (1-Bromo-2-hydroxyethane)

Ethylene dibromide — (EDB) (1, 2-Dibromoethane)

Ethylene dichloride — (1, 2-Dichloroethane)

Fluoroacetamide — (F-CH$_2$-$\overset{O}{\underset{C}{}}$-NH$_2$)

Fluoroacetanilides

Fluoroacetate — (F-CH$_2$-$\overset{O}{\underset{C}{}}$-ONa)

Folpet — (N-(Trichloromethylthio) phthalimide)

Griseofulvin — (7-chloro-4:6:2'-trimethoxy-6'-methylgris-2'-en-3:4'-dione) (7-chloro-4, 6-dimethyoxycou-maran-3-one-2-spiro-1'-(2'-methoxy-6'methylcyclohex-2'-en-4'-one))

Heptachlor — (Chlordane, Chlordene, Dihydrochlordene, Dihydroheptachlor, Heptachlor) (1,4,5,6,7,8 8-Heptachloro-3a,4,7,7a-tetrahydro-4, 7-methanoindene)

Ipazine® — (2-Chloro-4-diethylamino-6-isopropylamino-s-triazine)

Karsil — (N-(3, 4-Dichlorophenyl)-2-methylpentamide)

MCPA — (4-Chloro-2-methylphenoxyacetic acid)

MCPB — (4-(4-Chloro-2-methylphenoxy) butyric acid)

Methoxychlor — (1, 1, 1-Trichloro-2,2-bis-(p-methoxyphenyl)ethane)

2-Methoxy-3, 6-Dichlorobenzoic Acid

Methyl Bromide

Metobromuron — (Patoran) (3-p-Bromophenyl)-1-methoxy-1-methylurea)

Monuron — (N-(4-chlorophenyl)-N', N'-dimethylurea)

N-Serve — (2-Chloro-6-(trichloromethyl)pyridine)

OCS-21693 — (Methyl 2,3,5,6-Tetrachloro-N-Methoxy-N-Methylterephthalamate)

Paraquat — (1, 1'-Dimethyl-4, 4'-dipyridinium di(methyl sulfate))

PCNB — (Pentachloronitrobenzene)

PCP® — (Pentachlorophenol)

Phosalone — (O, O-Diethyl S-(6-Chloro-2-oxo-benzoxazol-3-ylmethyl) Phosphoro-dithioate)

Phosphamidon (N, N-diethyl 1-Chloro-2-dimethylphosphate-prop-2-enamide)

Picloram — (Tordon) (4-Amino-3, 5, 6-trichloropicolinic acid)

Prolan — (Bis(p-chlorophenyl)-2-nitropropane)

Propazine® — (2-Chloro 4, 6-bis(isopropylamino)-s-triazine)

Pyramin® — (Pyrazon, Pyrazonl, PCA, HS-119) (5-Amino-4-chloro-3-oxo-2-phenyl-(2H)-pyridazine)

Ronnel — (Trolene) (O, O)-Dimethyl O-(2, 4, 5-trichlorophenyl)phosphorothioate)

Ruelene® — (4-tert-butyl-2-chlorophenyl methyl methylphosphoroamidate)

SD 8447 — (2-Chloro-1-(2, 4, 5-trichlorophenyl)vinyl dimethyl phosphate)

Sesone® — (2, 4-Dichlorophenoxyethylsulfate)

Silvex® — (2, 4, 5-TP) (2-(2, 4, 5-Trichlorophenoxy)propionic Acid)

Simazine® — (2-Chloro-4, 6-bis (ethylamino)-s-triazine)

Sulfuryl Fluoride — ($SO_2 F_2$)

SWEP — (Methyl N-(3, 4-dichlorophenyl) carbamate)

2, 4, 5-T — (2, 4, 5-Trichlorophenoxyacetic Acid)

2,4,6-T — (2,4,6-Trichlorophenoxyacetic Acid)

2, 4, 5-TB — 3-(2, 4, 5-Trichlorophenoxy) butyric acid

TCA — (Trichloroacetate)

TCNB — Tetrachloronitrobenzene

Telodrin® — (1,3,4,5,6,7,8,8-Octachloro-3a, 4,7,7a-tetrahydro-4,7-methanophthalan)

TIBA — (2, 3, 5-Triiodobenzoic Acid)

TOK® — (2, 4-Dichloro-4'-nitrodiphenyl ether)

2, 3, 6-Trichlorobenzoic Acid

Trifluralin — (a, a, a-Trifluoro-2, 6-dinitro-N, N-di-n-propyl-p-toluidine)

Vapam® — (Metham Sodium) (Sodium N-methyldithiocarbamate)

VC-13® — (O-(2, 4-Dichlorophenyl)O,O-diethyl phosphorothioate)

Zytron® — (O-(2, 4-Dichlorophenyl) O-methyl N-isopropylphosphoramidothioate)

ARTICLES REVIEWED

Audus, L. J. and Symonds, K. V., Further studies on the breakdown of 2:4-dichlorophenoxyacetic acid by a soil bacterium, *Ann. Appl. Biol.*, 42, 174, 1955.

Bollen, W. B., Morrison, H. E., and Crowell, H. H., Effect of field treatments of insecticides on numbers of bacteria, Streptomyces, and molds in the soil, *Entomol.*, 47, 302, 1954.

Barker, P. S., Morrison, F. O., and Whitaker, R. S., Conversion of DDT to DDD by *Proteus vulgaris*, a bacterium isolated from the intestinal flora of a mouse, *Nature*, 205, 621, 1965.

Bartha, R., Pesticide interaction creates hybrid residue, *Science*, 66, 1299, 1969.

Bollag, J. M., Helling, C. S., and Alexander, M., 2,4-D metabolism: enzymatic hydroxylation of chlorinated phenols, *J. Agr. Food Chem.*, 16, 826, 1968.

Bollag, J. M., Briggs, G. G., Dawson, J. E., and Alexander, M., 2,4-D metabolism: enzymatic degradation of chlorocatechols, *J. Agr. Food Chem.*, 16, 829, 1968.

Bollen, W. B., Morrison, H. E., and Crowell, H. H., Effect of field and laboratory treatments with BHC and DDT on nitrogen transformations and soil respiration, *J. Econ. Entomol.*, 47, 307, 1954.

Braunberg, R. C. and Beck, V., Interaction of DDT and the gastrointestinal microflora of the rat, *J. Agr. Food Chem.*, 16, 451, 1968.

Bounds, H. C., Magee, L. A., and Colmer, A. R., The reversal of dalapon toxicity by microbial action, *Canad. J. Microbiol.*, 15, 1121, 1969.

Burge, W. D., Anaerobic decomposition of DDT in soil. Acceleration by volatile components in alfalfa, *J. Agr. Food Chem.*, 19, 375, 1971.

Chacko, C. I., Lockwood, J. L., and Zabik, M., Chlorinated hydrocarbon pesticides: degradation by microbes, *Science*, 154, 893, 1966.

Chacko, C. I. and Lockwood, J. L., Accumulation of DDT and dieldrin by microorganisms, *Canad. J. Microbiol.*, 13, 1123, 1967.

Collins, J. A. and Langlois, B. E., Effect of DDT, dieldrin, and heptachlor on the growth of selected bacteria, *Appl. Microbiol.*, 16, 799, 1968.

Cowly, G. T. and Lichtenstein, E. P., Growth inhibition of soil fungi by insecticides and annulment of inhibition by yeast extract or nitrogenous nutrients, *J. Gen. Microbiol.*, 62, 27, 1970.

Cox, J. L., DDT residues in marine phytoplankton: increase from 1955 to 1969, *Science*, 170, 71, 1970.

Cox, J. L., Low ambient level uptake of ^{14}C-DDT by three species of marine phytoplankton, *Bull. Environ. Contam. Toxicol.*, 5, 218, 1970.

Dalton, S. A., Hodkinson, M., and Smith, K. A., Interactions between DDT and river fungi. I. Effects of p, p'-DDT on the growth of aquatic hyphomycetes, *Appl. Microbiol.*, 20, 662, 1970.

Dougherty, E. M., Reichelderfer, C. F., and Faust, R. M., Sensitivity of *Bacillus* var. *thuringiensis* to various insecticides and herbicides, *J. Invert. Path.*, 17, 292, 1971.

Duxbury, J. M., Tiedje, J. M., Alexander, M., and Dawson, J. E., 2,4-D metabolism: enzymatic conversion of chloromaleylacetic acid to succinic acid, *J. Agr. Food Chem.*, 18, 199, 1970.

Eno, C. F. and Everett, P. H., Effects of soil applications of 10 chlorinated hydrocarbon insecticides on soil microorganisms and the growth of stringless black valentine beans, *Soil Sci. Soc. Am. Proc.*, 22, 235, 1958.

Faulkner, J. K. and Woodcock, D., Metabolism of 2, 4-dichlorophenoxyacetic acid ('2,4-D') by *Aspergillus niger* van Tiegh, *Nature*, 203, 865, 1964.

Fletcher, D. W. and Bollen, W. B., The effects of aldrin on soil microorganisms and some of their activities related to soil fertility, *Appl. Microbiol.*, 2, 349, 1954.

French, A. L. and Hoopingarner, R. A., Dechlorination of DDT by membranes isolated from *Escherichia coli*, *J. Econ. Entomol.*, 63, 756, 1970.

Focht, D. D. and Alexander, M., DDT metabolites and analogs: ring fission by *Hydrogenomonas*, *Science*, 170, 91, 1970.

Garretson, A. L. and San Clemente, C. L., Inhibition of nitrifying chemolithotrophic bacteria by several insecticides, *J. Econ. Entomol.*, 61, 285, 1968.

Gray, P. H. H. and Rogers, C. G., Effects of benzenehexachloride on soil microorganisms. IV. Benzenehexachloride-resistant bacteria from virgin soils, *Canad. J. Microbiol.*, 1, 312, 1954.

Guenzi, W. D. and Beard, W. E., Anaerobic biodegradation of DDT to DDD in soil, *Science*, 156, 1116, 1967.

Henzel, R. F. and Lancaster, R. J., Degradation of commercial DDT in silage, *J. Sci. Food Agr.*, 20, 499, 1969.

Hill, D. W. and McCarty, P. L., Anaerobic degradation of selected chlorinated hydrocarbon pesticides, *J. Water Pollut. Contr. Fed.*, 39, 1259, 1967.

Horvath, R. S., Cometabolism of the herbicide 2,3,6-trichlorobenzoate, *J. Agr. Food Chem.*, 19, 291, 1971.

Hurlbert, S. H., Mulla, M. S., Keith, J. O., Westlake, W. E., and Dusch, M., Biological effects and persistence of Dursban in freshwater ponds, *J. Econ. Entomol.*, 63, 43, 1970.

Johnson, B. T., Goodman, R. N., and Goldberg, H. S., Conversion of DDT to DDD by pathogenic and saprophytic bacteria associated with plants, *Science*, 157, 560, 1967.

Jones, L. W., Stability of DDT and its effect on microbial activities of soil, *Soil Sci.*, 73, 237, 1952.

Jones, L. W., Effects of some pesticides on microbial activities of the soil, Bulletin 390, Utah State Agricultural College, Division of Agricultural Sciences, Agricultural Experiment Station, 1956. 1956.

Kallman, B. J. and Andrews, A. K., Reductive dechlorination of DDT to DDD by yeast, *Science*, 141, 1050, 1963.

Kapoor, I. P., Metcalf, R. L., Mystrom, R. F., and Sangha, G. K., Comparative metabolism of methoxychlor, methiochlor, and DDT in mouse, insects, and a model ecosystem, *J. Agr. Food Chem.*, 18, 1145, 1970.

Kearney, P. C., Kaufman, D. D., and Beall, M. L., Enzymatic dehalogenation of 2,2-dichloropropionate, *Biochem. Biophys. Res. Commun.*, 14, 29, 1964.

Keil, J. E. and Priester, L. E., DDT uptake and metabolism by a marine diatom, *Bull. Environ. Contam. Toxicol.*, 4, 169, 1969.

Kim, S. C. and Harmon, L. G., Relationship between some chlorinated hydrocarbon insecticides and lactic culture organisms in milk, *J. Dairy Sci.*, 53, 155, 1970.

Ko, W. H. and Lockwood, J. L., Accumulation and concentration of chlorinated hydrocarbon pesticides by microorganisms in soil, *Canad. J. Microbiol.*, 14, 1075, 1968.

31

Ko, W. H. and Lockwood, J. L., Conversion of DDT to DDD in soil and the effect of these compounds on soil microorganisms, *Canad. J. Microbiol.*, 14, 1069, 1968.

Kokke, R., DDT: its action and degradation in bacterial populations, *Nature*, 226, 977, 1970.

Langlois, B. E., Collins, J. A., and Sides, K. G., Some factors affecting degradation of organochlorine pesticides by bacteria, *J. Dairy Sci.*, 53, 1671, 1970.

Langlois, B. E. and Collins, J. A., Growth of selected bacteria in liquid media containing DDT, dieldrin, and heptachlor, *J. Dairy Sci.*, 53, 1666, 1970.

Langlois, B. E. and Collins, J. A., Factors affecting the inhibition of growth of *Staphylococcus aureus* by heptachlor, *Pestic. Symp.*, 235, 1970.

Langlois, B. E., Reductive dechlorination of DDT by *Escherichia coli*, *J. Dairy Sci.*, 50, 1168, 1967.

Ledford, R. A. and Chen, J. H., Degradation of DDT and DDE by cheese microorganisms, *J. Food Sci.*, 34, 386, 1969.

Leshniowsky, W. O., Dugan, P. R., Pfister, R. M., Frea, J. I., and Randles, C.I., Aldrin: removal from lake water by flocculent bacteria, *Science*, 169, 993, 1970.

Lichtenstein, E. P. and Schulz, K. R., Volatilization of insecticides from various substrates, *J. Agr. Food Chem.*, 18, 814, 1970.

Lichtenstein, E. P., Anderson, J. P., Fuhremann, T. W., and Schulz, K. R., Aldrin and dieldrin: loss under sterile conditions, *Science*, 159, 1110, 1968.

MacRae, I. C. and Alexander, M., Microbial degradation of selected herbicides in soil, *J. Agr. Food Chem.*, 13, 72, 1965.

MacRae, I. C., Raghu, K., and Bautista, E. M., Anaerobic degradation of the insecticide lindane by *Clostridium* sp., *Nature*, 221, 859, 1969.

Malone, T. C., In vitro conversion of DDT to DDD by the intestinal microflora of the northern anchovy, *Engraulis mordax*, *Nature*, 227, 848, 1970.

Martin, J. P., Harding, R. B., Cannell, G. H., and Anderson, L. D., Influence of five annual field applications of organic insecticides on soil biological and physical properties, *Soil Sci.*, 87, 334, 1959.

Matsumura, F., Khanvilkar, V. G., Patil, K., Boush, G. M., Metabolism of endrin by certain soil microorganisms, *J. Agr. Food Chem.*, 19, 27, 1971.

Matsumura, F. and Boush, G. M., Dieldrin: degradation by soil microorganisms, *Science*, 156, 959, 1967.

Matsumura, F., Boush, G. M., and Tai, A., Breakdown of dieldrin in the soil by a microorganism, *Nature*, 219, 965, 1968.

Mendel, J. L. and Walton, M. S., Conversion of p,p'-DDT to p,p'-DDD by intestinal flora of the rat, *Science*, 151, 1527, 1966.

Menzie, C. M., Metabolism of pesticides, Bureau of Sport Fisheries and Wildlife, Special Scientific report, Wildlife No. 127, Washington, D.C., 1969.

Menzel, D. W., Anderson, J., and Randtke, A., Marine phytoplankton vary in their response to chlorinated hydrocarbons, *Science*, 167, 1724, 1970.

Miles, J. R. W., Tu, C. M., and Harris, C. R., Metabolism of heptachlor and its degradation products by soil microorganisms, *J. Econ. Entomol.*, 62, 1334, 1969.

Monheimer, R. H., Interaction between *Clostridium botulinum* Type E toxin and DDT in white mice, *Nature*, 222, 788, 1969.

Moore, R. B. and Dorward, D. A., Accumulation and metabolism of pesticides by algae, *J. Phycol.*, 4, 7(suppl.), 1968.

Okey, R. W. and Bogan, R. H., Apparent involvement of electronic mechanisms in limiting the microbial metabolism of pesticides, *J. Water Pollut. Contr. Fed.*, 37, 692, 1965.

Patil, K. C., Matsumura, F., and Boush, G. M., Degradation of endrin, aldrin, and DDT by soil microorganisms, *Appl. Microbiol.*, 19, 879, 1970.

Plimmer, J. R., Klingebiel, U. I., and Hummer, B. E., Photooxidation of DDT and DDE, *Science*, 167, 67, 1970.

Plimmer, J. R., Kearney, P. C., and von Endt, D. W., Mechanism of conversion of DDT to DDD by *Aerobacter aerogenes*, *J. Agr. Food Chem.*, 16, 594, 1968.

Ragland, W. L., Friend, M., Trainer, D. O., and Sladek, N. E., Interaction between duck hepatitis virus and DDT in ducks, *Res. Commun. Chem. Path. Pharmacol.*, 2, 236, 1971.

Schwartz, H. G., Microbial degradation of pesticides in aqueous solutions, *J. Water Pollut. Contr. Fed.*, 39, 1701, 1967.

Sethunathan, N., Bautista, E. M., and Yoshida, T., Degradation of benzene hexachloride by a soil bacterium, *Canad. J. Microbiol.*, 15, 1349, 1969.

Shaw, W. M. and Robinson, B., Pesticide effects in soils on nitrification and plant growth, *Soil Sci.*, 90, 320, 1960.

Shin, Y., Chodan, J. J., and Wolcott, A. R., Adsorption of DDT by soils, soil fractions, and biological materials, *J. Agr. Food Chem.*, 18, 1129, 1970.

Singh, J. and Malaiyandi, M., Dechlorination of p,p'-DDT in aqueous media, *Bull. Environ. Contam. Toxicol.*, 6, 337, 1969.

Sodergren, A., Uptake and accumulation of C^{14}-DDT by *Chlorella* sp. (Chlorophyceae), *Oikos*, 19, 126, 1968.

Stadnyk, L., Campbell, R. S., and Johnson, B. T., Pesticide effect on growth and ^{14}C assimilation in a freshwater alga, *Bull. Environ. Contam. Toxicol.*, 6, 1, 1971.

Stenersen, J. H. V., DDT-metabolism in resistant and susceptible stable-flies and in bacteria, *Nature*, 207, 660, 1965.

Sweeny, R. A., Metabolism of lindane by *Chlorella vulgaris* and *Chlamydomonas reinhardtii*, *J. Phycol.*, 4, 7(suppl.), 1968.

32

Tiedje, J. M., Duxbury, J. M., Alexander, M., and Dawson, J. E., 2,4-D metabolism: pathway of degradation of chlorocatechols by Arthrobacter sp., *J. Agr. Food Chem.*, 17, 1021, 1969.

Tu, C. M., Miles, J. R. W., and Harris, C. R., Soil microbial degradation of aldrin, *Life Sci.*, 7, 311, 1968.

Vance, B. D. and Drummond, W., Biological concentration of pesticides by algae, *J. Amer. Water Works Ass.*, 61, 360, 1969.

Voerman, S. and Tammes, P. M. L., Adsorption and desorption of lindane and dieldrin by yeast, *Bull. Environ. Contam. Toxicol.*, 4, 271, 1969.

Weber, J. B. and Coble, H. D., Microbial decomposition of diquat adsorbed on montmorillonite and kaolinite clays, *J. Agr. Food Chem.*, 16, 475, 1968.

Wedemeyer, G., Partial hydrolysis of dieldrin by *Aerobacter aerogenes*, *Appl. Microbiol.*, 16, 661, 1968.

Wheeler, W. B., Experimental absorption of dieldrin by *Chlorella*, *J. Agr. Food Chem.*, 18, 416, 1970.

Wilson, J. K. and Choudhri, R. S., Effects of DDT on certain microbiological processes in the soil, *J. Econ. Entomol.*, 39, 537, 1946.

Winely, C. L. and San Clemente, C. L., Effects of pesticides on nitrate oxidation by *Nitrobacter agilis*, *Appl. Microbiol.*, 19, 214, 1970.

Winely, C. L. and San Clemente, C. L., The effect of two herbicides (CIPC and eptam) on oxidative phosphorylation by *Nitrobacter agilis*, *Canad. J. Microbiol.*, 17, 47, 1971.

Wurster, C. F., DDT reduces photosynthesis by marine phytoplankton, *Science*, 159, 1474, 1968.

REFERENCES

1. Nakamura, S., General theory on the mechanics of enzyme action, *Enzymologia*, 25, 3, 1962.

2. Shattuck, G. E., Jr. and Alexander, M., A differential inhibitor of nitrifying microorganisms, *Soil Sci. Soc. Amer. Proc.*, 27, 600, 1963.

3. Alexander, M., Biodegradation: problems of molecular recalcitrance and microbial fallibility, *Advances Appl. Microbiol.*, 7, 35, 1965.

4. Menzie, C. M., Metabolism of Pesticides, Special Scientific Report, Wildlife No. 127, Washington, D. C., 1969.

5. Alexander, M., The breakdown of pesticides in soils, in *Agriculture and the Quality of Our Environment*, Brady, N. C., Ed., 331, 1967.

6. Bailey, G. W. and White, J. L., Review of adsorption and desorption of organic pesticides by soil colloids, with implications concerning pesticide bioactivity, *J. Agr. Food Chem.*, 12, 324, 1964.

7. Bollen, W. B., Interaction between pesticides and soil microorganisms, *Ann. Rev. Microbiol.*, 15, 69, 1961.

8. Marth, E. H., Residues and some effects of chlorinated hydrocarbon insecticides in biological material, *Res. Rev.*, 9, 2, 1965.

9. Martin, J. P., Influence of pesticide residues on soil microbiological and chemical properties, *Res. Rev.*, 4, 96, 1963.

10. Ware, G. W. and Roan, C. C., Interaction of pesticides with aquatic microorganisms and plankton, *Res. Rev.*, 33, 15, 1970.

BIOCHEMICAL TRANSFORMATION OF PESTICIDES BY SOIL FUNGI

Author: **J. -M. Bollag**
Department of Agronomy
The Pennsylvania State University
University Park, Pa.

INTRODUCTION

An increasing number of xenobiotic compounds such as pesticides are now used in agriculture and industry, but their fate in the various environments, which very often is influenced by biological activities, still leaves numerous questions unanswered. It is also necessary to keep in mind that in the various ecosystems, the added substances may be transformed by different groups of organisms and, consequently, the resulting products can vary and possess an unknown effect in the environment. It is essential to understand certain metabolic activities by specific groups of organisms from a fundamental standpoint, since most transformation products can be physiologically active toward other living organisms and cause potential hazard to life if they accumulate in extensive and consequently toxic amounts.

The most abundant microorganisms in soil are bacteria, but since the bacterial cell is quite small in size, the fungi account for the largest portion of the total microbial protoplasm in most well-aerated and cultivated soils.[1] Therefore, it is of considerable interest to examine the special fungal participation in the transformation of pesticides. Bacteria and fungi differ widely in

their reactions to foreign chemicals, and with the extensive use of synthetic organic pesticides, the need for unequivocal information on the transformation capacities of each group of organisms represents necessary knowledge which has to be acquired. Bacteria often can attack a certain pesticide and biodegrade it completely, i.e., decompose it to its mineral components. However, in many instances fungi provoke only minor chemical changes from the original compound.

Biotransformations of xenobiotic substances have usually been classified as 'detoxications' or 'detoxifications.' These expressions can be misleading without clarification. 'Detoxication' can refer to a self-protective action of the metabolizing organism or it can mean reduction or elimination of toxicity toward a defined target organism or organisms in general. In this review the described detoxication mechanisms are considered only as transformation processes by fungi.

There is little doubt that, especially among the fungi, the various detoxication mechanisms are not isolated biochemical curiosities, but represent common processes which also take place with naturally occurring substances. Very little is known and almost no research has been done on the fungal enzymes responsible for pesticide metabolism, but it appears to be a possible hypothesis that constitutive enzymes play an important role in the 'detoxifying' activity of fungi, which is in contrast to the apparent predominance of inducible enzymes in the metabolism of foreign compounds by bacteria.

This review summarizes known fungal detoxication reactions of pesticides to provide a general concept of the possible transformation reactions which can be expected by the interference of fungi with xenobiotic compounds and to stimulate interest for study in the widespread basic biochemical reactions. As Williams suggested in his book, *Detoxication Mechanisms,*[2] it was also attempted in this review to group the various reactions as oxidation, hydrolysis, reduction, and synthetic processes.

OXIDATION REACTIONS

Dealkylation

Although dealkylation reactions are truly one of the major detoxication activities of the soil fungi, no detailed knowledge on the dealkylation mechanism exists. There is no report available on the isolation of a specific fungal cell-free preparation for basic biochemical studies. A great number of pesticides from different groups are metabolized through dealkylation by fungi (Table 1)

and very often the resulting compounds can be isolated and even appear to be the final metabolic products.

N-Dealkylation

There is little doubt that N-dealkylation is the major initial route of phenylurea metabolism. Several studies with mixed cultures or in soil demonstrated the formation of dealkylated products. Geissbühler et al.[3] isolated and identified 3-(4-chlorophenoxy)phenyl-1-methylurea and 3-(4-chlorophenoxy)phenylurea from chloroxuron, 3-[4-(*p*-chlorophenoxy)-phenyl]-1,1-dimethylurea, in liquid cultures of mixed soil microbes. This indicated that the microbial attack of chloroxuron consisted of stepwise demethylation. A similar pattern of dealkylation was reported by Dalton et al.[4] for diuron, 3-(3,4-dichlorophenyl)-1,1-dimethylurea in soil samples from diuron-treated cotton fields. In this case, the products identified were 3-(3,4-dichlorophenyl)-1-methyl-urea and 3,4-dichlorophenylurea.

Consequently, a generalized pattern of stepwise dealkylation for dimethyl phenylureas can be assumed (Figure 1).

Reports on microbial degradation of the phenylureas have been primarily bacterial. Hill and McGahen[5] first reported on the involvement of fungi, *Penicillium* and *Aspergillus* spp., in the

TABLE 1

N-Dealkylation Reactions of Pesticides by Soil Fungi

Chemical group	Pesticide which is dealkylated	Fungus	Reference
Phenylurea	Metobromuron	*Talaromyces wortmanii, Fusarium oxysporum*	6
	Chlorbromuron	*Rhizoctonia solani*	9
	Monolinuron	*Aspergillus niger*	8
	Linuron	*Aspergillus niger*	8
Triazines	Atrazine	*Aspergillus fumigatus, A. ustus, A. flavipes, Fusarium moniliforme, F. oxysporum, F. roseum, Penicillium decumbens, P. janthinellum, P. luteum, P. rugulosum, Rhizopus stolonifer, Trichoderma viride*	11
	Simazine	*Aspergillus fumigatus*	12
Carbamates	Carbaryl	*Aspergillus terreus*	16
	Diphenamid	*Trichoderma viride, Aspergillus candidus*	18

35

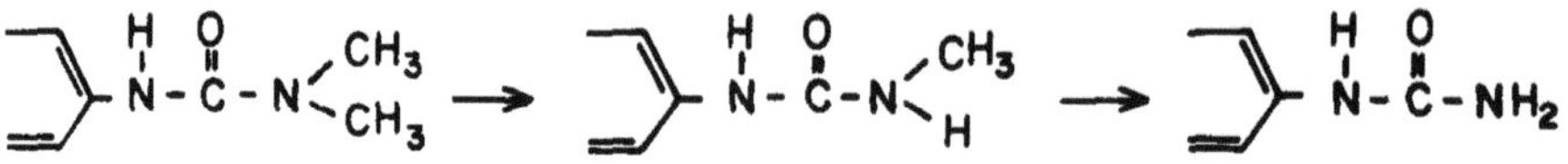

FIGURE 1. Stepwise dealkylation of dimethyl phenylureas.

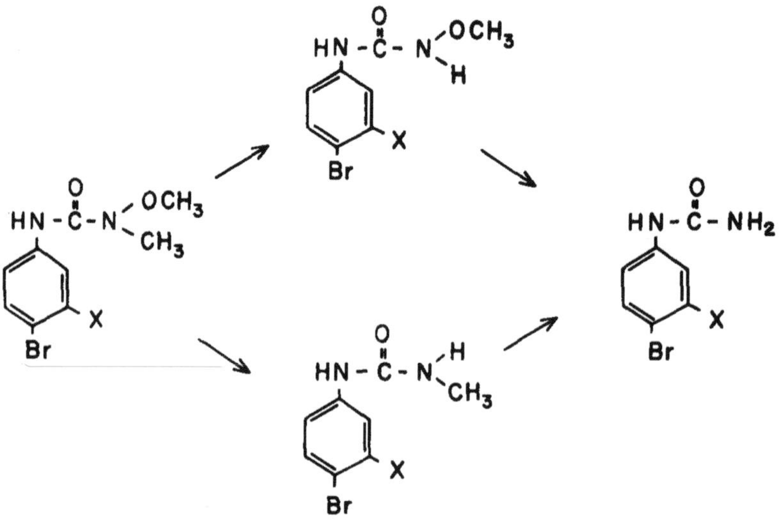

FIGURE 2. Dealkylation and dealkoxylation of halogen-substituted phenylurea pesticides (X represents H for metobromuron and Cl for chlorbromuron).

transformation of monuron, 3-(p-chlorophenyl)-1,1-dimethylurea. They claimed that it could be utilized as a carbon source in an agar medium, but they did not look for intermediates and consequently could not postulate any specific pathway.

However, meaningful data with fungi were obtained by Tweedy at al.,[6,7] Börner,[8] and Weinberger.[9] The fungus *Talaromyces wortmanii* metabolized the herbicide metobromuron, 3-(p-bromophenyl)-1-methoxy-1-methylurea, and the demethylated metabolite 1-(p-bromophenyl)-3-methoxyurea and the demethoxylated intermediate 1-(p-bromophenyl)-3-methylurea could be isolated.[6] It was also possible to identify p-bromophenylurea, which indicates that dealkylation and subsequent dealkoxylation, or vice versa, are involved in the stepwise degradation of the phenylurea compound (Figure 2). A similar mechanism could also be assumed for the fungus *Fusarium oxysporum.*

An analogous transformation for chlorbromuron, 3-(3-chloro-4-bromophenyl)-1-methoxy-1-methylurea, was demonstrated by Weinberger[9]

with *Rhizoctonia solani.* It was possible to isolate and identify 3-(3-chloro-4-bromophenyl)-1-methoxyurea, the demethylated herbicide (Figure 2). If *R. solani* was cultured in the presence of the demethoxylated herbicide, 3-(3-chloro-4-bromophenyl)-1-methylurea, it was partially transformed to the corresponding urea, indicating the general ability of this fungal species for dealkylation. Several other phenylureas (chloroxuron, diuron, fenuron, fluometuron, linuron, metobromuron, neburon, and siduron) were also attacked as indicated by thin-layer chromatography by *R. solani,* but no specific products were identified.

In his studies Börner[8] also found that monolinuron, 3-(p-chlorophenyl)-1-methoxy-1-methylurea and linuron, 3-(3,4-dichlorophenyl)-1-methoxy-1-methylurea, were dealkylated by an *Aspergillus niger* sp.; whereas the demethylated intermediate was the major product, the demethoxylated compound and the corresponding urea could be found sometimes as minor products. Börner et al.[10] also investigated a great number of bacteria and fungi, most of which appeared capable of transforming linuron, monolinuron, monuron, and diuron in a nutrient solution.

N-Dealkylation also is used by fungi as the initial attack on chloro-*s*-triazines. In studies with pure cultures of *Aspergillus fumigatus* Fres. it was possible to demonstrate that atrazine was dealkylated to 2-chloro-4-amino-6-ethylamino-*s*-triazine and 2-chloro-4-amino-6-isopropyl-amino-*s*-triazine[11] (Figure 3), whereas simazine was altered to 2-chloro-4-ethylamino-6-amino-*s*-triazine.[12] Further dealkylation of simazine and atrazine was observed, but in the case of atrazine the completely dealkylated product still could not be definitively identified. Kearney et al.[12] attempted to obtain an active preparation from fractionated fungal mycelia which would dealkylate simazine, but they were unsuccessful in this undertaking. Since the majority of the metabolites were found in the external solution, they speculated that an extracellular enzyme might catalyze the degradation of simazine or that the metabolism of the herbicide occurs on the surface of the cells with subsequent release of the dealkylated product to the external media.

An interesting observation is related to species specificity concerning the preferential removal of the ethyl side chain or the isopropyl group during atrazine metabolism.[11] All the fungi investigated were able to attack the herbicide by dealkylation of either alkylamino group, but quantitative and qualitative differences were observed in the activity of the various fungal species. *Aspergillus fumigatus*, for example, exhibited a greater tendency to remove the ethyl moiety than the isopropyl group, whereas *Rhizopus stolonifer*

metabolized the isopropyl rather than the ethyl moiety. Even in the same genus, differences could be established. *Fusarium moniliforme* dealkylated the ethyl group, whereas *F. roseum* showed a stronger activity in the removal of the isopropyl substitution.

It is worthwhile to note that dealkylation of *s*-triazine herbicides does not necessarily mean detoxication, if one considers the toxicity from a 'plant point-of-view.' It was often found that the dealkylated compound still has some phyto-toxicity.[13]

N-Dealkylation is also of great importance in the fungal attack on *N*-methylcarbamates. Since the *N*-hydroxymethyl derivatives of these compounds are relatively stable, it is possible to isolate them from the fungal growth media.[14-16] The identification of 1-naphthyl *N*-hydroxymethyl carbamate as well as the formation of the dealkylated intermediate, 1-naphthyl carbamate, could be demonstrated clearly with the fungus *Aspergillus terreus* (Figure 4) and also to a certain extent with *A. flavus*.[16] The *N*-hydroxymethyl intermediate also can be chemically decomposed to 1-naphthyl carbamate, but the study gave evidence that through the additional biological activity the formation of the dealkylated product was increased considerably.

The initial *N*-methyl hydroxylation can be assumed to be the first attack in dealkylation reactions, but since the hydroxylated intermediates are unstable chemical compounds in most cases, it is very hard to isolate them. The

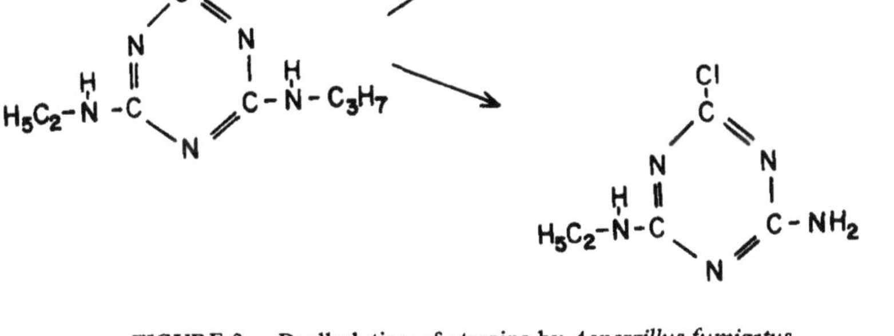

FIGURE 3.   Dealkylation of atrazine by *Aspergillus fumigatus*.

N-methylcarbamates provide, therefore, a kind of a model group in which the mechanism of dealkylation can be studied in more detail.

The exact mechanism of biological demethylation is still not clarified but from the chemical intermediates which could be isolated from various compounds, the following oxidation reactions can be anticipated:

$$\begin{array}{c}R\\R'\end{array}\!\!N - CH_3 \rightarrow \left[\begin{array}{c}R\\ \\R\end{array}\!\!\overset{\overset{\displaystyle O}{\|}}{N} - CH_3 \rightarrow \begin{array}{c}R\\ \\R\end{array}\right]\!\! NCH_2OH \rightarrow \begin{array}{c}R\\R'\end{array}\!\!NH + HCHO$$

and therefore it still has to be considered a hypothetical mechanism.

A thorough study combining hepatic microsomes and their dealkylating ability of pesticidal compounds was reported by Hodgson and Casida.[17] A large number of *N:N*-dialkyl carbamate pesticides were metabolized by rat liver microsomes after addition of NADPH and oxygen. With *N:N*-dimethyl carbamates, the formation of formaldehyde could be observed, but this was not found with *N*-methyl carbamates. The same microsomal enzyme system, however, showed a weak activity on urea herbicides such as fenuron, monuron, and diuron.

An interesting observation was made with a carbamate-related amide, namely the herbicide diphenamid, *N,N*-dimethyl-2,2-diphenylace-tamide.[18] It was found that diphenamid is far less toxic to tomato and certain grass seedlings than two dealkylated metabolites, and it could be shown that two fungi, *Trichoderma viride* and *Aspergillus candidus*, can be responsible for this "detoxication-toxication" reaction. Diphenamid is demethylated to *N*-methyl 2,2-diphenylacetamide and subsequently to 2,2-diphenylacetamide (Figure 5).

In some cases it appears that the first step of the methylated compound is the formation of an *N*-oxide which can represent an intermediate to the *N*-hydroxylated product; this in turn can be metabolized to the completely dealkylated compound under simultaneous release of formaldehyde. This entire sequence of products has not been isolated in any single dealkylation reaction,

The biochemical mechanism of dealkylation is by no means clarified for the various biological systems. There exist a large number of observations which do not appear to be of general significance. Brodie et al. claimed[19] that *N*-demethylation requires only oxygen and NADPH in liver microsomal system, but Nilsson and Johnson[20] claimed that for maximum rate and

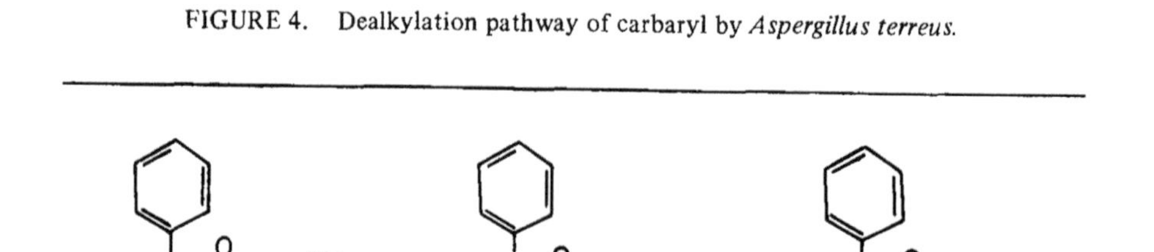

FIGURE 4. Dealkylation pathway of carbaryl by *Aspergillus terreus*.

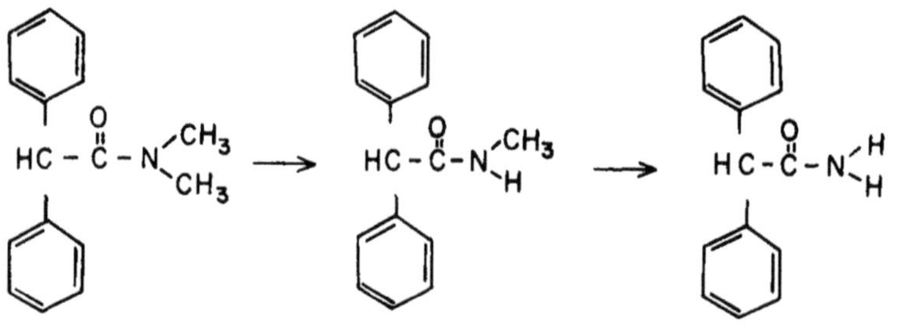

FIGURE 5. Stepwise dealkylation of diphenamid by *Trichoderma viride* and *Aspergillus candidus*.

activity of an *N*-demethylation and *O*-demethylation reaction, NADH is also required. A combination of these two cofactors was found more active than NADPH alone in the *N*-demethylation of 3-methyl-4-monomethyl-aminoazobenzene.

O-Dealkylation

In a recent review, Hollingworth[21] concluded that cleavage of the esterified alkyl group of organophosphorus pesticides may make a significant contribution to detoxication in microorganisms, plants, insects, and mammals. *O*-dealkylation may become a predominant detoxication mechanism in mammals and insects, since the removal of an alkyl group leads to a considerable reduction in anticholinesterase activity and consequently in toxicity. However, also in this case, very little is known on the specific mechanism. There are numerous reports with various liver microsome preparations; stimulation of dealkylation can be intermediated by glutathione;[22] one report describes the oxidative *O*-dealkylation of a diethyl phosphate triester in which the reaction was NADPH-dependent and resulted in the corresponding aldehyde and monodealkyl derivative as products.[23] In houseflies dealkylation of paraoxon may take place by an apparent hydrolytic mechanism which yields an alcohol.[21]

If *N*-demethoxylation involves the formation of the corresponding hydroxylamine or hydroxylamides, it would also constitute an *O*-demethylation reaction, but no hydroxylamine intermediate could be isolated from a methoxylated pesticide by fungal action.

However, it was possible to demonstrate that methoxylated aromatic compounds can be demethylated, e.g., *o*-, *m*-, and *p*-methoxybenzoic acid are converted to the corresponding hydroxybenzoic acids, and veratric acid is demethylated to vanillic acid by the soil fungus *Hormodendrum* sp., *Haplographium* sp., and a *Penicillium* sp. (Figure 6). The latter transformation also seems to indicate that a methoxyl group in the *p*-position is preferentially attacked.[24] This investigation was undertaken since decomposition of lignified material in soil is always indicated by reduction in methoxyl compounds.

O-Dealkylation occurs with methoxy-*s*-triazines in higher plants and it is a very likely reaction in soil fungi, but no conclusive observations have been reported.

An interesting observation of 'double-dealkylation' was made by Stenersen[25] during fungal metabolism of the insecticide bromophos, dimethyl 4-bromo-2,5-dichlorophenyl phosphorothionate. It was shown that *Alternaria tenius* and *Trichoderma lignorum* are particularly active in metabolizing bromophos and surprisingly it was possible to determine by thin-layer chromatography the bisdesmethylated bromophos, namely 4-bromo-2,5-dichlorophenyl phosphorothionate (Figure 7), but it was not possible to detect the monodesmethyl derivative. Several other fungi which were not identified showed a similar degradation pattern. The same author also described the simultaneous removal of both methyl groups from the bromophos by a mouse liver enzyme preparation which was glutathione-dependent.[26]

O-Dealkylation was also reported with anisole whose chlorinated derivatives were found occasionally as intermediates during the breakdown of chlorinated phenoxyacetic acids. By gas chroma-

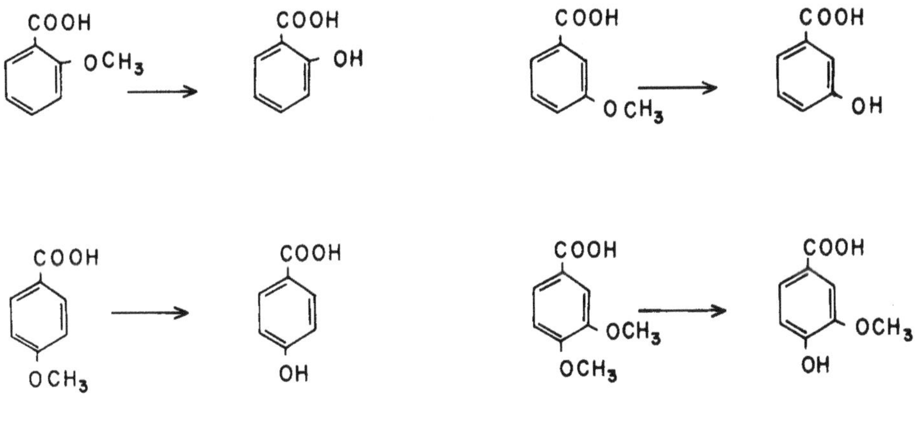

FIGURE 6.   Demethylation of methoxylated aromatic compounds.

39

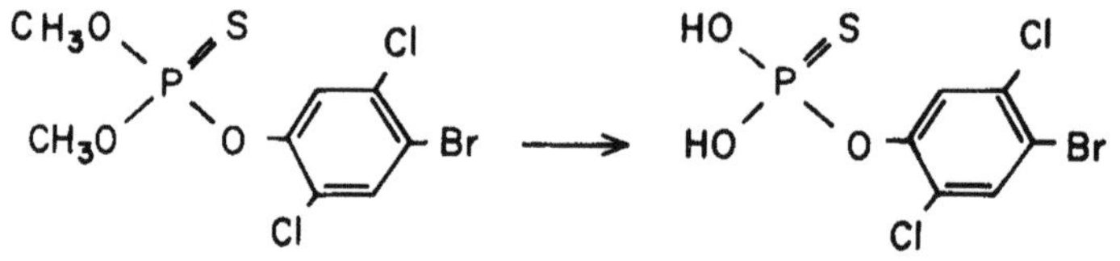

FIGURE 7.   'Double-dealkylation' of bromophos.

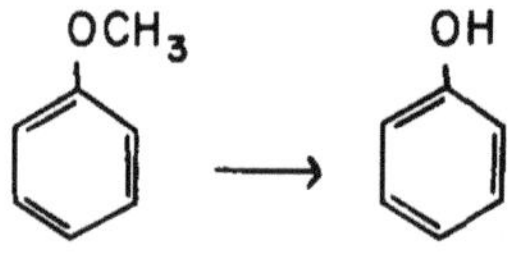

FIGURE 8.   Dealkylation of anisole by *Aspergillus niger*.

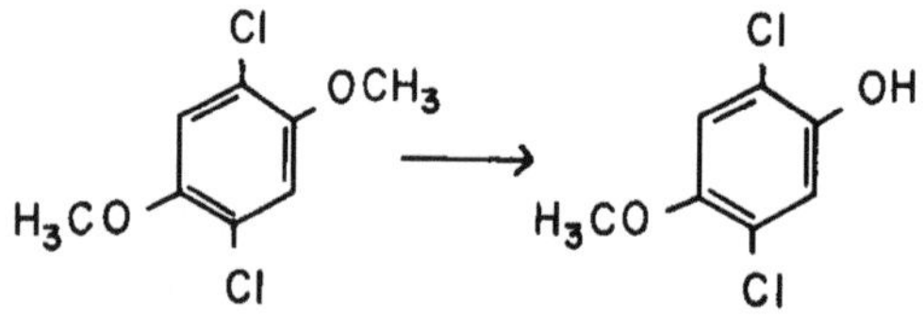

FIGURE 9.   *O*-Demethylation of chloroneb by *Rhizoctonia solani*.

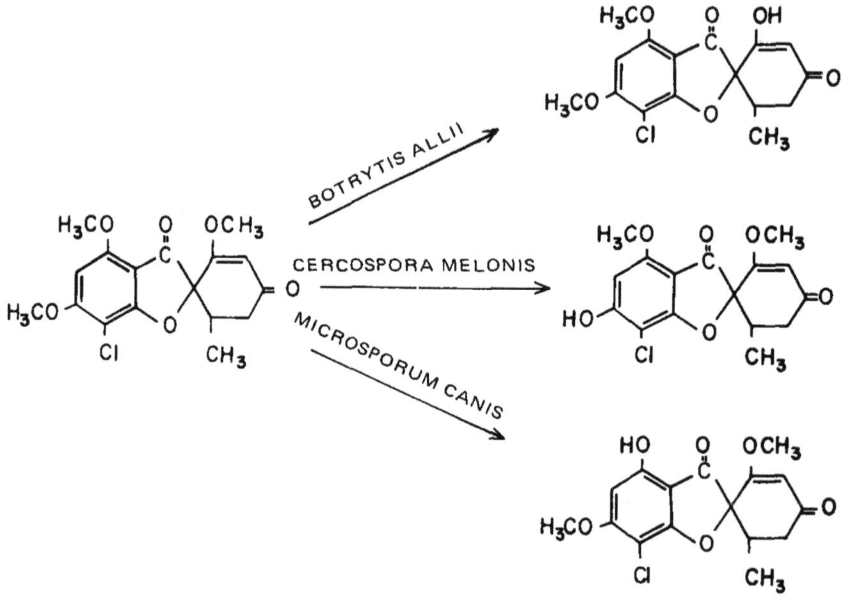

FIGURE 10.   Demethoxylation of griseofulvin by various fungi.

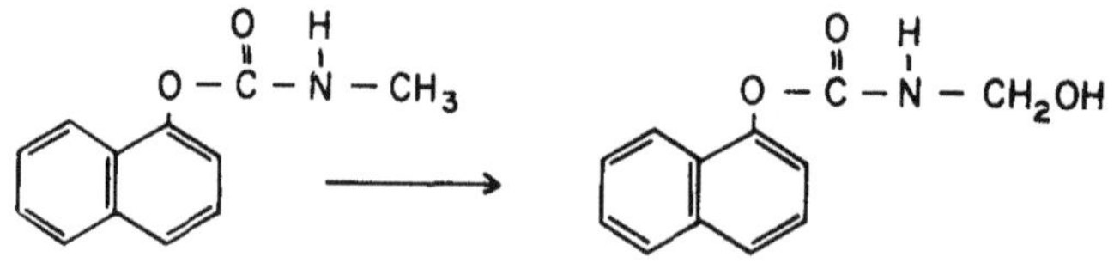

FIGURE 11.   Side-chain hydroxylation of carbaryl.

tography Bocks et al.[27] determined phenol as a product of anisole in replacement cultures of *Aspergillus niger* (Figure 8).

Chloroneb, 1,4-dichloro-2,5-dimethoxybenzene, which is used as a soil fungicide, can be converted by *Rhizoctonia solani* to 2,5-dichloro-4-methoxyphenol[28] (Figure 9). In this study it was clearly observed that the demethylation of a methoxy group to the hydroxy derivative could occur during mycelial growth. The metabolite formed by the dealkylation reaction inhibited the growth of the fungus at a concentration of 16 μg/ml. whereas chloroneb was toxic for fungal growth at 5 to 8 μg/ml.

In a study with the fungicide griseofulvin, 7-chloro-4:6:2'-trimethoxy-6'-methylgris-2'-en-3:4'-dione, it was observed that different fungi demethoxylate the fungicide at different methoxy groups which are attached to the molecule.[29] *Botrytis allii* formed the 2'-demethylgriseofulvin, *Cercospora melonis* produced the 6-demethylgriseofulvin, and *Microsporum canis* generated 4-demethylgriseofulvin (Figure 10).

Hydroxylation Reactions

The metabolism of xenobiotic compounds is often initiated by hydroxylation, but there is no clear physiological explanation for this reaction. It is possible that these substances are often oxygen deficient and need to be oxygenated to become biologically active or more soluble in water.[30] In plants or animals the introduction of a hydroxyl group into a compound can give a center at which conjugation can occur, but this does not seem to be the purpose during hydroxylation by microorganisms. The hydroxylation reaction usually takes place only in the presence of oxygen and with the participation of the cofactor NADPH, indicating that the process is catalyzed by a mixed function oxidase, and the enzyme is also referred to as an oxygenase. Hydroxylations can occur with aliphatic as well as aromatic compounds, and they can be involved in various more complex reactions such as dealkylations, decarboxylation, deamination, epoxide formation, etc.

It is important to emphasize that evidence exists that nonspecific hydroxylations can be caused by nonenzymic hydroxylating systems in vitro. In 1954 Udenfriend and his collaborators[31] developed a system, consisting of ferrous iron, ascorbic acid, a chelating agent (ethylenediaminetetraacetic acid), and oxygen, which under physiological conditions of temperature and pH hydroxylates aromatic substances in a manner shown to be closely analogous to in vivo hydroxylations.

Aliphatic Hydroxylation

As outlined previously under dealkylation, it is generally thought that N-demethylation reactions proceed through the intermediate formation of an N-hydroxymethyl intermediate. This mechanism could be observed especially well with various carbamate pesticides. Carbaryl, e.g., 1-naphthyl N-methylcarbamate is hydroxylated to the N-hydroxymethyl derivative (Figure 11) by a large number of fungi.[14-16,32] In a study with *Gliocladium roseum*, grown in a liquid growth medium, the side chain hydroxylated intermediate could be isolated and identified by ultraviolet, infrared, and mass spectroscopy.

During the fungal degradation of alkylated substituted ureas, the formation of a hydroxylated intermediate such as an N-hydroxymethylamide *could not be detected, but as outlined under dealkylation such a product is fairly unstable.*

The occurrence of N-hydroxylation caused by fungi has not been reported.

Aromatic Hydroxylation

The hydroxylation of aromatic compounds is a widespread detoxication reaction performed by soil fungi, and there exist a great number of pesticidal groups which are attacked through ring hydroxylation (Table 2). Like the process of N-dealkylation, ring hydroxylation reactions have been most extensively investigated with mammalian liver microsomal systems, and all observations of fungal hydroxylation so far have been made in growth cultures or in replacement culture studies. The isolation of the hydroxylated compounds from experiments with fungi also indicates that the products are quite stable and do not undergo immediate further degradation by fungal activity.

The aromatic hydroxylation of carbaryl could be shown unequivocally with *Gliocladium roseum*[15] and for a large number of fungi as indicated by thin-layer chromatography.[14] Carbaryl was hydroxylated to 4-hydroxy-1-naphthyl methylcarbamate and 5-hydroxy-1-naphthyl methylcarbamate (Figure 12).

As in other instances, it was interesting to observe that the hydroxylation in the side chain or

TABLE 2

Formation of Hydroxylated Metabolites from Methyl-[14]C Labeled Carbaryl by Various Soil Fungi[14]

(Radioactivity in dpm detected after TLC of ether extracts from five-day-old growth media)

	Naphthyl *N*-hydroxy-methylcarbamate	4-hydroxy-1-naphthyl methylcarbamate	5-hydroxy-1-naphthyl methylcarbamate
Aspergillus flavus Link ex Fries	4342	126	228
Aspergillus fumigatus Fresenius	0	0	0
Aspergillus niger Van Tieghem	739	140	305
Aspergillus terreus Thom.	4854	59	262
Aspergillus sp.	2257	186	334
Fusarium oxysporum Schlectendahl	0	0	0
Fusarium roseum Link	569	0	0
Fusarium sp.	0	0	0
Geotrichum candidum Link	0	0	0
Gliocladium roseum (Link) Thom.	829	254	408
Helminthosporium sp.	63	235	82
Mucor racemosus Fresenius	134	1275	834
Mucor sp.	65	406	666
Penicillium roqueforti Thom.	0	0	0
Penicillium sp. – isolate 1	291	0	0
Penicillium sp. – isolate 2	266	3342	1003
Rhizopus sp.	122	1400	719
Trichoderma viride Per. ex Fries	102	117	95
Control medium	0	0	0

related to the position in the ring varied qualitatively as well as quantitatively with the various fungal species which were investigated (Table 2). Considerable differences could be detected even within one genus. *Aspergillus flavus* showed strong hydroxylation of the side chain, whereas *A. fumigatus* did not hydroxylate at all; different *Penicillium* sp. showed great variability in their attack, one species hydroxylating essentially on the ring, and a second species showing preference for the side chain, whereas *P. roqueforti* did not indicate any hydroxylating activity.

Extensive work on ring hydroxylation of phenoxyacetic acid was undertaken with the fungus *Aspergillus niger*. Using a replacement culture technique, Byrde and Woodcock[33] showed that phenoxyacetic acid and related compounds hydroxylated predominantly in the *para*- and to a lesser extent in the *ortho*-position. In a later investigation it was also possible to demonstrate the formation of the *meta*-isomer[34] (Figure 13).

If 2-chlorophenoxyacetic acid and 4-chlorophenoxyacetic acid were incubated with *A. niger*, hydroxylation occurred in various ring positions.[35] From 4-chlorophenoxyacetic acid it was possible to isolate 4-chloro-2-hydroxy- and 4-chloro-3-hydroxyphenoxyacetic acid after incubation with *A. niger* mycelium (Figure 14).

Exposure of 2-chlorophenoxyacetic acid to *A. niger* resulted in a much more interesting hydroxylation pattern, since all possible hydroxy derivatives were formed; hydroxylation in the 4- and 5-positions produced the major metabolites, whereas hydroxylation in the 3- and 6-positions resulted in minor products. It was even possible to isolate a fifth product from 2-chlorophenoxyacetic acid which was identified as 2-hydroxyphenoxyacetic acid, demonstrating the replacement of a chlorine by a hydroxyl group.[35]

The compound 2,4-dichlorophenoxyacetic acid, the most important of the phenoxyalkanoates, was metabolized by *A. niger* essentially to 2,4-dichloro-5-hydroxyphenoxyacetic acid and a product which was identified by infrared spectroscopy and mixed m.p. determination as 2,5-dichloro-4-hydroxyphenoxyacetic acid.[36,37] The latter product, also generated by the fungal

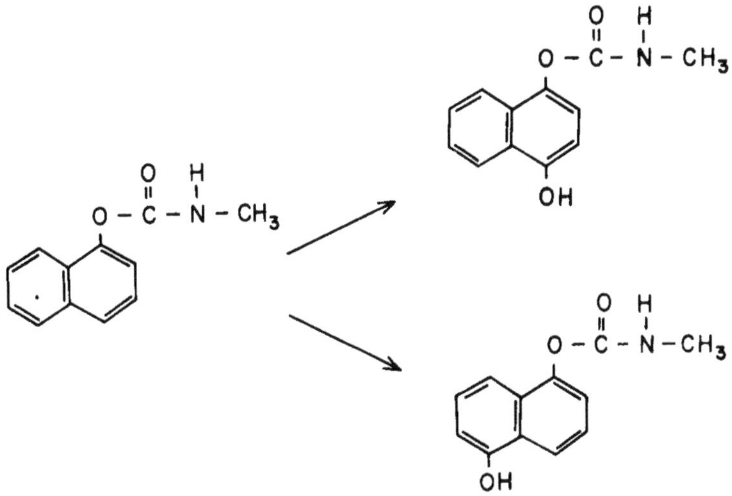

FIGURE 12.   Ring hydroxylation of carbaryl.

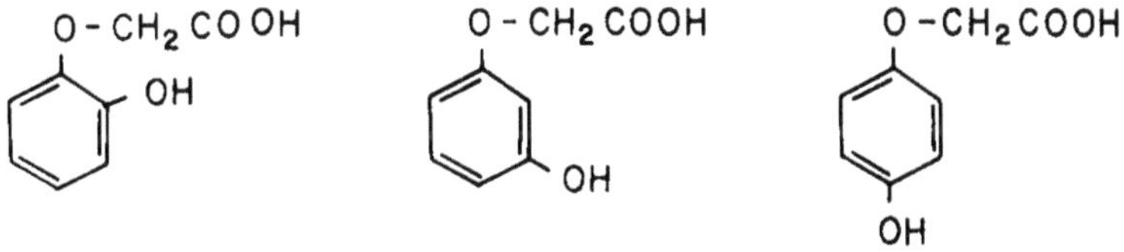

FIGURE 13.   Ring hydroxylation products of phenoxyacetic acid by *Aspergillus niger*.

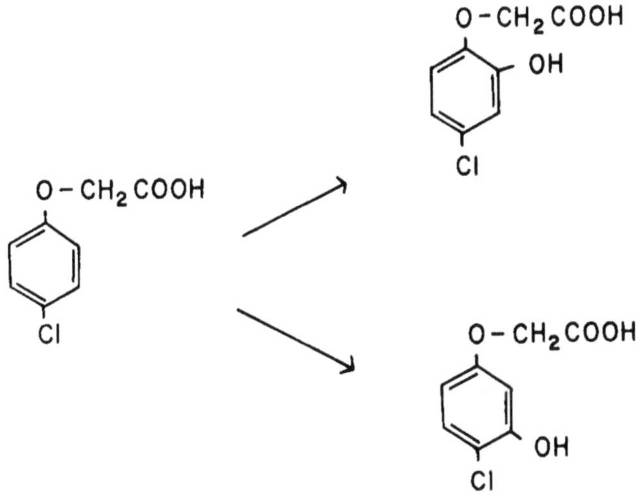

FIGURE 14.   Hydroxylation of 4-chlorophenoxyacetic acid.

43

activity, indicated a shift of a chlorine atom which was coupled with the introduction of a hydroxyl group in its place (Figure 15). However, it was not possible to find 3- or 6-hydroxy-2,4-dichlorophenoxyacetic acid, which might have been expected by analogy with previous results with 2-chlorophenoxyacetic acid and the same fungus.

In a similar investigation with the herbicide MCPA, 2-methyl-4-chlorophenoxyacetic acid, it was only possible to identify 2-methyl-5-hydroxy-4-chlorophenoxyacetic acid as a major metabolite.[36]

A highly selective nuclear hydroxylation system appeared to be active in the metabolism of ω-(2-naphthyloxy)-n-alkylcarboxylic acids. *Aspergillus niger*, e.g., converted γ-(2-naphthyloxy)-n-butyric acid to 6-hydroxy-2-naphthyloxyacetic acid[38] (Figure 16). The 6-hydroxy product was the only hydroxylation intermediate which could be isolated. Subsequently, the side chain was shortened by β-oxidation, but no evidence of ring fission by the fungal activity was obtained.

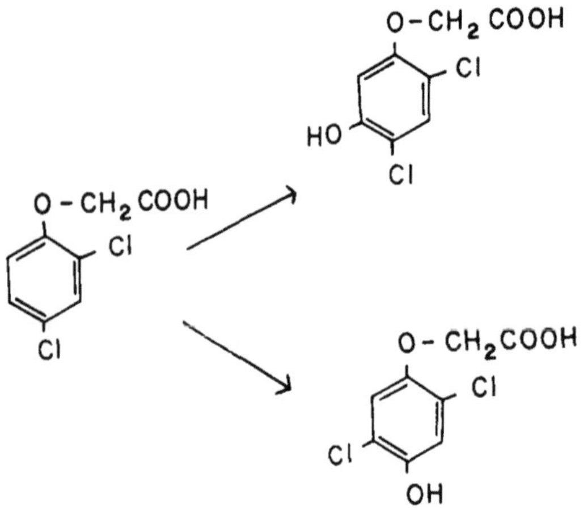

Attack on Aromatic Ring

While a great number of soil bacteria are capable of attacking and cleaving the aromatic ring and causing complete breakdown to carbon dioxide and water, there is only limited detailed knowledge of this mechanism occurring with soil fungi. Many pesticides are aromatic compounds and consequently their complete biodegradation is achieved only after cleavage of the ring. A mere transformation of the aromatic molecule easily can produce a less degradable compound than the original pesticide, and this 'detoxication reaction' therefore has to be investigated extensively if one is concerned with the fate of aromatic substances in the environment.

The metabolic pathways of the breakdown of aromatic substances were fairly well demonstrated with a number of various bacterial species,[39,40] but the fungi have received little detailed attention at the enzyme level despite the fact that they are probably the principal agents of lignin decomposition and consequently also of other aromatic compounds. Practically all bacterial enzymes involved in the pathway of ring fission could be isolated and characterized for the specific mechanism in which they participate, and it was possible to identify the various intermediate products. It is generally accepted that dihydroxylation is a prerequisite for enzymatic cleavage of the benzene ring. The hydroxyl groups may be in the *ortho*-position to each other, as in catechol or protocatechuic acid, or in *para*-position as in gentisic and homogentisic acids, and this configuration is considered a precondition prior to ring fission for mononuclear or polynuclear aromatic compounds.

The formation of an *ortho*- or *para*-dihydroxylated compound by a fungus was found, but the actual ring cleavage mechanism of fungi usually was indicated only by the trapping of C^{14}-labeled carbon dioxide from radiolabeled ring molecules and the detection of considerable amounts of oxalic acid.

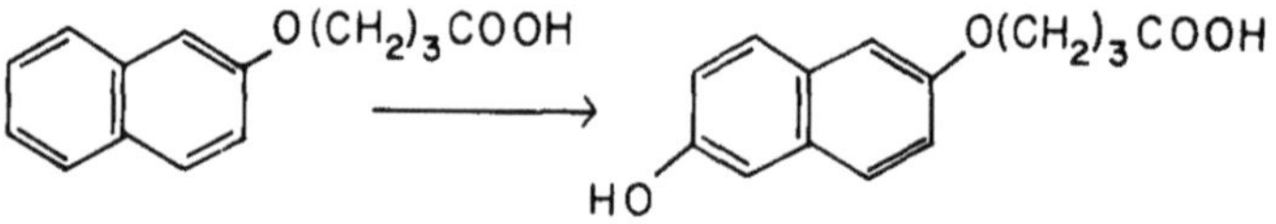

FIGURE 15. Metabolism of 2,4-dicholorophenoxyacetic acid by *Aspergillus niger*.

FIGURE 16. Hydroxylation of γ-(2-naphthyloxy)-n-butyric acid.

In 1950[41] it was reported that the fungi *Aspergillus niger, Penicillium notatum,* or *P. chrysogenum* convert tyrosine to homogentisic acid, which means the formation of a *para*-dihydroxylated product (Figure 17).

Using a replacement culture technique with *A. niger,* Kluyver and Van Zijp[42] claimed that both benzoic and salicylic acid were rapidly "consumed" and degraded and oxalic acid was the only detectable product. However, in studies with phenylacetic acid they were able to identify homogentisic acid as an oxidation product (Figure 18). Since they did not find additional intermediates, they concluded that it "is warranted that the greater part of phenylacetic acid had been completely oxidized to CO_2 and H_2O." This statement and conclusion can be found in numerous investigations, but this warranty would look much more convincing if some proof of the fungal participation in the aromatic breakdown could be provided on an enzymatic and biochemical basis.

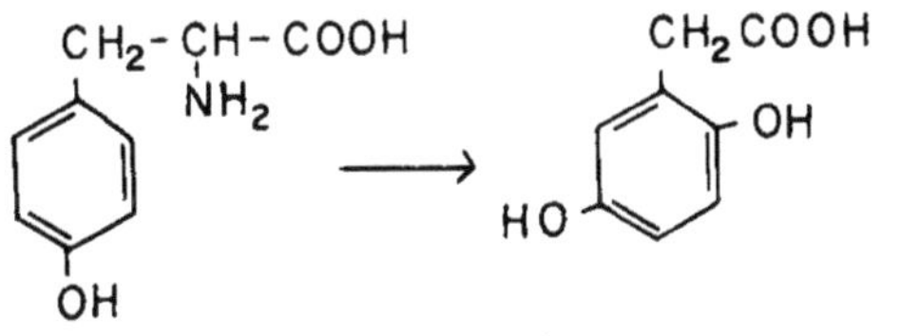

Several other studies were made with phenylacetic acid, but the attack of the intact ring appeared to be quite different with various fungal species.

Hockenhull et al.[43] demonstrated that the initial oxidation caused by *Penicillium chrysogenum* also took place in the side chain, but phenylacetic acid was converted to benzaldehyde (Figure 18), and it was assumed that mandelic acid and benzylalcohol were probable intermediates.

Isono[44] used a different strain of *P. chrysogenum* and isolated 2-hydroxyphenylacetic acid as an intermediate (Figure 18). In a subsequent investigation with manometric techniques, there was good evidence that the aromatic ring of phenylacetic acid had to be cleaved. However, using the 'sequential induction' technique, Isono[45] concluded that the cleavage of the ring follows neither the homogentisic acid nor the catechol pathway.

The various attacks on the phenylacetic acid molecule leave much room for speculation on the mechanism of ring cleavage, since even with the formation of a hydroxylated intermediate, the regular and known pathway does not appear to be the only possibility.

Ferulic acid, *p*-hydroxybenzaldehyde, syringaldehyde, and vanillin, which are believed to be

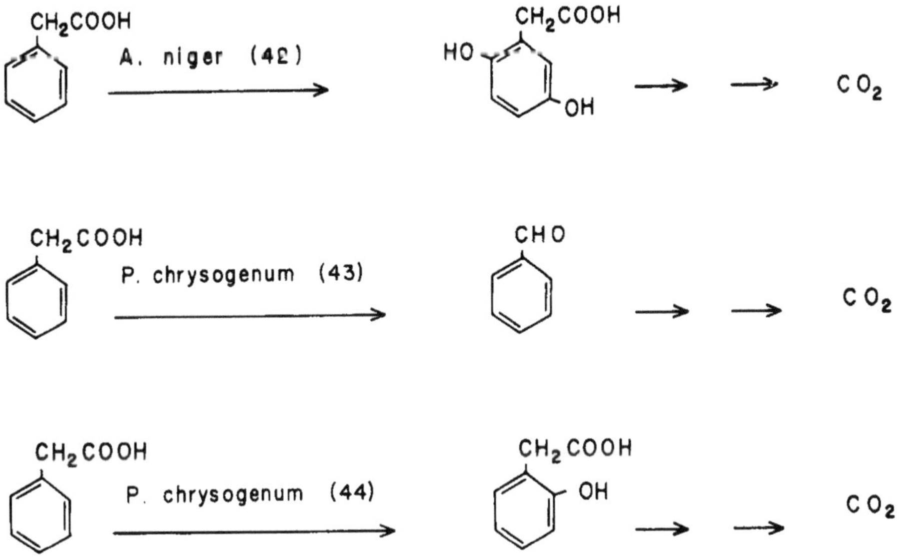

FIGURE 17. Conversion of tyrosine to homogentisic acid.

FIGURE 18. Oxidation of phenylacetic acid by fungi.

related structurally to the lignin molecule, can be utilized by some 60 fungi as sole carbon source.[46] From absorption spectrometry and paper chromatography, it was concluded that ring fission took place with all chemicals investigated. The benzene ring was completely degraded by species of *Alternaria, Aspergillus, Chaetomella, Coniothyrium, Cylindrocarpon, Hormiscium, Hormodendrum, Penicillium, Pyrenochaeta, Sphaeronaema,* and *Torula.* Several fungi converted vanillin and ferulic acid to vanillic acid, and syringaldehyde to syringic acid (Figure 19), and most fungi attacked *p*-hydroxybenzaldehyde most readily, but no intermediate was traced from this compound.

It was also claimed that complete disappearance of the benzene ring could be demonstrated by the applied spectrophotometric technique. In several instances, metabolites could be detected after analysis with paper chromatography, but they were not identified.

In a subsequent study these results were confirmed using spore suspensions of *Haplographium* sp., *Hormodendrum* sp., *Penicillium* sp., and *Spicaria* sp.[47] The previously described intermediate products, *p*-hydroxybenzoic acid, vanillic acid, and syringic acid, were found to be attacked by inducible enzymes, but Warburg studies did not reveal more details on the formation of products or ring cleavage. In the reported studies it was found that *p*-hydroxybenzaldehyde was more rapidly oxidized than the other compounds which were substituted with one or two methoxyl groups.

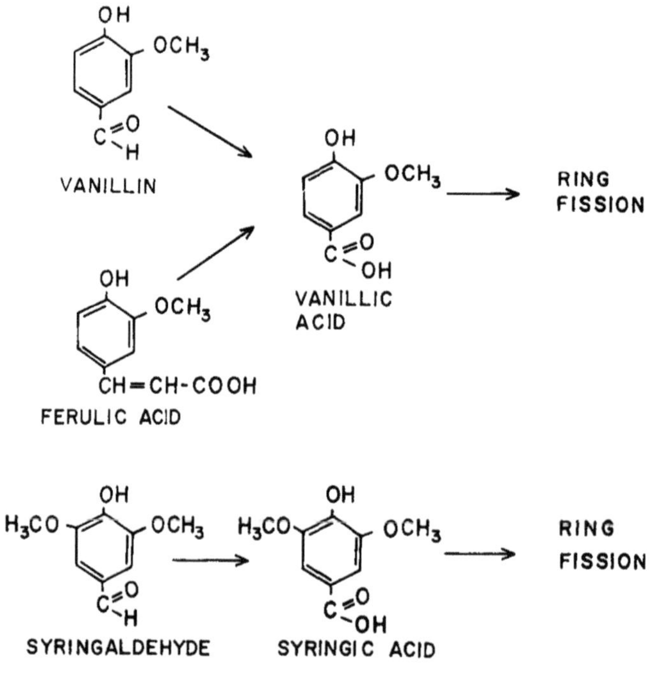

FIGURE 19. Transformation of vanillin, ferulic acid, and syringaldehyde by various soil fungi.

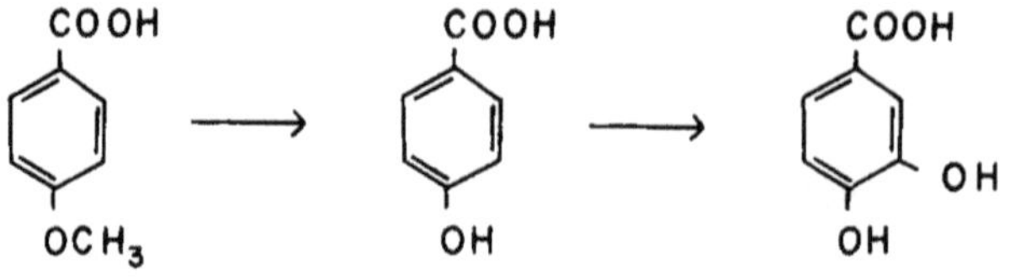

FIGURE 20. Dihydroxylation of *p*-methoxybenzoic acid by *Hormodendrum* sp. and *Penicillium* sp.

Henderson[24] also succeeded in isolating proto-catechuic acid, which was generated from p-methoxybenzoic acid via p-hydroxybenzoic acid (Figure 20), but a respiration experiment with six various dihydroxybenzoic acids gave only small increases in oxygen consumption during incubation with mycelia of Hormodendrum sp. and Penicillium sp.

It is also possible to conclude from these experiments that there still exists a vast field for speculations concerning the rupture and subsequent pathway of the benzene ring.

Rich and Horsfall[48] tested 42 different phenols and quinones on their degradability by mycelial extracts of Stemphylium sarcinaeforme and Monilinia fructicola. The indication used for transformation of the compounds by mycelial extracts was a color change in the reaction mixture which appears to be connected with polyphenol oxidase activity. If this change occurred it was concluded that the tested compound was oxidized and polymerized. The mycelial extract from S. sarcinaeforme contained a polyphenol oxidase capable of acting upon hydroquinone and catechol, but not tyrosine, i.e., a laccase, whereas the extract from M. fructicola gave color reactions with tyrosine and catechol but not with hydroquinone, therefore representing a tyrosinase type enzyme. The laccase oxidized and polymerized 41% of the test compounds, whereas the tyrosinase acted on only 17% of the investigated substances. This investigation also seems to indicate that fungi possess enzymes of a constitutive character which attack many aromatic compounds. However, the resulting transformation products remain unknown.

Lyr[49] stated that chlorophenols are oxidized by fungal oxidases such as laccase, tyrosinase, and peroxidase, but the chemical changes occurring are not clear and polymerization appears probable. After incubation of chlorophenols, a release of some of the bound chlorine was detected and, as second step of oxidation, ring cleavage "was expected."

In a recent publication, Cain et al. surveyed[50] a number of fungi in relation to their metabolic and ring-fission ability and they claimed that most of the fungi examined were able to convert proto-catechuate to β-ketoadipate with β-carboxy-muconolactone as a reaction intermediate. The fungi which were grown on p-hydroxybenzoate rapidly oxidized protocatechuate, but not

catechol. They concluded also that since the fungi did not degrade β-oxoadipate enol-lactone, the ortho-fission route for the investigated fungi is quite different from the bacterial pathway in which β-oxoadipate enol-lactone and γ-carboxy-muconolactone are established as intermediates. The following mycelia-forming fungi which originated from soil were investigated in this study: Penicillium spinulosum, Aspergillus niger, Fusarium oxysporum, Cephalosporium acremonium, a Cylindrocephalum sp., and a Phoma sp.

All the studies described above on aromatic attack by fungi did not concern actual pesticides, but they related to compounds with a similar chemical configuration. In the numerous investigations of the fungal attack on pesticidal compounds, and their transformation, no conclusive report on the breakage of an aromatic ring exists.

Several phenylurea herbicides which are attacked by fungi such as Penicillium and Aspergillus are used as a carbon source in an agar medium[5] or disappear from a liquid growth medium, as indicated by loss of radioactivity.[9,10]

Ring cleavage of triazine herbicides does not appear to be caused by soil fungi. First there was a report that Aspergillus fumigatus reduces the amount of radioactivity in culture solutions containing C^{14}-chain-labeled simazine more rapidly than in culture solutions containing C^{14}-ring-labeled simazine.[51] However, in a subsequent study it was claimed that $C^{14}O_2$ was evolved only from the chain-labeled simazine, which then leads to the conclusion that the ring portion of the molecule was still intact.[52]

The ring of atrazine also is not cleaved by fungi. Experiments with C^{14}-ring-labeled atrazine gave no indication that ring fission occurs in the growth medium, although transformations in the side chains took place.[11] The fungi investigated included Aspergillus fumigatus, A. flavipes, A. ustus, Rhizopus stolonifer, Fusarium moniliforme, F. roseum, F. oxysporum, Penicillium decumbens, P. janthinellum, P. rugulosum, P. luteum, and Trichoderma viride.

Since there was little evidence of ring fission in the metabolism of various phenoxyalkanoates, the group at Long Ashton selected[53] a nonpolar source to examine the effect of this chemical configuration on ring cleavage. In this case it was possible to demonstrate that 2-methoxynaphthalene was degraded in the presence of *A. niger* mycelium to 4-methoxysalicylic acid (Figure 21).

The presence of the hydroxyl group in the final product suggests that this substitution must occur after the initial ring fission, otherwise the methoxyl-substituted ring would have been the more active one during ring cleavage. Even though an intermediate could be isolated after ring cleavage, the attack on the aromatic ring remains nebulous in this case.

Epoxidation

The direct addition of oxygen to a double bond presents an epoxidation reaction and it is now recognized as a normal process for the metabolism of xenobiotic compounds. Chlorinated hydrocarbon pesticides can undergo epoxidation provoked by microbial activity to yield products which possess an increased toxicity in the environment. Especially, fungi were found to convert the insecticides heptachlor and aldrin to their corresponding epoxides.

Various epoxidized products from the insecticide heptachlor (1,4,5,6,7,8,8-heptachloro-3a,4,7,7a-tetrahydro-4,7-methanoindene) could be isolated after incubation with numerous fungal organisms.[54] Heptachlor was directly oxidized to heptachlor epoxide by cultures of *Rhizopus, Fusarium, Penicillium,* and *Trichoderma* species. Another epoxidation reaction occurred after an initial chemical hydrolysis of heptachlor to 1-hydroxychlordene. In this case, the chemical intermediate product was converted by the fungi to 1-hydroxy-2,3-epoxychlordene.

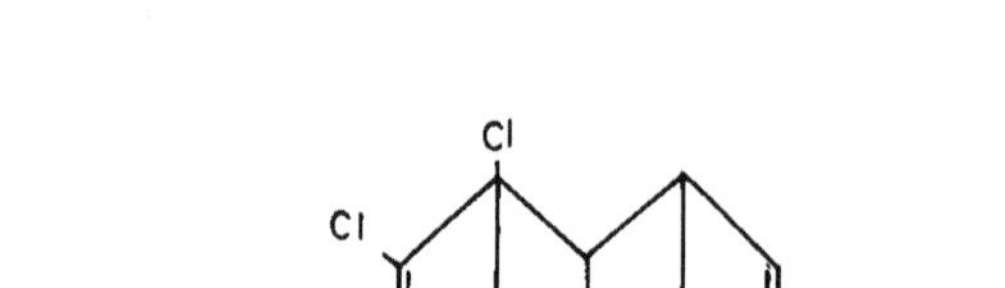

A similar epoxidation was found with aldrin which is transformed to dieldrin (Figure 22). Ninety-two different strains of bacteria, actinomycetes, and fungi were tested for their ability to transform aldrin and most of them could epoxidize the original pesticide to dieldrin.[55]

The most active isolates in this conversion were species of *Trichoderma,* but other fungi like *Fusarium, Penicillium,* and *Aspergillus* were also quite effective in the epoxidation process. In these studies it was evident that transformation of the insecticide in liquid medium took place in the exponential growth period of the fungi, but the insecticide alone was considered as an inadequate sole source of carbon.

β-Oxidation

β-Oxidation of an aliphatic side chain proceeds by the removal of two-carbon fragments from a fatty acid; the shortened fatty acid can be further degraded in the same manner until the chain length is four or two carbons. This reaction can be observed among the pesticides, especially with phenoxyalkanoic acids which possess essentially fatty acids linked by an ether bridge to an aromatic body.

In replacement cultures, *Aspergillus niger* was capable of degrading the side chains of various phenoxyalkanoic acids by β-oxidation and ring hydroxylation.[33] The fungus transformed phenoxybutyric acid to 4-hydroxyphenoxyacetic acid (Figure 23) and phenoxyvaleric acid to 3-(4-hydroxyphenoxy)-propionic acid. β-Oxidation by *A. niger* was also found with various members of the ω-(2-naphthyloxy)-*n*-alkyl carboxylic acid series,[38] whereas all compounds having an even number of carbon atoms in the side chain are all

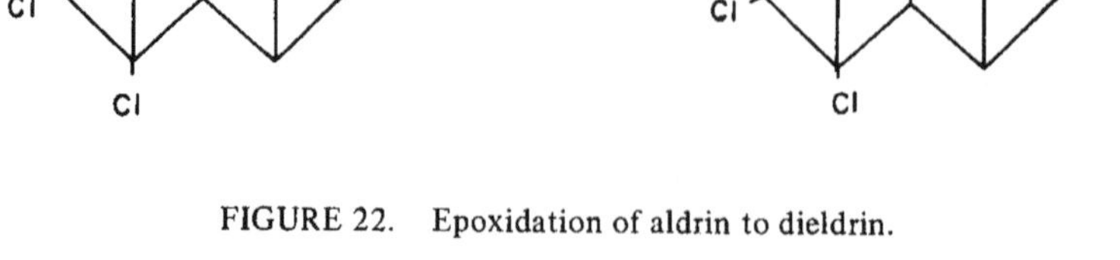

FIGURE 21. Ring cleavage of 2-methoxynaphthalene by *Aspergillus niger*.

FIGURE 22. Epoxidation of aldrin to dieldrin.

48

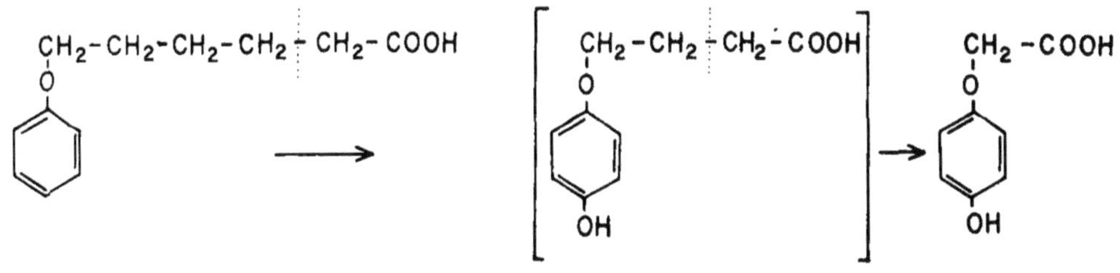

FIGURE 23.   β-Oxidation of phenoxybutyric acid by *Aspergillus niger.*

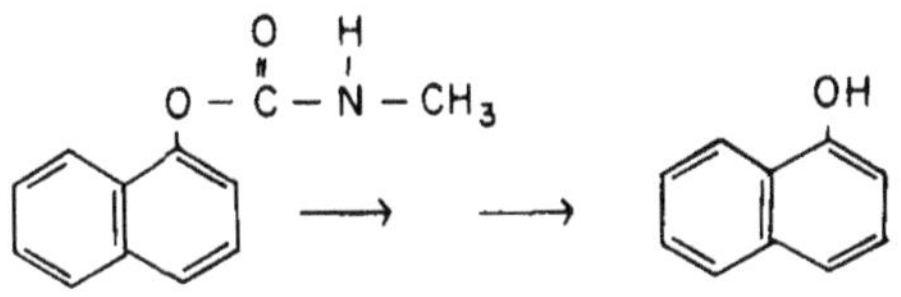

FIGURE 24.   Hydrolysis of carbaryl.

successively β-oxidized to 6-hydroxy-2-naphthyloxyacetic acid and the compounds with an odd number of carbon atoms are degraded only to 6-hydroxy-2-naphthyloxypropionic acid.

Even though there are several reports about β-oxidation of chlorinated phenoxyalkanoates by bacteria and actinomycetes as well as in natural soil,[56] no fungal β-oxidation was described with the chlorinated compounds. β-Oxidation of phenoxy herbicides with long aliphatic side chain moieties results in no loss of phytotoxicity and may even lead to an increase.

HYDROLYTIC REACTIONS

Since a great number of pesticides are esters, hydrolysis is a very likely degradative reaction. This process which can be provoked by enzymes may also occur chemically by exposing a certain substance to alkaline or acidic conditions. Therefore, sometimes it may be difficult to decide if hydrolysis is a *primary* biological reaction catalyzed by an enzyme or if it is a *secondary* caused by biological changes in pH, thereafter inducing a chemical change.

Fungal hydrolysis was found in various groups of herbicides and insecticides and it was reported often that the hydrolytic product could be isolated easily and identified in pure culture studies; this indicates that the hydrolytic process may often constitute the only detoxication reaction used by the fungus.

Carbamates are esters which are relatively easily hydrolyzed by chemical as well as biological processes. However, in this group, it is sometimes difficult to establish which one of these processes is predominant, since through biological hydroxylation or dealkylation, an unstable compound can be produced whose spontaneous decomposition results in a hydrolysis product.

Carbaryl, e.g., is hydrolyzed to 1-naphthol, but it is difficult to establish to what extent the pathway of degradation is enzymatic or chemical (Figure 24).[57]

The only report which describes the activity of microbial hydrolytic activity on a carbamate at a subcellular level was performed with a bacterial cell-free extract.[58] A partially purified enzyme from a *Pseudomonas* sp. hydrolyzed various phenylcarbamates with the appearance of the corresponding aniline, but methylcarbamates were not transformed.

As in the case of phenylcarbamates, it appears that hydrolysis is also the major metabolic reaction for acylanilides. Sharabi and Bordeleau[59] isolated two species of *Penicilllum* and one species of *Pulullaria* capable of hydrolyzing the herbicide karsil [*N*-(3,4-dichlorophenyl)-2-methylpentanamide] from soil. One *Penicillium* species produced 3,4-dichloroaniline as well as 2-methylvaleric acid (Figure 25). The latter compound disappeared during additional incubation or it could serve as the sole carbon source for growth, whereas 3,4-dichloroaniline was not metabolized further and accumulated in the medium. A cell-free extract was prepared and partially purified from the *Penicillium* sp., and it was concluded that the active enzyme was inducible. The specificity of the cell-free extract was tested on a large number of structurally related compounds. Increasing activity was established with increasing chain length, whereas activity was best with four-carbon side chain compounds. In addition, the substitution of the N-acyl group or the phenyl ring also influenced enzyme activity. However, diuron and isopropyl N-(3-chlorophenyl)carbamate (CIPC), a phenylurea and phenylcarbamate herbicide, respectively, were not attacked. It is also interesting to point out that not only the inducing substrate but also the composition of the growth medium seems to influence the formation of the active enzyme.

A similar hydrolysis reaction of an acylanilide herbicide was reported with *Fusarium solani*.[60] This fungus could use propanil, 3',4'-dichloropropionanilide, as a sole carbon source in pure culture studies and transformed it to 3,4-dichloroaniline and propionic acid (Figure 26). It appeared that 3,4-dichloroaniline accumulated in the medium and was not further degraded; at a certain concentration it even inhibited further utilization of the added herbicide.

An acylamidase was isolated[61] from the fungal mycelium which hydrolyzed propanil to 3,4-dichloroaniline, when acetanilide was used as the inducing substrate. The active enzyme was concentrated by salt precipitation and characterized with acetanilide as substrate in relation to the influence of pH, temperature, and substrate concentration. Hydrolysis rates were decreased by various *para*-substitutions of acetanilide; chloro-substitution reduced the rate of hydrolysis relative to acetanilide to 78% and methoxy-substitution to 31%. The described fungal acylamidase was highly specific for N-acetylarylamines and was unable to hydrolyze phenylurea herbicides (monuron and fenuron). Other acylanilides such as dicryl and karsil appeared unaffected by this enzyme system.

Matsumura and Boush[62] demonstrated that *Trichoderma viride* hydrolyzes malathion [O,O-dimethyl S-bis (carboethoxy)ethyl phosphorodithioate]. Soluble carboxyl esterase enzymes were isolated which could hydrolyze malathion to various carboxylic acid derivatives; hydrolysis of the O-methyl group phosphate esters also occurred (Figure 27). There is one report claiming that atrazine is hydrolyzed by *Fusarium roseum* to its corresponding hydroxy analogue,[63] but the identification was only assured by paper-chromatography and co-chromatography with the authentic compound. However, this hydrolysis is not surprising, since several similar hydrolysis reactions are known with higher plants.[13]

REDUCTION PROCESSES

Several reductive processes in the fungal degradation of pesticides appear to be of great importance also. One example is the reduction of nitro-substituents to amines. Madhosingh[64] showed that 2,4-dinitrophenol, which is used in fungicidal preparations, was reduced by *Fusarium oxysporum* to 2-amino-4-nitrophenol and 4-amino-2-nitrophenol in a liquid basal medium (Figure 28). It was anticipated that this reduction took place in stages, involving the intermediate formation of the nitroso and hydroxyamino groups. Formation of the 4-aminonitrophenol compound appeared to be favored in acid cultures, whereas the 2-amino isomer dominated at a higher pH value.

A large number of microorganisms appear to be capable of reducing pentachloronitrobenzene, a

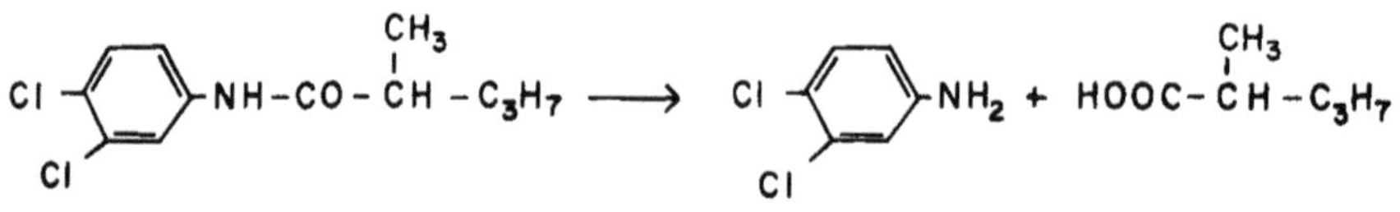

FIGURE 25.   Hydrolysis of karsil by a *Penicillium* sp.

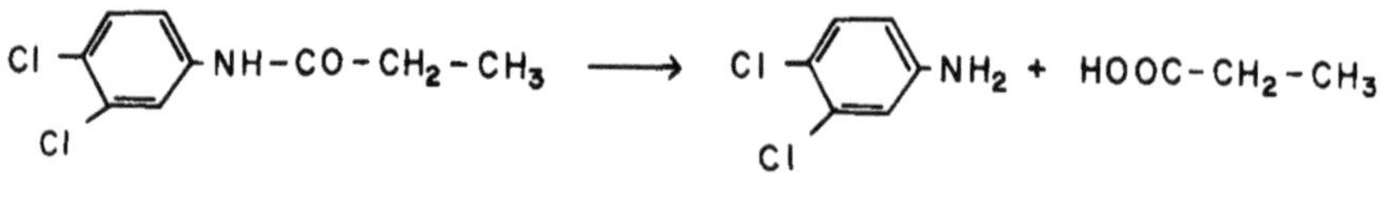

FIGURE 26.   Hydrolysis of propanil by *Fusarium solani*.

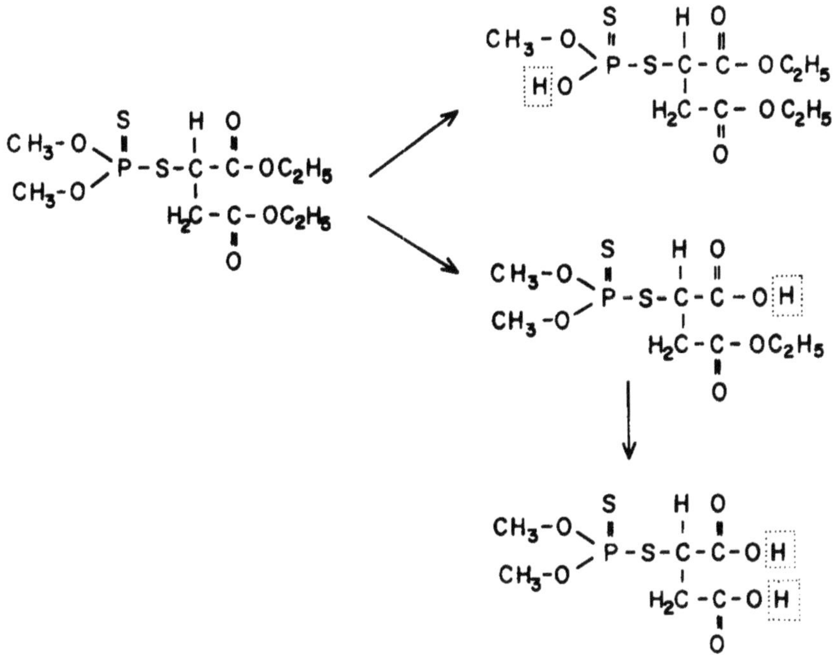

FIGURE 27. Hydrolytic transformation of malathion by *Trichoderma viride*.

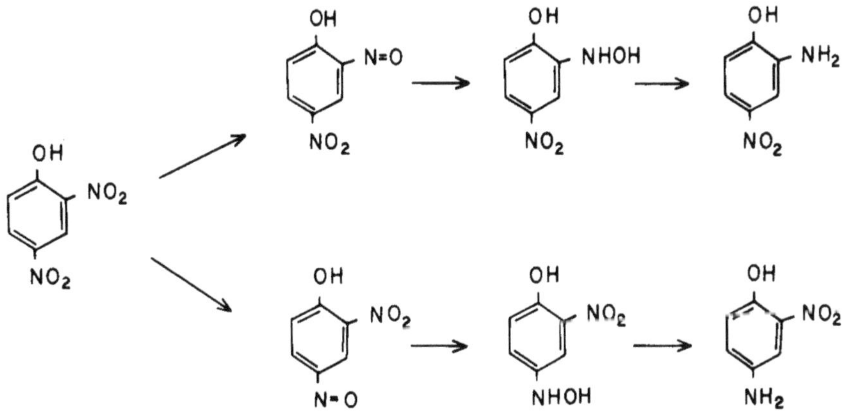

FIGURE 28. Suggested pathway of reduction of 2,4-dinitrophenol by *Fusarium oxysporum*.

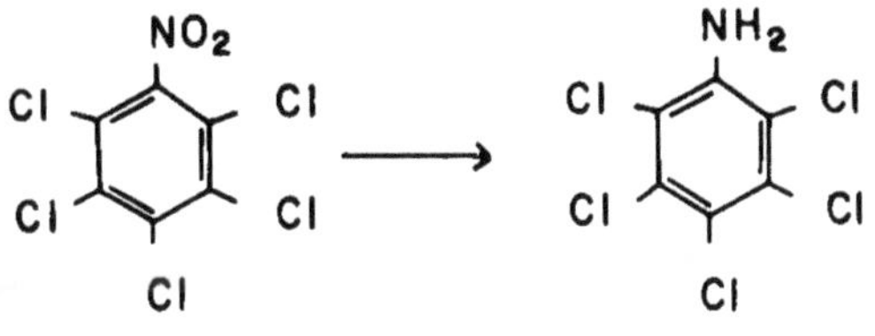

FIGURE 29. Conversion of pentachloronitrobenzene to pentachloroaniline.

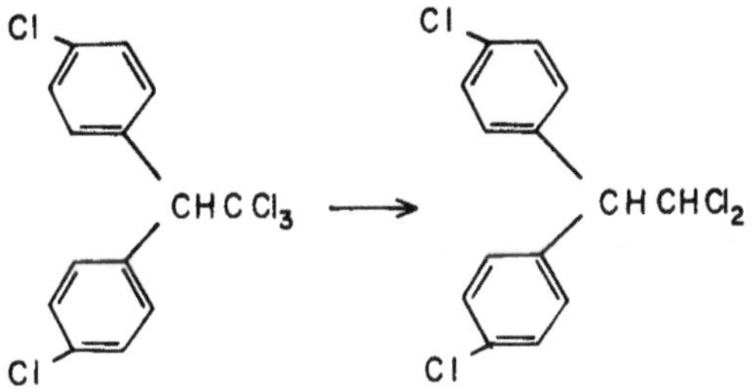

FIGURE 30. Reductive dehalogenation of DDT to DDD.

fungicidal compound, to pentachloroaniline (Figure 29).[65] All fungi tested participated in this reduction, namely *Aspergillus niger*, *Fusarium solani*, *Glomerella cingulata*, *Helminthosporium victoriae*, *Mucor ramannianus*, *Myrothecium verrucaria*, *Penicillium frequentans*, and *Trichoderma viride*.

Reduction of the insecticide DDT is caused by various organisms. The conversion of DDT, 2,2-bis(*p*-chlorophenyl)-1,1,1-trichloroethane, to DDD, 2,2-bis(*p*-chlorophenyl)-1,1-dichloroethane, represents a reductive dehalogenation (Figure 30). This process was demonstrated with yeast[66] under anaerobic conditions in soil,[67] but there are contradictory findings with fungi. Chacko et al.[65] claimed that a dechlorination of DDT by the tested soil fungi was not detected in liquid growth cultures, but Matsumura and Boush[68] received evidence that some species of *Trichoderma viride* reduced DDT to DDD. *Mucor alternans* partially metabolized DDT in shake cultures to various metabolites, but these products could not be identified as any known metabolites.[69] If spores of *M. alternans* were added to DDT-contaminated soil, there was no indication that the fungus degraded the insecticide.

SYNTHETIC MECHANISMS

Synthetic reactions are concerned with conjugation processes, i.e., the reaction between a xenobiotic compound or its metabolite(s) with an endogenous substrate resulting in the formation of glycosides, amino acid conjugates, acetylated amines, methylated substances, etc. All these mechanisms have still not been found extensively in the fungal metabolism of pesticides, but they occur frequently in higher organisms.

Formation of glucoside of pesticides is common in plants and insects, but has not yet been described for fungi.

A few reports exist on fungal methylation and acetylation. According to Williams,[2] methylation is a normal process of metabolism, which also occurs with xenobiotic organic molecules. Methylation is known with nitrogenous compounds, and it was also found in connection with phenolic hydroxyl groups.

Recently, it was claimed that pentachlorophenol, a substance of fungicidal and various other pesticidal activities, can be methylated by *Trichoderma virgatum* in liquid cultures,[70] and the

resulting product was identified as pentachloroanisole by melting point determination and infrared spectroscopy (Figure 31). It is still not clear whether methylation is the only initial attack of the pentachlorophenol molecule or rather if it presents a side reaction. In any case, it was found that pentachloroanisole is more resistant to further chemical and biological degradation than the parent chlorinated phenol, but it appears less toxic to various organisms.

Acetylation has been mostly observed as a reaction of foreign amines, when the amino group is attached to an aromatic ring. Enzymic acetylation usually requires coenzyme A for its activity and it was observed in most species of animals, where it is considered as the most common conjugation reaction for aromatic amines.[2]

Many herbicides which belong to the groups of phenylacylanilides, phenylcarbamates, or substituted phenylureas are degraded by a variety of microorganisms to the corresponding halogen-substituted anilines. These intermediates are biologically active and may undergo various transformation reactions. Tweedy et al.[6,7] demonstrated that the herbicide metobromuron, 3-(*p*-bromophenyl)-1-methoxy-1-methylurea, is demethylated and demethoxylated by the fungi *Talaromyces wortmanii* and *Fusarium oxysporum* and isolated *p*-bromoacetánilide as a product which indicates the occurrence of an acetylation reaction (Figure 32). *p*-Bromoaniline could not be

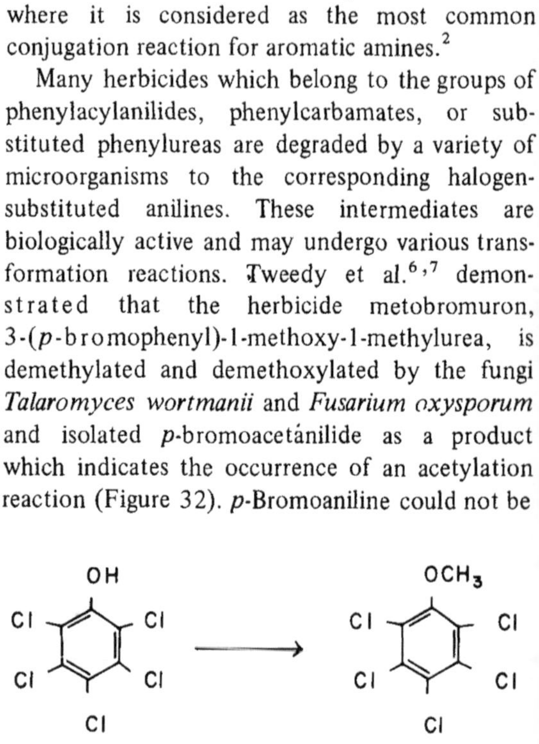

FIGURE 31. Methylation of pentachlorophenol to pentachloroanisole by *Trichoderma virgatum*.

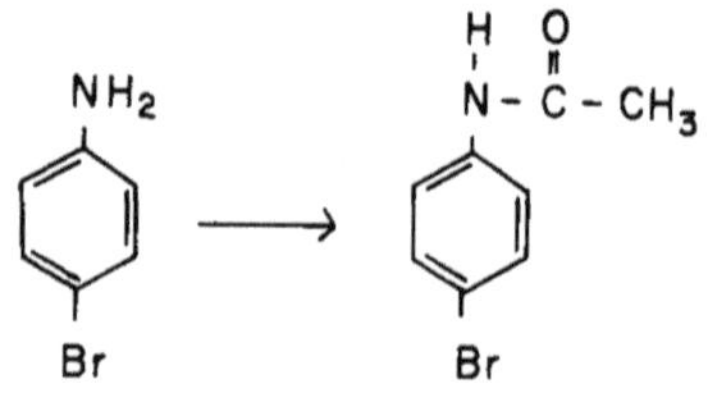

FIGURE 32. Acetylation of *p*-bromoaniline by *Talaromyces wortmanii* and *Fusarium oxysporum*.

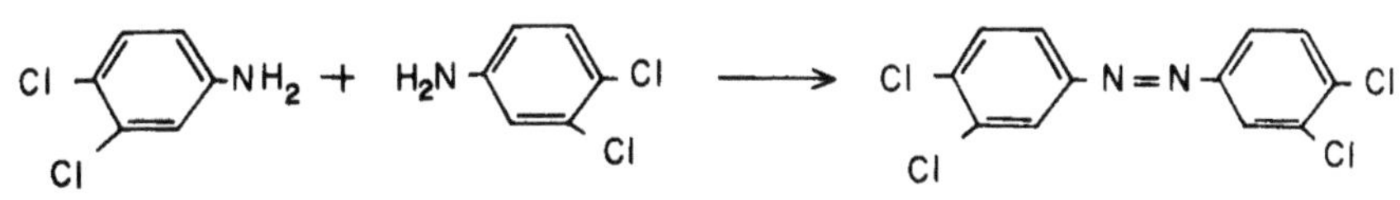

FIGURE 33. Dimerization reaction of 3,4-dichloroaniline by *Geotrichum candidum.*

found as an intermediate, but it can be assumed that acetylation of the aniline is a fast process and consequently it did not accumulate in the culture medium. However, if *p*-bromoaniline was used as a substrate, the brominated aniline was converted completely to *p*-bromoacetanilide after incubation for some days.

Lyr[71] concluded from UV absorption spectra that transformation of chlorinated phenols which are often used as fungicides resulted in the inactivation of phenolic hydroxyl-groups; no dihydroxylation or ring-cleavage could be found, but various observations indicated that oxidative polymerizations or other synthetic mechanisms lead to more complex compounds. It appears that this unexplored area may still reveal numerous unknown products by the fungal metabolism.

The actual participation of a fungus in polymerizing an aniline compound to an azo-derivative was described recently with *Geotrichum candidum* (Figure 33).[72] The acylanilide herbicide propanil, *N*-(3,4-dichlorophenyl)-propionamide, can be hydrolyzed by various fungi and other microorganisms with the release of the corresponding aniline, 3,4-dichloroaniline.[73] It was previously shown that 3,4-dichloroaniline can be transformed biochemically to 3,3',4,4'-tetrachloroazobenzene, but this transformation was not assigned to any specific microbe. Bordeleau and Bartha[72] demonstrated that *Geotrichum candidum* was capable of performing this reaction in nutrient solution as well as in sterilized soil. They stipulated that the ability of the fungus to produce considerable amounts of peroxidase is the cause of the polymerization reaction. The role of peroxidase in the conversion of anilines to azo-derivatives was previously established with pure peroxidases and in soil systems containing the enzyme.[74]

CONCLUSIONS

Fungi are an important group of microorganisms which participate actively in the biotransformation of pesticides, but the resulting products

in laboratory experiments often show only a slight change from the parent substance. With respect to pollution, it is of great importance to be aware of the possible toxicity of metabolic products toward the various organisms, and it is not justified to be concerned mainly with the contamination of our environment by the original form of the applied pesticides.

Although biochemical changes of pesticides appear to be a consequence of a few general biological mechanisms catalyzed by enzymes, basic differences undoubtedly exist between the detoxication reactions of filamentous fungi and bacteria. Pure culture experiments seem to indicate that the ability to transform or degrade pesticides is much more limited for fungi than for bacteria.

In the study of drug metabolism it has been previously pointed out[75] that most compounds are metabolized in higher organisms by nonspecific enzymes that catalyze relatively few general reactions. A similar statement seems appropriate for the detoxication reactions of fungi. Turner[76] also stated in his book, *Fungal Metabolites*, that in spite of the diversity of their chemical structures and their biological activities, "the fungal metabolites are derived from some half dozen basic biosynthetic pathways."

How a fungus reacts in a soil ecosystem toward a xenobiotic compound is almost impossible to determine, since a natural habitat is so complex in relation to the existing microflora, the influence of the chemical composition of the environment and the numerous competing reactions. Therefore, it is necessary to study the basic concepts of biochemical transformations in pure or model culture systems. Ideally it should be demanded that a metabolic process should be proven by the isolation of the enzyme system involved, but this objective may not be achieved easily because of great lability of certain enzymes or other inherent biochemical problems.

It is assumed[1,77] that filamentous fungi possess high biochemical activity upon a wide variety of simple and very complicated organic substances to convert them into cell material. It is even assumed that the transformation of complex organic materials such as lignin, cellulose, and hemicellulose can be performed better by fungi than by bacteria. However, this assumption often cannot be verified with model substrates where only a partial transformation took place. In his book, Foster[77] asks why fungi produce metabolic products other than cell material and CO_2 from sugar when cultivated under laboratory conditions. He suggests that the influence of abnormal environmental conditions provokes "deranged or pathological behaviour" of the fungi. The introduction in nature of xenobiotic compounds such as pesticides can also be considered as 'abnormal,' and could cause — in analogy to Foster's explanation — only partial decomposition of the chemicals.

From studies with chlorinated phenols, Lyr[49,71] concluded that fungal oxidases such as laccase, tyrosinase, and peroxidase play a major role in the dechlorination of these compounds without ring fission, while assuming that polymerization reactions or other synthetic mechanisms lead to detoxication or removal of these substances. It appears that the metabolic activity of fungal oxidases, which are often constitutive or easily induced in fungi, is an area of research which was very much neglected and which might reveal frequent transformation processes occurring with many xenobiotic compounds.

Fungi, in particular, produce enzymes which are secreted into the environment and clearly function in making outside nutrients available to the cell.[78] Pathogenic fungi, for example, excrete pectinases and cellulases which loosen cell walls of plants and facilitate penetration of the parasite into the plant. It is conceivable that such a mechanism may also be active in the attack on toxic pesticidal compounds, but no specific research has been devoted to the activity of fungal extracellular enzymes.

Whereas pesticidal molecules seem little changed, often only dealkylated, hydroxylated, etc., by fungal interference when observed under the minimal, artificial conditions of a synthetic growth medium, contrary results were obtained with a purified enzyme. In a study on the decomposition of 1-naphthol, a hydrolysis product of the insecticide carbaryl, it was shown[79] that little radioactivity of C^{14} labeled 1-naphthol

disappeared in the growth medium of *Fusarium solani,* whereas a cell-free preparation from the same fungus completely degraded the 1-naphthol molecule, and it was possible to trap the radioactivity as $C^{14}O_2$. This case demonstrates that an enzyme may behave differently when free in solution, or when adsorbed on certain surfaces or associated with other colloidal systems.[80] In a natural environment, lysis of fungal cells occurs and fungal enzymes are released which may be responsible for the transformation of a pesticidal chemical.

Fungi appear in different morphological forms such as spores and mycelia, but little work has been done to compare the biochemical activity of the various morphological stages; however, in the few cases investigated, a surprising similarity exists. Comparing the activity of fungal mycelium and spores on *p*-hydroxybenzaldehyde, ferulic acid, syringaldehyde, and vanillin, Henderson and Farmer[46] and Henderson[47] found basically the same transformation ability with each of the two different morphological forms. In a recent publication,[79] isolated enzymes from spores and mycelium were compared. The results showed that cell-free extracts had similar activity in the transformation of 1-naphthol which served in these experiments as substrate.

As previously outlined, one of the major deficiencies in the understanding of the biochemical activities of fungi is the lack of isolated and purified enzyme preparations which would allow detailed examination of the various reactions. Basic knowledge is necessary if one is concerned with the fate of pesticides in the environment, and there is no doubt that fungi participate in the transformation of these compounds by particular reactions which produce specific products. In vitro studies with isolated enzymes can shed light on the conditions and rates of transformation of certain molecules, as well as on the relationships between enzymatic activity and the possible attacks on compounds of various chemical structures.

The available techniques and the actual possibilities for studying xenobiotic transformations in a natural soil ecosystem are not satisfactory and it is conceivable that results from in vitro experiments may provide initial indications of a dangerous product which may cause serious problems in the environment.

REFERENCES

1. Alexander, M., *Introduction to Soil Microbiology*, John Wiley & Sons, New York, 1961, 63.
2. Williams, R. T., *Detoxication Mechanisms*, 2nd ed., Chapman and Hall, Ltd., London, 1959.
3. Geissbühler, H., Haselbach, C., Aebi, H., and Ebner, L., The fate of N'-(4-chlorophenoxy)-phenyl-N, N-dimethylurea (C-1983) in soils and plants. III. Breakdown in soils and plants, *Weed Res.*, 3, 277, 1963.
4. Dalton, R. L., Evans, A. W., and Rhodes, R. C., Disappearance of diuron in cotton field soils, *Weeds*, 14, 31, 1966.
5. Hill, G. D. and McGahen, J. W., Further studies on soil relationships of the substituted urea herbicides for pre-emergence weed control, *Proc. South. Weed Conf.*, 8, 284, 1955.
6. Tweedy, B. G., Loeppky, C., and Ross, J. A., Metabolism of 3-(p-bromophenyl)-1-methoxy-1-methylurea, *J. Agr. Food Chem.*, 18, 851, 1970.
7. Tweedy, B. G., Loeppky, C., and Ross, J. A., Metobromuron: Acetylation of the aniline moiety as a detoxification mechanism, *Science*, 168, 482, 1970.
8. Börner, H., Der Abbau von Harnstoffherbiziden im Boden, *Z. Pflanzenkr. Pflanzenpathol. Pflanzenschutz*, 74, 135, 1967.
9. Weinberger, M., Transformation of substituted phenylurea herbicides by the fungus *Rhizoctonia solani*, M. S. Thesis, Pennsylvania State University, 1972.
10. Börner, H., Burgemeister, H., and Schroeder, M., Untersuchungen über die Aufnahme, Verteilung und Abbau von Harnstoffherbiziden durch Kulturpflanzen, Unkräuter und Mikroorganismen, *Z. Pflanzenkr. Pflanzenpathol. Pflanzenschutz*, 76, 385, 1969.
11. Kaufman, D. D. and Blake, J., Degradation of atrazine by soil fungi, *Soil Biol. Biochem.*, 2, 73, 1970.
12. Kearney, P. C., Kaufman, D. D., and Sheets, T. J., Metabolites of simazine by *Aspergillus fumigatus*, *J. Agr. Food Chem.*, 13, 369, 1965.
13. Knuesli, E., Berrer, D., Dupuis, G., and Esser, H., s-Triazines, in *Degradation of Herbicides*, Kearney, P.C. and Kaufman, D. D., Eds., Marcel-Dekker, New York, 1969, 51.
14. Bollag, J.-M. and Liu, S.-Y., Hydroxylations of carbaryl by soil fungi, *Nature*, 236, 177, 1972.
15. Liu, S.-Y. and Bollag, J.-M., Metabolism of carbaryl by a soil fungus, *J. Agr. Food Chem.*, 19, 487, 1971.
16. Liu, S.-Y. and Bollag, J.-M., Carbaryl decomposition to 1-naphthyl carbamate by *Aspergillus terreus*, *Pestic. Biochem. Physiol.*, 1, 366, 1971.
17. Hodgson, E. and Casida, J. E., Metabolism of *N,N*-dialkyl carbamates and related compounds by rat liver, *Biochem. Pharmacol.*, 8, 179, 1961.
18. Kesner, C. D. and Ries, S. K., Diphenamid metabolism in plants, *Science*, 155, 210, 1967.
19. Brodie, B. B., Axelrod, J., Cooper, J. R., Gaudette, L., La Du, B. N., Mitoma, C., and Udenfriend, S., Detoxication of drugs and other foreign compounds by liver microsomes, *Science*, 121, 603, 1955.
20. Nilsson, A. and Johnson, B. C., Cofactor requirements of the *O*-demethylating liver microsomal enzyme system, *Arch. Biochem. Biophys.*, 101, 494, 1963.
21. Hollingworth, R. M., The dealkylation of organophosphorus triesters by liver enzymes, in *Biochemical Toxicology of Insecticides*, O'Brien, R. D. and Yamamoto, I., Eds., Academic Press, New York, 1970, 75.
22. Fukami, J. and Shishido, T., Nature of a soluble, glutathione-dependent enzyme system active in cleavage of methyl parathion to desmethyl parathion, *J. Econ. Entomol.*, 59, 1338, 1966.
23. Donninger, C., Hutson, D. H., and Pickering, B. A., Oxidative cleavage of phosphoric acid triesters to diesters, *Biochem. J.*, 102, 26, 1966.
24. Henderson, M. E. K., The metabolism of methoxylated aromatic compounds by soil fungi, *J. Gen. Microbiol.*, 16, 686, 1957.
25. Stenersen, J., Degradation of P^{32}-bromophos by microorganisms and seedlings, *Bull. Environ. Contam. Toxicol.*, 4, 104, 1969.
26. Stenersen, J., Demethylation of the insecticide bromophos by a glutathione-dependent liver enzyme and by alkaline buffers, *J. Econ. Entomol.*, 62, 1043, 1969.
27. Bocks, S. M., Lindsay-Smith, J. R., and Norman, R. O. C., Hydroxylation of phenoxyacetic acid and anisole by *Aspergillus niger* van Tiegh, *Nature*, 201, 398, 1964.
28. Hock, W. K. and Sisler, H. D., Metabolism of chloroneb by *Rhizoctonia solani* and other fungi, *J. Agr. Food Chem.*, 17, 123, 1969.
29. Boothroyd, B., Napier, E. J., and Somerfield, G. A., The demethylation of griseofulvin by fungi, *Biochem. J.*, 80, 34, 1961.
30. Hayaishi, O., Enzymic hydroxylation, *Annu. Rev. Biochem.*, 38, 21, 1969.
31. Udenfriend, S., Clarke, C. T., Axelrod, J., and Brodie, B. B., Ascorbic acid in aromatic hydroxylation. 1. A model system for aromatic hydroxylation, *J. Biol. Chem.*, 208, 731, 1954.
32. Bollag, J.-M. and Liu, S.-Y., Fungal transformation of the insecticide carbaryl, *Bacteriol. Proc.*, A94, 16, 1971.
33. Byrde, R. J. W. and Woodcock, D., Fungal detoxication. 2. The metabolism of some phenoxy-*n*-alkylcarboxylic acids by *Aspergillus niger*, *Biochem. J.*, 65, 682, 1957.
34. Clifford, D. R. and Woodcock, D., Metabolism of phenoxyacetic acid by *Aspergillus niger* van Tiegh, *Nature*, 203, 763, 1964.

35. Faulkner, J. K. and Woodcock, D., Fungal detoxication. Part V. Metabolism of *o*- and *p*-chlorophenoxyacetic acid by *Aspergillus niger, J. Chem. Soc.,* 5397, 1961.

36. Faulkner, J. K. and Woodcock, D., Fungal detoxication. Part VII. Metabolism of 2,4-dichlorophenoxyacetic and 4-chloro-2-methylphenoxyacetic acids by *Aspergillus niger, J. Chem. Soc.,* 1187, 1965.

37. Faulkner, J. K. and Woodcock, D., Metabolism of 2,4-dichlorophenoxyacetic acid by *Aspergillus niger* van Tiegh, *Nature,* 203, 865, 1964.

38. Byrde, R. J. W., Harris, J. F., and Woodcock, D., Fungal detoxication. 1. The metabolism of ω-(2-naphthyloxy)-*n*-alkylcarboxylic acids by *Aspergillus niger, Biochem. J.,* 64, 154, 1956.

39. Rogoff, M. H., Oxidation of aromatic compounds by bacteria, *Adv. Appl. Microbiol.,* 3, 193, 1961.

40. Gibson, D. T., Microbial degradation of aromatic compounds, *Science,* 161, 1093, 1968.

41. Utkin, L. M., Homogentisic acid in the metabolism of molds, *Biokhimiia,* 15, 330, 1950.

42. Kluyver, A. J. and Van Zijp, J. C. M., The production of homogentisic acid out of phenylacetic acid by *Aspergillus niger, Antonie van Leeuwenhoek,* 17, 315, 1951.

43. Hockenhull, D. J. D., Walker, A. D., Wilkin, G. D., and Winder, F. G., Oxidation of phenylacetic acid by *Penicillium chrysogenum, Biochem. J.,* 50, 605, 1952.

44. Isono, M., Oxidative metabolism of phenylacetic acid by *Penicillium chrysogenum* Q176. III. Intermediary production of 2-hydroxyphenoxyacetic acid, *J. Agr. Chem. Soc. Jap.,* 27, 255, 1953.

45. Isono, M., Oxidative metabolism of phenylacetic acid by *Penicillium chrysogenum* Q176. II. Disappearance of phenylacetic acid in Penicillin culture broth, and manometric estimation of phenylacetic acid oxidation, *J. Agr. Chem. Soc. Jap.,* 27, 198, 1953.

46. Henderson, M. E. K. and Farmer, V. C., Utilization by soil fungi of *p*-hydroxybenzaldehyde, ferulic acid, syringaldehyde and vanillin, *J. Gen. Microbiol.,* 12, 37, 1955.

47. Henderson, M. E. K., A study of the metabolism of phenolic compounds by soil fungi using spore suspensions, *J. Gen. Microbiol.,* 14, 684, 1956.

48. Rich, S. and Horsfall, J. G., Relation of polyphenol oxidases to fungitoxicity, *Proc. Natl. Acad. Sci. USA,* 40, 139, 1956.

49. Lyr, H., Detoxification of heartwood toxins and chlorophenols by higher fungi, *Nature,* 195, 289, 1962.

50. Cain, R. B., Bilton, R. F., and Darrah, J. A., The metabolism of aromatic acids by micro-organisms. Metabolic pathways in the fungi, *Biochem. J.,* 108, 797, 1968.

51. Kaufman, D. D., Kearney, P. C., and Sheets, T. J., Simazine: degradation by soil microorganisms, *Science,* 142, 405, 1963.

52. Kaufman, D. D., Kearney, P. C., and Sheets, T. J., Microbial degradation of simazine, *J. Agr. Food Chem.,* 13, 238, 1965.

53. Byrde, R. J. W., Downing, D. F., and Woodcock, D., Fungal detoxication. 4. Metabolism of 2-methoxynaphthalene by *Aspergillus niger, Biochem. J.,* 72, 344, 1959.

54. Miles, J. R. W., Tu, C. M., and Harris, C. R., Metabolism of heptachlor and its degradation products by soil microorganisms, *J. Econ. Entomol.,* 63, 1334, 1969.

55. Tu, C. M., Miles, J. R. W., and Harris, C. R., Soil microbial degradation of aldrin, *Life Sci.,* 7, 311, 1968.

56. Loos, M. A., Phenoxyalkanoic acids, in *Degradation of Herbicides,* Kearney, P. C. and Kaufman, D. D., Eds., Marcel-Dekker, New York, 1969, 1.

57. Bollag, J.-M. and Liu, S.-Y., Degradation of Sevin by soil microorganisms, *Soil Biol. Biochem.,* 3, 337, 1971.

58. Kearney, P. C., Purification and properties of an enzyme responsible for hydrolyzing phenylcarbamates, *J. Agr. Food Chem.,* 13, 561, 1965.

59. Sharabi, N. El-D. and Bordeleau, L. M., Biochemical decomposition of the herbicide *N*-(3,4-dichlorophenyl)-2-methylpentanamide and related compounds, *Appl. Microbiol.,* 18, 369, 1969.

60. Lanzilotta, R. P. and Pramer, D., Herbicide transformation. I. Studies with whole cells of *Fusarium solani, Appl. Microbiol.,* 19, 301, 1970.

61. Lanzilotta, R. P. and Pramer, D., Herbicide transformation. II. Studies with an acylamidase of *Fusarium solani, Appl. Microbiol.,* 19, 307, 1970.

62. Matsumura, F. and Boush, G. M., Malathion degradation by *Trichoderma viride* and a *Pseudomonas* species, *Science,* 153, 1278, 1966.

63. Couch, R. W., Gramlich, J. V., Davis, D. E., and Funderburk, H. H., The metabolism of atrazine and simazine by soil fungi, *Proc. South. Weed Conf.,* 18, 623, 1965.

64. Madhosingh, C., The metabolic detoxification of 2,4-dinitrophenol by *Fusarium oxysporum, Can. J. Microbiol.,* 7, 553, 1961.

65. Chacko, C. I., Lockwood, J. L., and Zabik, M. L., Chlorinated hydrocarbon pesticides: Degradation by microbes, *Science,* 154, 893, 1966.

66. Kallman, B. J. and Andrews, A. K., Reductive dechlorination of DDT to DDD by yeast, *Science,* 141, 1050, 1963.

67. Guenzi, W. D. and Beard, W. E., Anaerobic biodegradation of DDT to DDD in soil, *Science,* 156, 1116, 1967.

68. Matsumura F. and Boush, G. M., Degradation of insecticides by a soil fungus, *Trichoderma viride, J. Econ. Entomol.,* 61, 610, 1968.

ACTIVITY, ECOLOGY, AND POPULATION DYNAMICS OF MICROORGANISMS IN SOIL

Author: **G. Stotzky**
Department of Biology
New York University
New York, N.Y.

I. INTRODUCTION

The soil was probably the first microbial ecosystem to be studied in any detail. Whether this was a result of the early practical development of microbiology (e.g., to increase food production through the understanding of nitrogen fixation, prevention of plant diseases and food spoilage) or of the, still apparent, perversity of microbiologists — especially soil microbiologists — is not clear. What is abundantly clear, however, is that the soil is undoubtedly the most complex of all microbial habitats. As such, it also offers one of the greatest challenges to microbial ecologists. Despite the plethora of books, reviews, and papers dealing with soil microbiology, understanding of the activity, ecology, and population dynamics of microorganisms in soil is still scant.

Soil, which is found almost everywhere on Earth, can best be defined as a tripartite system composed of finely divided minerals (both primary and secondary); animal, plant, and microbial residues in various stages of decay; and a living and metabolizing microbiota. Consequently, microbes not only conduct their activities in soil but also constitute an integral part of soil. Furthermore, the other two components are, to a great degree, the result of microbial conversion of parent rock materials and living matter. In addition to the solid phase, soil also contains a liquid and a gaseous phase. The results of interactions between these phases are a complex of chemical, physical, and biological interfaces, which must be recognized and made intelligible if the ecology of microbes in soil is to be unraveled and understood.

Although individual chemical (e.g., organic and inorganic nutrients, growth factors, ionic composition), physical (e.g., moisture, temperature, pressure, radiation, atmospheric composition, pH, oxidation-reduction potential, particulates), and biological (e.g., characteristics of the organisms, positive and negative interactions between organisms) factors that influence microbial ecology in soil can be easily listed, their relative importance and influence cannot be assumed nor studied as easily, inasmuch as their expression usually does not result from the action of one or several variables, but rather from the sum of numerous interactions. For example, the water content of soil cannot be altered without concomitantly influencing the atmospheric composition; changes in pH are accompanied by changes in oxidation-reduction potentials and ionic composition; pressure and temperature are interrelated. Consequently, an alteration in one environmental factor causes a change not only in that factor but simultaneously and subsequently in several, with the result that the entire environment is changed.

Obviously, many of the factors that influence microbes in soil also influence microorganisms in other habitats. Attempts are usually made, in all habitats, to study each factor individually, primarily for reasons of simplicity in experimentation and presentation of data. However, the essentially unlimited permutations between factors probably exert a greater influence in soil than does an individual factor in less complex habitats.

Another consideration in the study of microbes in soil is that microbial life in soil exists in discrete microhabitats, the chemical, physical, and biological characteristics of which undoubtedly differ in both time and space. Although knowledge of these microhabitats is sparse and primarily deductive, the concept of discrete microhabitats is a necessary assumption at this stage of development of the art of soil microbial ecology. The concept is derived primarily from the fact that soil is a heterogeneous, discontinuous, and structured environment, dominated by a solid phase varying in size from less than 0.2 μm to greater than 2 mm. This discontinuity and variability in particle size results in soil being a composite of enumerable small microbial communities, each circumscribed by its own immediate environment.

The description of the microhabitat, as that of the soil itself, is difficult, and each soil microbiologist undoubtedly has his or her own visual conception. Many would agree, however, that the basic components are sand- and silt-sized particles coated, completely or partially, with skins or cutans of clay minerals and organic matter[1] and surrounded by water films of various thicknesses. These water-coated particulates can probably exist individually but, most likely, occur as aggregates, with water films surrounding the entire aggregate and forming bridges with adjacent aggregates (i.e., microhabitats) close enough to retain water by surface tension. In this manner, each microhabitat is an independent entity, whose interchange with adjacent worlds is limited by the energy within the shared water bridges. This energy influences, among other phenomena, the movement of soluble nutrients and the exchange of volatiles and gases, and it imposes an energy threshold which microbes must exceed if they are to move from one

microhabitat to another. The remaining space between microhabitats, the pore space, is filled with gases, the amounts of which depend on the water content of the soil at any given time.

Also inherent in the concept of the microhabitat is that microbes are aquatic creatures, and, therefore, even in a terrestrial habitat, such as soil, are restricted to the aqueous phase. Although some microbes, primarily filamentous forms, can grow through the voids between water films, their structures are coated with water, and nutrient uptake probably occurs only in sites of accumulated water.

If the concept of discrete microhabitats is accepted, even if only temporarily, it is possible also to accept the concept of the diversity of microhabitats and, therefore, the variability in the microbial composition between even closely adjacent microhabitats. For example, if the clay cutans around the sand grains composing an aggregate are predominantly of one mineralogical type, whereas the cutans of an adjacent aggregate are predominantly of another, the cation exchange capacity, the exchangeable cations, the pH, the thickness of the water films, and most other physicochemical characteristics will differ, and, in turn, the types of microbes capable of invading and persisting in these two microenvironments will differ.

A further assumption to be derived from the microhabitat concept is that, because of the relatively small size of the microhabitats and their essential isolation from one another, only small environmental changes are necessary to alter their microbiological composition. Consequently, whereas the soil as a whole is generally considered to be a "well-buffered system" in "dynamic equilibrium," the sites where the action is, i.e., the microhabitats, may, in fact, be less buffered and either undergo rapid and extreme fluctuations or, depending on their degree of isolation, they may attain something akin to an adapted climax population until the isolation is ended and the environment again is changed. These aspects may be magnified even more, as the size of individual microbes, and certainly of microcolonies, may not be much smaller than that of their microhabitats.

These concepts, in concert, probably explain, at least in part, why it is more difficult to evaluate the influence of various environmental factors on the activity, ecology, and population dynamics of microorganisms in soil than in more homogeneous and less compartmentalized microbial habitats. For example, the oxidation-reduction potential of soil is relatively simple to determine, but its interpretation is difficult and, often, meaningless. The measured potential is a composite of all the half-cells present, and, as these are usually not known, no indication is obtained of the relative states of oxidation and reduction of components between and within individual microhabitats. Because nutrient supply, water content, aeration, mineral species, pH, composition of the microbiota, etc. influence the metabolism within the microhabitats, the redox potential and, in turn, the microbial composition of each will differ.

Consequently, the soil microbiologist has enlisted techniques and concepts of methodology from all branches of science. Analytical (autecological) and synthetic (synecological) approaches to enumerate and define the soil microbiota are used; metabolic studies are conducted with pure cultures of microbes originally isolated from soil as well as with mixed populations either in laboratory culture or in the soil itself; analytical techniques, from the simplest spot test to the most sophisticated instrument currently available, have been invoked. Despite the wealth of information that has been derived from these and other approaches, too infrequently have several been applied simultaneously to the same question. Because of the complexity of soil as a microbial environment, the methodology must, obviously, also be complex and multipronged. Furthermore, the soil microbial ecologist must be not only a microbiologist but also a soil scientist, a physical chemist, a mineralogist, an ecologist; in fact the "compleat scientific man."

In an attempt to simplify at least some of the complexities of the ecology of microorganisms in soil, attention here is focused on (1) the factors that influence microorganisms in soil; (2) some of the methodology used to study these factors; and (3) an example of how one factor interacts with others to influence the activity, ecology, and population dynamics of microbes in soil. This presentation stresses not so much what is known about the ecology of microbes in soil, but rather attempts to emphasize that which is known about microbes in general but which has not been examined and related sufficiently in soil.

To facilitate both the writing and reading of this presentation, references to original sources have been kept at a minimum. For greater detail and differing points of view, the attention of the interested reader is directed to various reviews and texts dealing with microbial ecology in soil.[2-21]

II. FACTORS AFFECTING MICROORGANISMS IN SOIL

A. Substrates

Most soils probably provide an array of carbonaceous substrates as vast as the microbial types present therein. All possible organic materials (e.g., the complex residues of plants and animals; the products of these residues transformed by microbes, by man, or abiologically; materials synthesized by man from primary elements) eventually find their way into the soil, where they serve as substrates for one or another group of microorganisms. In addition to these sources, microbes themselves serve, either alive or dead, as substrate for other microorganisms. A recent discussion[22] on the biological transformation of microbial residues in soil is most informative on this important, but often unappreciated, source of substrates in soil.

Although the types and sources of substrates in soil are relatively well defined, considerably less is known about their distribution in soil. It is a simple matter to determine the total organic matter content of soil (which may vary from less than 1% in some mineral soils to more than 90% in some organic soils) and somewhat more difficult to define the individual chemical components of the organic matter, but very little is known about where these components are located in soil. A sizable fraction of the soil organic matter is present as partially decomposed residues, which may be relatively unavailable − structurally, chemically, and geographically − to the bulk of the soil microbiota, and the remainder, which has undergone more extensive microbial degradation, is probably distributed throughout soil microhabitats. Even the latter fraction, however, is apparently not distributed uniformly but may accumulate at specific loci, determined by the geometry and the chemical, physical, and mineralogical properties of the microhabitats. This discrete and discontinuous distribution may account, in part, for a lack in uniformity in the distribution of the soil microbiota.

In many microbial habitats, both natural and in the laboratory, the biochemical complexity of the interior of the microbial cell is in marked contrast to the chemical simplicity of its environment, and few materials are assimilated without further extensive modifications. This relationship does not pertain to the soil environment, wherein the chemical composition of the organic matter ranges from the simple to the highly complex, and a variety of materials are assimilated. Most of these materials, undoubtedly, are used as energy sources, although some may be incorporated, without marked alteration, into structural components of the cell. The latter is especially true for fastidious heterotrophs incapable of synthesizing required growth factors, such as certain amino acids and vitamins. The soil also harbors and nourishes autotrophs and hypotrophs, as well as heterotrophs. Auxotrophic strains probably also arise, and recent studies[23] indicate that these can proliferate in soil.

Although the presence of microbes of this nutritional spectrum can be easily demonstrated, their location and distribution with respect to each other are not known. For example, autotrophic nitrifiers can grow in the laboratory in the presence of low concentrations of some organic compounds, but higher levels are usually inhibitory.[24] These autotrophs are easily isolated from nature, especially from soil after enrichment with amino acids, which are heterotrophically deaminated to provide the source of NH_4^+ on which growth of the nitrifiers depends. Nothing is known, however, of the spatial basis of this interaction. Where and how do the heterotrophs and autotrophs reside that enables them to interact metabolically in soil? Similarly, what is the relative competitiveness of a fastidious hetero troph (or an auxotroph) with respect to a non-fastidious heterotroph (or prototroph) in a nutrient-poor microhabitat and in one that is more luxuriantly endowed with a broad spectrum of soluble carbonaceous compounds? Do some microhabitats, because of the nutrients that they provide, encourage vegetative growth of spore-forming bacteria, whereas, in other microhabitats, the same species reside only as spores?

Because of its apparent omnivorousness, soil has been compared to a digestive system composed of cells differing only in their degree of nutritional specificity. This analogy, however, is restrictive, even if the ability of soil to degrade and assimilate substrates is the sole consideration. The soil microbiota is composed of individual species, each with unique biochemical potentialities and environmental requirements, which operate sometime in concert but probably as often in discord. The types of substrate, the spatial relations between substrates and microbial cells and their enzymes, and the conditions prevailing in the microhabitats probably dictate which substances are attacked by which microbes and at what rate. Little is known in soil about the relative mobility of substrates to sites of microbial concentrations, of the effective distribution radius of exoenzymes, of the functioning of the various control mechanisms (e.g., induction and repression) that are observed in the laboratory, of chemotactic or other movements of microbes to sites of substrate accumulation, of the nutritional importance of inorganic gases and volatile organics, or of the passage of carbonaceous materials through a sequence of species that comprise a foodweb and the importance of this in succession.

Nor is the soil now considered to be as omnivorous as was once assumed. Not only do some natural products (e.g., "humus," porphyrins, some D-amino acids) accumulate under certain conditions, but the persistence of some man-made molecules (e.g., certain detergents, plastics, pesticides) has cast doubt on the classical concept that, to paraphrase Alexander,[25] the soil is the all-powerful and final incinerator that eventually converts all organic substances to their theoretical end-products by virtue of its "microbial infallibility." There exist recalcitrant molecules, which even the most adaptable of microbial communities – the soil – does not decompose.[26]

Recalcitrance may also have an environmental basis. The low solubility of some organics in the soil solution may prevent their assimilation by microbial cells. The adsorption and binding of some materials (e.g., proteins) by certain clay minerals (e.g., montmorillonite) and, possibly, by soil organic matter may render these materials secure from enzymic attack. Similarly, the complexing of materials (e.g., proteins) by some components of soil organic matter (e.g., lignin) will decrease their availability to microbes. Substrates occluded within aggregates or narrow capillary openings will be physically inaccessible to microbes and their exoenzymes until these aggregates are disrupted, perhaps by the soil micro- and macrofauna or by environmental events such as freezing and thawing or wetting and drying. Environmental recalcitrance may be less permanent than chemical recalcitrance, but both serve to restrict the availability to microbes of some carbonaceous substrates in soil.

Although it is doubtful that substrates ever become limiting in the soil as a whole, it is conceivable that they may become depleted within individual microhabitats. Because the majority of microbes is dependent upon carbonaceous substrates for subsistence, such depletion can markedly alter their ecology and population dynamics in soil.

B. Mineral Nutrients

As with carbonaceous substrates, most soils contain, somewhere in their profile, most, if not all, of the noncarbonaceous nutrients necessary for microbial life. Nevertheless, despite the predominantly inorganic nature of soil, mineral nutrition (and mineral is used here to encompass all except organic materials) can limit the development of indigenous microorganisms. This limitation, which can greatly influence the microbial ecology of soil, is usually not caused by a paucity of a specific nutrient but by its unavailability.

For example, most soils usually contain adequate total nitrogen, in both the combined and gaseous states, to support maximum microbial development, but seldom contain enough nitrogen in a form readily available to the microbes. In addition to gaseous nitrogen, most combined soil nitrogen is present in organic form, in either primary residues or microbial products and protoplasm,[27] and some may be fixed between lattices of expanding clay minerals.[28] Some of the organic nitrogen may be utilizable by some species, but, until it is converted into inorganic forms, it may be inaccessible to many others. Species possessing proteolytic and deaminating enzyme systems should, therefore, have an ecological advantage over species not containing one or both of these systems in microhabitats containing only organic forms of nitrogen. Similarly, in microhabitats low in or devoid of available combined forms of nitrogen, nitrogen-fixing species would have an ecological advantage.

Phosphorus and sulfur, even though required in smaller quantities than nitrogen, may also be present in forms not readily assimilable by an actively developing microbial population.[29,30] The availability of other nutrients is usually adequate to sustain high levels of microbial activity, although some soils will respond (e.g., respiratory rates are increased) to additions of potassium (unpublished results).

The limiting effects of inorganic nutrients in soil are usually only apparent when substantial amounts of readily oxidizable carbonaceous substrates are introduced.[29-32] The concentration of the limiting nutrients then determines the rate of substrate oxidation but, given adequate time, not the total amount of oxidation. This "turn-over rate" of the nutrients is determined by the chemical structure of the nutrient-containing compound, by the rate at which microbial protoplasm is decomposed, by the ambient environment within the microhabitats, etc., and, in most cases, probably follows the pathways of the classical biogeochemical cycles.[33,34]

Little is known, however, about which mineral nutrients are absolutely required for microbial life in soil. (In fact, this is not even well defined for microbes grown in pure culture.) Similarly, knowledge about ion specificity and antagonism in soil is scant. Although it may appear, on first inspection, that such information about an environment containing a vast array of minerals is academic, these phenomena could have considerable relevance to the ecology and population dynamics of microbes in soil microhabitats. For example, molybdenum appears to be required by *Azotobacter* sp. for nitrogen fixation, and fixation is competitively inhibited by tungsten.[35] Deficiencies of some minerals can affect the synthesis of enzymes and other biopolymers, the stabilization of cell walls and tertiary structures of DNA and RNA, phage attachment, cell division, mobility, host-parasite interactions, and a variety of other physiological and biochemical processes.[36,37] It is also important to note that mineral requirements change with changes in temperature, and, for some phototrophs, with the concentration of organic carbonaceous substrates. Most of these responses have been observed in pure culture, but similar effects undoubtedly occur in soil and other natural habitats, where they remain unstudied.

In addition to their role as structural components of protoplasm, some mineral forms (e.g., NH_4^+, NO_2^-, H_2S, S) also provide the energy for lithotrophs to live autotrophically. The presence of suitable energy sources for these organisms in microhabitats is necessary not only for their survival, but the conversion of these sources into other forms is necessary for the survival of many heterotrophs. Consequently, the types of minerals present in the microhabitats profoundly influence the ecological successions they may sustain.

Mineral nutrients present as water-soluble compounds and on exchange complexes are usually readily available to microbes. Some essential nutrients, however, may become inaccessible as the environment changes and their solubility decreases. For example, under conditions of low dissolved oxygen, reduced sulfur may be precipitated by iron and other metals, thereby making both the metal and sulfur less available. Similarly, extremes in pH may decrease the availability of phosphates. Although microbial metabolism in the microhabitats can elicit these and other alterations in the ambient environment, some products of microbial metabolism (e.g., citric, oxalic, 2-ketoglutaric acids) are excellent chelating agents, able to maintain mineral nutrients at levels that are available but not toxic to many species.

The great bulk of mineral nutrients in the soil, however, is present in insoluble forms as primary and secondary minerals. The solubilization of these materials is caused, in great part, by the metabolic activities of some microbial species, especially those capable of producing large quantities of acid, both as degrading agents and as chelators.[33,34] Not only are these biogeochemical processes necessary for soil formation, but the presence of microbes with such capabilities may increase the mineral nutrient status of microhabitats that would otherwise be deficient in essential nutrients.

Obviously, much more needs to be learned about the impact of the noncarbonaceous portion of soil on the ecology and population dynamics of microbes in this habitat.

C. Growth Factors

Growth factors, as defined here, are organic substances, other than major carbonaceous substrates, that are either essential or stimulatory to the growth of microorganisms and are required in only minute amounts. Various vitamins, amino acids, purines, pyrimidines, and other miscellaneous substances have been identified in pure culture as being growth factors. These materials are usually required in concentrations ranging from a fraction of 1 mg to 100 mg/l.[38] Although

their biochemical roles (e.g., synthesis of macromolecules and pigments, cell division, sporulation, germination) are relatively well defined, not much is known of the importance of growth factors in the ecology and population dynamics of the heterogeneous microbiota in soil. Based on studies in pure cultures and in other microbial habitats, there is little doubt, however, that growth factors are important in soil and that their requirement by different species is dependent upon environmental conditions, such as pH, E_h, temperature, ionic composition, and the level of nutrition.

Although it is difficult to establish growth factor requirements for individual species in the laboratory, it is much more difficult to do so for the soil microbiota. The most direct evidence for such requirements is the spectrum of microbes having different nutritional requirements — ranging from autotrophs to highly fastidious heterotrophs and auxotrophs — that can be readily isolated from soil. Lochhead and co-workers[39,-44] have demonstrated not only the stimulatory effects of various vitamins and amino acids, but also the absolute requirement of some soil bacteria for specific amino acids (e.g., the sulfur-containing group) and vitamins (e.g., B_{12}). In general, bacteria having requirements for growth factors are isolated more frequently from the rhizosphere than from soil some distance from plant roots. Furthermore, organisms requiring amino acids are more abundant in June than in October when plant growth has decreased. These workers also found that, whereas fertile soils, in general, contain most of the vitamins and other growth factors presumably required by microbes, 27% of soil bacteria isolated from such soils required one or more vitamins. Those found essential were, in order of frequency: thiamine, biotin, B_{12}, pantothenic acid, folic acid, nicotinic acid, and riboflavin.

Because of the complexity of soil, it is difficult to demonstrate the need for growth factors other than by testing for the growth factor requirements of isolated organisms. There is, however, an inherent bias in this procedure: namely, the organisms are initially isolated on nutritionally complex media, which, as will be discussed later, reflect only a portion of the soil microbiota. However, the addition of yeast extract or mixtures of vitamins to soils amended with glucose and mineral nutrients resulted in a significant decrease in the time required to reach the respiratory maximum (Stotzky and Norman, unpublished results).

In addition to vitamins and amino acids, a water-soluble, heat-stable soil substance, termed the terregens factor,[45] is required for the growth of some soil bacteria, primarily by the genus *Arthrobacter*. This factor can apparently be replaced by some iron-containing porphyrins.[46] The need for agarized soil or "fulvic acids" extracts for the isolation of unusual soil forms[47] further emphasizes the importance of growth factors in evaluating the composition of the soil microbiota.

Aside from this fragmentary information, detailed aspects of growth factors in soil are unknown. Essentially no work has been done to determine whether the effects and mechanisms elucidated in the laboratory function in soil, especially in the microhabitats. Does a deficiency in vitamins (e.g., in B_6, B_{12}, para-aminobenzoic acid) influence the distribution of microbes that require these vitamins for the synthesis of essential amino acids?[38] What are the antagonistic and inhibitory aspects between amino acids; e.g., if an excess of one amino acid increases the requirements for another essential amino acid which is not present in sufficient concentrations in the microhabitat, is the growth of some species thereby prevented? How important is end product repression in amino acid utilization?[48] What is the relative importance of growth factors as mediators of the environment through chelation, buffering, and altering the permeability of cell membranes? How important are growth factors in the various symbiotic, pathogenic, and parasitic relationships that occur in soil? Do growth factors in soil influence spore formation, spore germination, and morphological changes, all phenomena that have differential survival values in the inhospitable soil environment? How many other growth factors and their functions are yet to be discovered in soil?

Although growth factors are provided by various microorganisms, by root excretions, and by organic residues, and most are required in only minute quantities, their distribution is probably not uniform throughout the soil. Consequently, in some microhabitats, the rate of synthesis and release of growth factors probably mediates the sequential development of the resident microbes. Species not requiring such factors, or able to make use of those already present, probably develop rapidly after the introduction of readily utilizable substrates, followed by species dependent on growth factors synthesized by these primary populations. Unfortunately, evidence for such succession is sorely lacking. Perhaps all that can be stated with any certainty about growth factors in soil is that CO_2 is probably uniformly distributed throughout the microhabitats and is never limiting.

D. Ionic Composition

The soil solution is essentially a weak electrolyte composed of a variety of organic and inorganic cations and anions of different valencies. Microbial cell membranes and many soil particulates are charged and, therefore, interact with these ions. Consequently, the ionic composition of the microhabitats must be considered in any discussion of microbial ecology in soil, not only for the nutritional aspects of the ions but for their physicochemical effects. Unfortunately, however, not as much is known about the influence of ionic composition on microbes in soil as is known about their effects in pure culture and in aquatic habitats, especially marine.

The single most important parameter of the ionic environment is probably the ionic strength (μ):

$$\mu = 1/2 \ \Sigma \ CZ^2$$

where C = concentration of the ionic species and Z = valency of the ionic species. The ionic strength influences, among other factors, the solubility of salts, the ionization constants of weak electrolytes, the oxidation-reduction potential, the denaturation of proteins, the electrokinetic potential of cell surfaces and clay minerals, and, therefore, adsorption and flocculation. Some of the specific effects of ionic strength will be discussed later. Because of the ionic complexity of the soil solution, however, it is not possible to determine accurately the concentration and valence of the component ionic species and, therefore, to calculate the ionic strength.

In addition to the overall ionic strength of the environment, the type, charge, valence, and size (i.e., the hydrated ionic radius) of the predominant ions present probably influence microbial events. Ions appear to be important in maintaining the integrity of cells, both on the gross morphological level (e.g., cell membranes) and on the molecular level (e.g., macromolecules such as proteins, nucleic acids, and ribosomes). Chemical, morphological, and physiological differences between species (e.g., gram-negative vs.

gram-positive bacteria, halophilic vs. non-halophilic bacteria, filamentous vs. single cell forms) have been attributed to the ions in the ambient environment.[49-52] Although these aspects have not been explored extensively in soils, it is possible that the differences observed in morphology of microbes in soil and on media after isolation from soil[47,53-56] reflect differences in the ionic composition of these environments.

The ionic composition also appears to influence toxin production by some bacteria,[57-59] cell division,[60,61] formation and stabilization of osmotically fragile cells,[62,63] induction of permeases,[63] spore germination,[64,65] inhibition of nitrification,[66] and attachment, penetration, and multiplication of bacteriophages.[67-70] In some instances, the effects of the ions are not specific, and one ion can be replaced by another. In other instances, the effect is highly specific as to the type of ion, and ion antagonisms can be demonstrated. Generally, the cation is the important species, although, in some cases (e.g., spore germination, nitrification), the effects appear to be anion-specific.

The permeability to and uptake of ions and water by cells, whether by osmosis, diffusion, facilitated diffusion, Donnan equilibrium, or active transport, are also greatly influenced by the ionic composition of the medium. Essentially nothing, however, is known about these phenomena in the soil environment.

Because the cell-soil solution interface is essentially an ionic one, there can be little doubt that small changes in the ionic composition induce large changes in the physiology and, therefore, in the ecology of microbes. These changes would be magnified in the microhabitats, as the concentration of ions is probably greater here than in the rest of the soil solution, because of the charged surfaces of both the microbes and some soil particulates. Furthermore, changes in ionic composition are directly influenced by other physicochemical factors (e.g., pH, E_h, temperature, atmospheric composition) as well as by the metabolic activity of the microbes themselves. Obviously, therefore, much more study is needed on this aspect of the soil environment if the ecology of microbes in soil is to be better comprehended.

E. Water

Microbes, regardless of their habitat, are aquatic creatures. Not only is water necessary for intra-cellular metabolism, but an adequate layer of external water is necessary for maintenance of turgor, movement of substrates, toxic products, cells, etc. Many microbes can persist for long periods of time in the absence of unbound water, usually by virtue of specialized structures, such as spores, but free water is necessary for metabolism.

Most of the interactions of water with microbes and their environment are related to the physicochemical properties of water. The water molecule is V-shaped and polar, which enables it to undergo H-bonding with other water molecules, to form a tetrahedral structure, and with adjacent charged surfaces. In ice, the tetrahedral bonding becomes more regular, and the density is low because of the rigidity and the large hexagonal spaces within the crystal structure. In solution, the structure becomes more random and the density greater (being greatest at 4°C) because of the electrostatic attraction of charged particles, which become surrounded by an oriented sphere of water molecules. The degree of organization within this sphere of hydration varies inversely with the distance from the particle, and the disruption of the structure of the remaining water is proportional to the charge, size, and concentration of the charged particles. The intermolecular attraction between water molecules also imparts a high degree of thermal stability, as shown by the high specific heat, high latent heat of vaporization, and high heat of fusion of water.

The availability of water to microbial cells is more important than the total water content of soil. Availability is usually expressed as the activity of water (a_w):

$$a_w = \frac{P_s}{P_w} = \frac{RH}{100}$$

where P_s = vapor pressure of the solution, P_w = vapor pressure of pure water at the same temperature as the solution, and RH = relative humidity. Considerable data have been accumulated, in pure culture studies, on the a_w requirements for various microbial functions.[71] The limits in a_w range, in general, from 0.999 to 0.63, with bacteria requiring a range from 0.99 to 0.93. The minimal a_w requirements for yeasts range from 0.91 to 0.88 and, for some filamentous fungi (e.g., *Aspergillus glaucus*), from 0.70 to 0.65, although growth of germ tubes requires a range from 0.999 to 0.90. Most microbial activities are greatest at higher a_w, and reductions in a_w increase the length of the lag phase and decrease rates of growth and synthesis. As other environmental factors (e.g., temperature, pH, pO_2, ionic and nutritional composition) deviate from their optima, the range of tolerance of organisms to a_w decreases.

The survival of microbes at low a_w differs with species, type of cell, and other environmental conditions. Various vegetative bacterial cells can survive an a_w as low as 0.30, with reduced temperatures increasing viability at this a_w. The survival of endospores is maximal at a_w values ranging from 0.80 to 0.22.

The relevance of these pure culture studies to the natural soil environment is not sufficiently clear.[72] Most soils are seldom devoid of water, and, even in desert soils, the particles retain some water against the extreme dryness of the atmosphere above the soil. For example, the relative humidity in desert soils seldom ranged below 85%, even at the surface and when the relative humidity of the air was less than 15%.[73] This water, however, was probably not available to most microbes, as the soil water tension at 85% relative humidity is approximately 218 atmospheres.[74]

The retention of water by soil against adverse concentration gradients and its unavailability at low concentrations to biological systems result, in part, from the energy with which water is held by some particulates. The charged surfaces of clay minerals induce a rearrangement of adjacent water molecules, resulting in a quasicrystalline structure within the surrounding water films.[75,76] This water exhibits greater viscosity than normal water, and temperatures exceeding $500°C$ are necessary to remove it.[77] To date, the occurrence of "polywater" in soil has not been reported.

The mineralogical composition of soil influences the extent to which water is retained. For example, the addition of 2% water to 50 g oven-dried fine sand resulted in a relative humidity and a soil water tension of 99.7% and 4.0 atm, respectively; in 64.5% and 589 atm when added to an oven-dried sandy loam; in 30% and 1617 atm when added to a clay loam; and in 10% and 3093 atm when added to a silty clay soil. When 4% water was added to 50 g of the latter soil, the values were 53% and 853 atm, and the addition of 64% water was necessary to bring the values to 99.9% relative humidity and 1.7 atm.[74]

These values indicate that the water content of soils markedly influences their microbial composition. However, little direct evidence for this is available, especially at the level of the microhabitats. The respiration of soil decreases proportionally as the soil water content is reduced below the optimum level (approximately 0.3 bar tension) (unpublished). It is not clear, however, whether this decrease in respiration is caused by a uniform reduction in available water in all microhabitats or whether some microhabitats remain at or near the optimum level and others dry below this, and, therefore, the decrease in respiration reflects only fewer microhabitats capable of sustaining active metabolism. Nevertheless, the lower numbers of microorganisms usually isolated from dry and stored soils and the usual overgrowth of such dilution plates by *Bacillus* sp. would indicate that the drying of soils does result in a killing of many vegetative forms and in an increase in sporulation. Furthermore, in desert soil, for example, specific microbial groups may develop in water pockets formed from water retained by soil particulates and deliquescent salts and from condensation resulting from shadows, diurnal temperature cycles, and different opacities of surface crusts.[73] Limitations in available water may also select for genetic types capable of persisting under conditions of drought and of rapidly exploiting sporadic additions of water,[73] similar to the apparent selection of some soil microbes capable of rapidly exploiting carbonaceous substrates by virtue of having a higher metabolic rate than other species.[78,79]

Cycles of wetting and drying, as well as of freezing and thawing, of soil markedly stimulate metabolic activities in soil.[80-83] This stimulation probably results from the increase in substrates released from clay surfaces and killed cells, rather than from changes in water structure.

The crystalline nature of water also appears to influence a variety of metabolic processes.[76,84] The presumed subtle structural changes in water at specific temperatures (e.g., at approximately 15, 30, 45, and $60°C$) have been correlated in pure culture with decreases in growth and other microbial activities.[85-90] Unfortunately, no comparable studies have been conducted with microbes in soil. Although the occurrence of discrete thermal discontinuities in water has been questioned[91,92] the biological data suggest that comparable studies should be conducted in soil, especially in view of the presumed high degree of orderliness of water in the microhabitats.

The retention of water in soil is also a function of its high heat of vaporization, high dielectric constant, and high surface tension. These characteristics are modified in the soil, especially in the microhabitats, by the ions present and by the metabolites produced by the microbes. In addition, some metabolites will reduce the surface tension of the water films, thereby enhancing entrance of nutrients, gas exchange, and motility of cells. The relative hydrophilic and hydrophobic nature of the cells surfaces will influence the location of the cell in the water films and, therefore, their access to nutrients and gases. To what extent these phenomena are countered by the influence of the charged clay surfaces is not known.

The microbial cell itself retains water films, the thickness and other physicochemical characteristics of which are probably influenced by the chemical nature of and the charge on the cell membranes. The source of this water is not only from the ambient environment but also is a product of cellular metabolism. The degree of interaction and exchange between water films on the surfaces of cells and inanimate soil particulates is probably dependent on the solute and temperature gradients between them. These gradients are also probably responsible, in part, for the movement of nutrients, toxic substances, and gases through the water phase, as diffusion would be too slow to sustain active microbial life (e.g., at $25°C$, O_2 diffuses at a rate of 1.7×10^{-5} cm^2/sec, 0.01 M NaCl at 1.5×10^{-5} cm^2/sec, 0.39% glucose at 0.673×10^{-5} cm^2/sec).

Although much is known about the water relations of microbes in pure culture, relatively little is known about the reactions of water in soil microhabitats where the presence of a solid phase markedly alters its characteristics. Inasmuch as microbes reside in this water, its physicochemical characteristics will influence the ecology and population dynamics of the resident species.

F. Temperature

The activities of microbial cells, as those of other biological systems, are governed, to varying degrees, by the laws of thermodynamics. Consequently, temperature affects not only the overall physiological reaction rates of cells,[93] but also most of the physicochemical characteristics of the environment (e.g., volume, pressure, oxidation-reduction potentials, diffusion,

Brownian movement, viscosity, surface tension, water structure) that impinge on the cells. It is not surprising, therefore, that changes in soil temperature influence the activity, ecology, and population dynamics of microbes in soil.

There appear to be no known natural areas on the Earth's surface, with the exception of active volcanoes, where microbial life is absent as a result of extremes in temperature. Microbes are present in the Gobi Desert, where diurnal temperature fluctuations can attain $50°C$; at the polar caps, where temperatures may remain as low as $-23°$ to $-40°C$; in composts with temperatures as high as $60°C$; in hot springs, where temperatures may reach $90°C$; in deserts; in frozen soils; and in mountain snowbanks.[6,73,94,95]

Populations of microorganisms survive extremely low temperatures, although their numbers are reduced: e.g., cells and spores of bacteria can survive diurnal fluctuations from $-60°$ to $30°C$;[96,97] numerous fungal species can grow in soil exposed to diurnal extremes of $-94°$ to $23°C$;[98] spores of various bacterial and fungal species, many of which are commonly isolated soil forms, can survive a 2-hr exposure to temperatures as low as $0.0047°K$;[99] some fungal spores have been successfully germinated after storage at $-196°C$;[100] a bacteriophage (T1) of *Escherichia coli*, poliovirus type III, and *Penicillium roqueforti* survived temperatures ranging between $-75°$ and $-45°C$.[101]

Freezing of microbes in soil does not appear to be particularly detrimental, even though temperatures of $-20°$ to $-30°C$ may persist for several months. In fact, some yeasts, *Azotobacter* sp., and root-nodule bacteria (which also survived one month at $-180°C$) exhibited more vigorous metabolism and reproduction after a 3-week exposure to $-15°$ to $-20°C$.[12] Aerobes, in general, appear to be more tolerant to diurnal freeze-thaw cycles than organisms capable of growing in a CO_2 atmosphere without O_2.[102]

The critical factors influencing survival of microbes in low temperatures appear to be the rates at which cells are frozen and subsequently thawed. The reasons for this are not completely clear. Originally it was speculated that the formation of ice crystals within the cell during freezing caused derangement of cellular organelles. However, examination of frozen cells does not indicate such damage, and fast freezing, which would reduce the formation of individual crystals, accelerates death. Survival is increased if cells are frozen slowly and thawed rapidly. Another theory suggests that, as the ambient solution begins to freeze, the osmotic pressure in the unfrozen portion increases and damage results from cellular dehydration. The data for these hypotheses have been derived from pure culture studies, and it is interesting to speculate how the quasicrystalline water in soil microhabitats affects these phenomena.

Similarly, the phenomenon of "cold shock," whereby more cells are killed by rapid cooling to nonlethal temperatures than by gradual cooling to the same temperature,[103] should be investigated in soil. The sensitivity to cold shock differs with species, and temperatures in many parts of the world have diurnal fluctuations similar to those used to demonstrate cold shock in the laboratory.

At high temperatures, some microbes probably survive in soil, as they do in pure culture and in other natural habitats, predominantly by the formation of spores. Most vegetative cells are killed by temperatures approaching 85°C, although the thermal death point of some thermophilic bacteria in pure culture may be as high as 110° to 120°C. Bacterial spores, however, can, in pure culture, resist temperatures exceeding 100°C for extended periods, and spores of *Bacillus anthracis* have been recorded to survive a 30-sec exposure to 400°C.[104] The temperature maxima for survival of vegetative cells and spores in soil have apparently not been determined. Although the microhabitats may be buffered somewhat against temperature extremes by the surrounding water films, the detrimental effects of this water are not known. In pure cultures, cells and spores are more susceptible to elevated temperatures when the a_w is high (e.g., the heat resistance of spores is greatest between an a_w of 0.40 and 0.22).

Despite the apparent knowledge of the effects of temperature on microbes in soil, most of this knowledge is of a general or gross nature. Little is known of the effects, and even less of the mechanisms, of temperature fluctuations in the microhabitats. Although the high specific heat of water provides some buffering of the microhabitats, ambient temperature fluctuations, both seasonally and diurnally, undoubtedly exert a selective effect on the soil microbiota. Temperature limits have been defined for individual species in pure culture but not for species comprising the heterogeneous populations in soil or for the populations themselves. These limits are probably modified by the physicochemical characteristics of the microhabitats. Pure culture studies have also demonstrated that there are different temperature optima for various physiological activities (e.g., total cell yield; rates of respiration, fermentation, and solute transport; length of the lag period; synthesis of flagella, pigments, polysaccharides, fatty acids; saturation of fatty acids).[93,104] If similar differential optima are necessary in soil, it is conceivable that, at a given temperature, the growth of a species may be impaired, but its competitiveness and survival may be enhanced by the production of substances inhibitory to other species or by other mechanisms. Conversely, extremes in temperature may not inhibit growth, but the microbe may become auxotrophic for some nutrient not present in its microhabitat, and, therefore, its population density would decline. Such temperature-induced auxotrophy has been demonstrated in pure culture.[93,105]

The adaptability or selection of the soil microbiota to the prevailing temperature regimes is demonstrated by the preponderance of mesophilic species in temperate zone soils and the higher percentage of psychrophiles present in Antarctic soils. Although it is possible to isolate psychrophiles, mesophiles, and thermophiles from soils in all parts of the world, little is known of their location and importance within the microhabitats and even less of the mechanisms whereby they compete with forms more suited to the prevailing temperature conditions.

Another area that requires study in soil is the effects of temperature fluctuations on various microbiological rhythms. In pure culture, temperature effects on rhythms in growth patterns, sporulation, spore dispersal, etc. have been demonstrated.[105,107] Are such rhythms significant in the colonization, ecology, and population dynamics of microbes in soil, especially in locations where diurnal and seasonal changes in temperature are large?

More studies in soil are obviously needed on the effects of temperature changes on the activity and shifts of individual species comprising the heterogeneous microbial population within the microhabitats. These studies should include the effects of temperature on other physicochemical characteristics of the microhabitats and vice versa. These studies will not be easy to perform, inasmuch as the effects of temperature on microbes in even pure culture are difficult to determine. Even more difficult will be the delineation of mechanisms, as the many pure culture studies based on the concept of the temperature coefficient (Q_{10}) have shown.

G. Pressure

Changes in atmospheric pressure are probably too small to influence microbes in soil: e.g., the atmospheric pressure at sea level is 1.000 atm; at an elevation of 5,000 ft, it is 0.832 atm. Similarly, the solute concentration of normal soil solutions is usually not high enough to suggest that osmotic pressure inhibits the soil microbiota: the solute concentration seldom exceeds approximately 0.05%, with osmotic pressures ranging from 0.2 to 1 atm.[19]

These average values, however, probably do not reflect the situation within the soil microhabitats where the solute concentration is probably higher than that measured in the soil solution, inasmuch as solutes concentrate at particle-water interfaces, especially at charged surfaces. Furthermore, during periods of reduced moisture, the solute concentration and, therefore, the osmotic pressure may increase sufficiently to affect some microbes, especially in saline and alkaline soils. For example, at $25°C$ and at the a_w of a 1 N salt solution (a_w = 0.9828), the osmotic pressure is approximately 25 atm. Although such solute concentrations may seldom, if ever, occur in soil, bacteria isolated from soil begin to show reduced metabolic activity at osmotic pressures approaching 5 atm.[108,109]

Inhibition of microbial activity in soil by hypertonic osmotic pressures has been demonstrated[29-31] (Figure 1), and a spectrum of organisms isolated from soil are sensitive to elevated solute concentrations (unpublished). Fungi are much more resistant than bacteria to hypertonic osmotic pressure in pure culture, but this differential effect has not been studied in soil.

Although most soil microorganisms are probably euryhaline, the limits of their sensitivity in situ to hypertonic osmotic pressures have not been defined. Similarly, the interaction between osmotic pressure and other physicochemical changes in the microhabitats requires investigation. In pure culture studies, an increase in atmospheric or hydrostatic pressure permits organisms to grow at temperatures that are lethal at normal pressure, presumably because of compensatory volume changes in cell constituents. In soil, this inter-

action becomes more complex, as an increase in temperature also causes a reduction in soil water, thereby increasing the relative solute concentration and the osmotic pressure.

Elevated pressures induce a variety of changes in microbes in pure culture: e.g., decreases in cell division and mobility; numerous metabolic changes, such as decreases in the reduction of nitrate and nitrite and in other enzymic activities; changes in virulence of pathogens. However, it is not known whether such pressure-induced changes — changes which would significantly affect microbial ecology — occur in soil.

Although there appear to be relative differences in the salt-tolerance and salt-dependence of microbes isolated from saline and nonsaline habitats,[110-113] the adaptability and enrichment of osmotolerant strains in soil have not been resolved. The possible evolution of marine microbes from soil forms[112,114,115] also needs further study. Similarly, the role of the soil environment must be defined, especially as some clay minerals appear to protect microorganisms from the detrimental effects of hypertonic osmotic pressures (unpublished).

Microbes, in general, appear to be tolerant to extremes in pressure: e.g., bacterial and fungal spores can survive vacuums ranging from 1 x 10^{-6} to 2 x 10^{-10} torr[102,116-118] and hydrostatic pressures as high as 1500 atm.[115,119-121] Although such extremes in pressure are not encountered in soil, the extremes tolerated by the soil microbiota should be determined and related to pressure changes that actually occur in soil.[122] Interesting speculations on the evolution of microbes on Earth and on other planets and on their ecology in soil could result from such data.

H. Radiation

Radiation is probably the most important ecological factor on Earth, inasmuch as all life is ultimately dependent upon the energy (i.e., heat and driving force for photosynthesis) derived from the sun. Essentially no studies have been conducted on the effects of biologically important radiation (i.e., from approximately 200 to 1000 nm) on the ecology of microbes in soil. This is not surprising, however, as the opacity of soil restricts the penetration of light (i.e., the soil is dark!). Any significant effects of radiation are probably restricted to microbes on the soil surface, although infrared radiation influences the temperature of

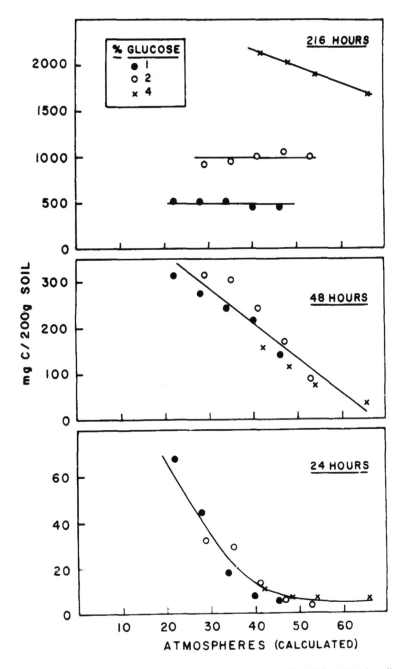

FIGURE 1. Effect of hypertonic osmotic pressure on microbial activity in soil. Soils were amended with three concentrations of glucose and with four or five concentrations of mineral salts (Hoagland solution No. 1) at each of the glucose concentrations. Data expressed as the cumulative C evolved as CO_2 from 200 g soil after 1, 2, and 9 days. Osmotic pressure of amendments calculated using the ideal vant Hoff factor. After 9 days of incubation, recovery of glucose carbon was essentially complete with the 1 and 2% glucose additions. (Stotzky and Norman, unpublished)

soil, and radionuclides, either natural or introduced by man, may be present in soil and affect microbes within their range. Cultivation, heaving, and other "soil mixing processes" may sporadically expose microbes from depths below the surface to solar radiation.

Much is known about the effects of biologically important radiation on microbes in pure culture and in some natural habitats, notably aquatic ones. The action spectra for photosynthesis of algae and photosynthetic bacteria are well defined, as are the radiation requirements for a variety of photodependent movements (e.g., phototropism, topophototaxis, phobophototaxis, photokinesis). The inhibitory and mutagenic effects of ultraviolet radiation are well known, and the mechanisms of these effects and of photoreactivation are under intensive study.

In the absence of any data, the importance of these radiation-dependent phenomena on the ecology of microbes in soil is difficult to assess. Photosynthesis on the soil surface undoubtedly adds carbonaceous substrates to the underlying microhabitats, and in these habitats, in turn, metabolites may be produced which may affect the primary producers. The amount of photosynthesis will be dependent on the type of plant cover and the intensity of light impinging on the soil surface. Most of the photosynthesis is probably by algae, inasmuch as the photosynthetic bacteria are anaerobes (with the possible exception of the nonsulfur purple bacteria), and it is doubtful if many microhabitats in the photic zone of soil are devoid of oxygen.

Light intensity, however, may have an indirect influence on the soil microbiota by affecting the qualitative nature of root exudates, which, in turn, influence the selection and/or physiology of the rhizosphere inhabitants. For example, approximately 30% of the bacteria isolated from the rhizosphere of the wheat grown with 300 ft-candles (f-c) of light required amino acids, whereas, from wheat grown with 1,000 f-c, approximately 65% required amino acids. The total numbers of bacteria isolated from the rhizospheres of wheat grown with 300 and 1,000 f-c were 3×10^9 and 7.9×10^9, respectively. The fungi in the rhizosphere of wheat were not significantly affected by light intensity.[123]

Many fungi, however, are phototropic, and spore formation and release, as well as other morphogenetic changes, are light-dependent and,

in some species, rhythms are apparent. Consequently, the dispersal and, therefore, the ecological distribution of light-sensitive fungi growing or sporulation in the photic zone will be influenced by the incident radiation.

In aquatic environments, radiation — both directly and indirectly through its effects on the microbes — diurnally affects other characteristics (e.g., pH, E_h, pO_2, pCO_2) of the environment. Whether such marked changes occur in soil and what effects such changes have on the ecology of the soil microbiota warrant study.

I. Atmospheric Composition

The inner layer of the Earth's atmosphere, the troposphere, is composed of approximately 78% N_2, 21% O_2, 0.9% Ar, 0.03% CO_2, plus variable trace amounts of water vapor, Ne, Kr, CH_4, N_2O, SO_2, NH_3, O_3, HCL, HF, H_2S, dust, pollen, and microorganisms. The composition of the soil atmosphere, however, is considerably different: the CO_2 content, even in well-aerated soil, is 10 to 100 times higher, the O_2 content may decrease to less than 5% of that found in the atmosphere above the soil, and more H_2S, CH_4, NH_3, H_2, and organic volatiles are present. These changes are the result, primarily, of microbial metabolism within the soil.

Compounds comprising the soil atmosphere can be present in the gaseous form, dissolved in the soil solution, or adsorbed on the solid phase. Because liquid and gas compete for the same space (i.e., the voids between the particulates), the contents of soil atmosphere and solution are inversely proportional. Regardless of the form in which a gas is present, its assimilation by the microbes is probably in the dissolved state, inasmuch as the gas must pass through the water films that surround the cells.

The solubility of gases in the soil solution is governed by the gas laws and, therefore, is markedly affected by other physicochemical characteristics of the microhabitats (e.g., temperature, ionic strength, pH, pressure). In addition, solubility is influenced by the extent to which a gas reacts chemically with water. Several biologically important gases, such as O_2 and N_2, are relatively insoluble in water (e.g., the adsorption coefficients are 0.049 and 0.024, respectively), whereas those that form compounds with water, such as NH_3 and CO_2, are highly soluble (e.g., the adsorption coefficients are 1.300 and

1.713, respectively). The solubility of the latter types of gases is greatly influenced by the pH of the soil solution. For example, CO_2 is present primarily as HCO_3^- at pH 7 to 9, as a mixture of CO_2 and HCO_3^- between pH 5 to 7, and as CO_2 below pH 5. In most soils, therefore, it would appear that CO_2 is present primarily as HCO_3^-. However, as the pH in the microhabitats adjacent to charged surfaces may be several pH units lower than that measured for the soil as a whole (see Section J), much of the CO_2 in the immediate environment of the microbiota may, in fact, be in the form of CO_2.

The rate of gas exchange between the soil and the overlying atmosphere is also governed by a variety of environmental factors. Although diffusion probably occurs, it is too slow a process to account for much of the gas exchange. The structure of the soil, which is dependent primarily on the particle size distribution, mineralogical composition, and organic matter content, is a major factor in gas exchange, as is the amount of water present. Fluctuations in temperature (especially diurnal), barometric pressure, wind velocity, relative humidity, and concentration gradients (resulting from the differential utilization and production of various gases by the microbiota) influence the rates of gas exchange. Within the microhabitats, the surface area, surface tension, and orderliness of the structure of the water films influence the rate at which gases pass through the air-water interface. This rate can be approximated by the Bohr invasion coefficient:

$$C = \frac{V_m P}{(P_a - P_s)A}$$

where V_m = volume of gas entering the liquid phase in one minute; P = atmospheric pressure; P_a = partial pressure of the gas in air; P_s = partial pressure of the gas in solution; and A = area of the interface.

In waterlogged soils, dissolved O_2 is quickly consumed by plant roots and the microbiota, and, because replenishment of O_2 is slow, the soils rapidly become anaerobic. This results in a marked change in the dominant microbiota from aerobic to anaerobic and microaerophilic forms and also in the production of metabolites which further influence the composition of the active components of the soil microbiota. Many of the heterofermentative products (e.g., short-chain organic acids, alcohols, aldehydes, esters, methane) as well as reduced inorganic products (e.g., NH_4^+, H_2S, H_2, Fe^{+2}) may accumulate to concentrations that become toxic to some species.

The incomplete oxidation of substrates under anaerobic conditions also influences the energetics and successions within the microhabitats. Although anaerobic metabolism is less efficient than aerobic metabolism (e.g., aerobic metabolism yields approximately 690 kcal/mol glucose, of which approximately 55% is lost as heat and the remainder retained in the net 38 moles of ATP formed; anaerobic metabolism yields approximately 58 kcal/mol glucose, of which approximately 75% is lost as heat and the remainder retained in the net 2 moles of ATP formed), fermentation results in the production of a variety of endproducts, which serve as substrates for the growth of many different forms and increase the species diversity of the soil microbiota. Although some intermediates are also produced in aerobic metabolism, the yield is relatively low; most of the energy contained in the initial substrate is utilized by the primary populations, and most of the substrate carbon is lost from the system as CO_2. Because of this rapid depletion of energy and carbonaceous substrates, the diversity and successions of populations may not be as great in the presence of oxygen as in its absence.

Even in well-aerated soils, the CO_2 content is high and the O_2 content low, as a result of microbial activity and the impairment of gas exchange. This imbalance is accentuated in microhabitats, especially when readily oxidizable substrates are introduced and, even in the laboratory, cannot be corrected by rapid aeration. For example, when soils to which glucose was added were aerated with CO_2-free air at flow rates exceeding 15 l./hr, the respiratory quotients during the period of active metabolism were higher than the theoretical quotient of 1.[124] With large additions of glucose (e.g., 4% w/w), quotients higher than 10 have been observed, even in soil containing 96% sand (unpublished). Once the peak in respiratory activity passed and the glucose was oxidized to CO_2 and intermediary metabolites, the quotients decreased to 1 and below. Metabolism in soil apparently switches from predominantly aerobic to anaerobic pathways when the O_2 concentration drops below approximately $3 \times 10^{-6} M$.[125,126]

The widespread distribution of facultative and obligate anaerobes in soil further indicates that

O_2-deficient loci occur commonly, even in well-aerated soils, and that anaerobic microhabitats coexist with aerobic ones. The atmosphere within these microhabitats probably fluctuates between extremes of high and low tension of both CO_2 and O_2, depending on the levels of microbial activity and the physicochemical characteristics that govern gas exchange. The tolerance of the soil microbiota to these fluctuations appears to be greater than that of less complex and more specialized microbial populations of other habitats.

Soil fungi are more tolerant, in general, of high CO_2 and low O_2 tensions than are bacteria and actinomycetes, even though bacteria, as a group, are most tolerant of low O_2 tensions. The inhibition by high CO_2 concentrations seems to be a function of the pCO_2 and not of the low pO_2. More bacteria, fungi, and actinomycetes are capable of developing under N_2 than under CO_2 tensions approaching 100%, and the presence of even a small amount of O_2 mitigates the inhibitory effects of CO_2. Soil microbes also appear to be capable of "adapting" to high CO_2 and low O_2 tensions, although it has not been established whether this adaptation is physiological or represents only an enrichment of segments within the heterogeneous populations already tolerant of these conditions. Regardless of the mechanisms involved, this adaption is caused by elevated CO_2 concentrations rather than by reductions in O_2.[127,128]

The O_2/CO_2 ratio also influences the vertical distribution of microbes in soil. The proportion of microbes capable of growing under reduced O_2 concentrations increases with depth to approximately 30 cm and then decreases at lower depths. The zone of most rapid depletion of O_2 is usually 10 to 20 cm below the soil surface. Even though O_2 replenishment is slower at depths greater than 20 to 30 cm, the microbial activity is less due to substrate deficiencies, and the rate of O_2 utilization is reduced. The microorganisms found at lower depths are primarily bacteria, although some species of fungi and actinomycetes can be isolated, probably because their spores are carried downward by gravitational water.[7] Because most soil actinomycetes and filamentous fungi are aerobes, their activity and competitiveness in zones of reduced O_2 tension are probably low.

Some soil fungi, however, appear to have an ecological advantage in habitats low in O_2 and

high in CO_2: for example, in pure culture, the growth of some soil-borne plant pathogens (e.g., *Aphanomyces euteiches*, *Fomes annosus*, some *Fusarium* sp.) and the germination of spores of others (e.g., some *Phytophthora* sp.) are greater under these conditions.[129-132]

The distribution of clones of *Rhizoctonia solani* also appears to be related to the pCO_2 of their environment; e.g., clones isolated from lower soil depths are more tolerant to high pCO_2 than clones isolated from aerial plant parts and from the soil surface.[133,134]

The morphological changes induced in some organisms by altered atmospheres could also have ecological consequences, if these occur in soil as they do in pure culture. For example, *Mucor rouxii* shifts from its usual filamentous form to a yeast-like form when the pCO_2 is increased. This shift in morphology apparently results from an increased synthesis and accumulation of cell wall components and can be prevented by the presence of even small amounts of O_2.[135] Similar morphological changes have been observed in some *Fusarium* sp.[127] The metabolism and morphology of *Blastocladiella emersonii*, an aquatic phycomycete, are also altered by increased concentrations of HCO_3^- in the medium.[136] A similar response in soil-borne phycomycetes has not been reported. Some enterobacteria form fimbriae when the O_2 content of the medium decreases, presumably because these appendages enable the cells to form a pellicle on the surface and, therefore, enhance their exposure to atmospheric O_2.[137] The sporulation of some *Bacillus* sp. and of some fungi also appears to be dependent on minimum concentrations of O_2,[138,139] as does the production of some antibiotics.[140] Increased pCO_2 can induce the encystment of some protozoa.[141]

Although there appear to be no reported studies of morphological changes of microbes in soil induced directly by altered atmospheres, it is probable that such occur. The differences in morphology between microbes observed directly in soil and on media[47,53-56] may reflect such differences in atmospheric composition. Furthermore, the distribution of cells within the microhabitats relative to the air-water interface may be influenced by the requirement of and adaptability to the prevailing atmospheric composition.

In addition to an imbalance in the O_2/CO_2 ratio, an increase in other components of the soil atmosphere, such as NH_3 and organic volatiles, could induce morphological changes.[142-144] Volatile metabolites, which are formed more under anaerobic conditions but are also present in aerobic systems, have been shown differentially to inhibit or stimulate growth, sporulation, and spore germination, as well as influence the morphology, of various fungi.[145,146] These volatiles, which are produced by microbes and by germinating seeds,[147] can also serve as substrates and chemotactic stimuli for microbes (unpublished). Because the diffusion radius of a volatile is greater than that of a substance in solution, the sphere of influence of volatile metabolites in soil would be greater, and even very small amounts could affect numerous microhabitats.

The apparent current increase in air pollution indicates that more studies should be conducted on the influence of such gases on microbes in soil. Although these gases are present in considerably lower concentration than when volatile pesticides are applied to soil, air pollutants are chronically present, in contrast to the latter which are usually added as acute doses. Consequently, some air pollutants could accumulate in soil to toxic levels, especially those that have high solubility in the soil solution (e.g., SO_2, H_2S, HCl, HF, oxides of nitrogen). Most air pollutants are toxic to microbes in pure culture, but usually at concentrations higher than those present in ambient air.[148] At sublethal concentrations, some (e.g., nitrous acids, bisulfite, peroxyacetyl nitrate) are mutagenic agents. The susceptibility of microbes to air pollutants differs with the type of pollutant and with the species and type of propagule of the organism.[118,142,148,149] In general, microbes appear to be more resistant to air pollutants than other forms of life. Little is known about the effects of these pollutants on the ecology and population dynamics of microbes in soil nor about the ability of microbes to metabolize these materials and, therefore, for soil to act as a detoxifying medium.[149]

Soil can apparently serve as a biotic and abiotic sink for some air pollutants.[148,150,151] For example, ethylene[150] and CO[152] are degraded by microbes in soil, and both chemical and microbiological processes in soil transform SO_2 and NO_2.[150] Fungi in soil metabolize and, therefore, remove from the atmosphere organic volatiles arising from plant foliages,[153] as well as methane, ethane, propane, and butane.[154] Consequently, because of the apparent ability of soils to detoxify the atmosphere, considerably more emphasis should be directed toward studying the effects of air pollutants on the activity, ecology, and population dynamics of microbes in soil.

J. pH

Few environmental factors have been invoked as often as has pH in explaining variations in microbial activity and ecology in soil. There is, however, little definitive evidence for the direct and indirect effects of pH nor of the mechanisms involved, especially at the level of soil microhabitat.

On a gross level, the measured pH of soil appears to influence the predominance of one group of organisms over another, at least as determined primarily by dilution plating. In acidic soils (below approximately pH 5.5), fungi appear to predominate, whereas, in near-neutral or moderately alkaline soils (between approximately pH 6 and 8), bacteria and actinomycetes predominate. The reduction in fungi in the latter soils is not because most fungi are intolerant of these pH values but probably because bacteria and actinomycetes are efficient competitors and prevent the establishment and proliferation of fungi. Conversely, the lower numbers of many bacteria and actinomycetes in acidic soils are apparently in direct response to their inability to tolerate these H^+ concentrations, whereas most fungi can tolerate them and, therefore, in the absence of intense competition, proliferate.

Exceptions to this general scheme, of course, exist. For example, iron- and sulfur-oxidizing autotrophs are tolerant of extremely low pH values (e.g., < pH 1.0), and bacteria of the lactic and acetic acid groups are tolerant of less extreme, but still low, pH values.[104] Some actinomycetes require pH values as high as 8.5 for optimum growth and are more prevalent in alkaline than in acid soils,[155,156] but other species of actinomycetes appear to be acid-tolerant.[157]

Although some of the direct effects of pH on microbial metabolism in pure culture have been defined (e.g., on enzyme activity and synthesis), and the mechanisms that some species use to survive at extreme pH values have been studied, very little is known about these effects and mechanisms in soil.

In addition to its direct effects on metabolism, pH also influences morphology and other characteristics of microbes (see Section N), interactions between microbes (see Section O), availability of nutrients, and toxicity of various compounds. For example, at low pH values, H^+ predominates on the exchange complex, and replaced cations may be leached from the microhabitat, thereby increasing its acidity and decreasing its nutritive capacity. Reductions in pH also increase the solubility of some ions (e.g., Fe^{+++}, Al^{+++}, Mn^{++}), which may either be toxic or remove $PO_4^=$ from the soil solution.[158] At high pH values, essential minor elements and $PO_4^=$ may become limiting, primarily as a result of decreased solubility.[19] When solubilized, some ions are chelated by microbial metabolites, which both decreases their toxicity and makes them available for metabolism. However, the chelating ability of many metabolites is also pH-dependent, and the persistence of these metabolites in soil microhabitats has not been determined.

At pH values below their isoelectric point, some amphoteric substrates may be sorbed to negatively charged clay minerals, but whether such sorption influences substrate availability has not been resolved for an adequate spectrum of compounds. The entrance into cells of both nutrients and toxic organics is also influenced by pH, inasmuch as entrance is greater when compounds are undissociated (i.e., at pH values below their pK values).

Fluctuations in soil pH are caused, to a large extent, by the metabolic activities of microbes, which, in turn, are dependent on the types of substrate present. For example, the addition of carbohydrates results in an initial decrease in pH, caused by the production of CO_2 and acidic intermediary metabolites,[29-31] whereas the addition of amino acids causes an initial increase in pH, primarily as a result of the release of NH_4^+ (unpublished). These changes in pH, however, are transitory, and their duration is determined by the rate at which metabolic products are further metabolized (e.g., to CO_2, which is evolved and the pH increases; or the oxidation of NH_4^+ to NO_3^-, which lowers the pH). These transitory changes in pH, however, probably have a marked influence on the succession of populations within microhabitats.

In general, the pH of the soil solution is dependent primarily on the type of parent material, amount of rainfall, and other soil-forming factors.[19] The existence and importance of various buffering systems (e.g., the CO_2-HCO_3^--$CO_3^=$ system) are also not well defined, either for soil as a whole or in microhabitats. Inasmuch as production of respiratory CO_2 is highest in the microhabitats, any buffering capacity present may be overcome more rapidly and the pH may be reduced more than is apparent on a gross soil level.

Although not enough is known about the gross effects of soil pH on the activity, ecology, and population dynamics of microbes in soil, even less is known about pH and its effects on microbes in microhabitats. The pH of soil, as usually determined (e.g., on a 2:1 water:soil suspension), reflects predominantly the pH of the bulk soil solution and a multitude of microhabitats, rather than the pH within individual microhabitats and near the surface of the component particles, especially clay minerals. Because of the increased concentration of cations within the double layer surrounding electronegative particulates,[75,76,159] the pH at the charged interface, the surface pH (pH_s), may be several times more acid than that of the adjacent soil solution (pH_b = bulk pH).[159-166] The magnitude of this ΔpH ($\Delta pH = pH_b - pH_s$) has not been experimentally determined in soil, or even in aqueous systems, but has been deduced and calculated from a variety of indirect methods.[159-162,167] For example, the pH of montmorillonite saturated with H^+ and Al^{+++} has been calculated from infrared studies to be lower than 0.8.[167] The ΔpH presumably extends 50 to 150 A from the surface of the particle into the surrounding solution phase, as based on the distance that the electrokinetic potential of particles presumably exerts its effect.[168]

The concept of ΔpH has been used to explain the requirement for higher pH values for microbial activity in soil than in liquid culture. For example, the optimum pH for oxidation of NO_2^- to NO_3^- is approximately 0.5 pH units higher in soil than in solution;[169] maximal oxidation of various substrates by bacteria presumably adsorbed on anionic exchange resins (used as model soil systems) occurs at pH values that are at least one pH unit higher than those required by nonsorbed cells;[170,171] enzymes adsorbed on clay minerals have a higher pH optimum than enzymes in solution.[168,172] Because these activities are presumably occurring at or near the particle surface, where the pH_s is lower than the pH_b, the necessary increase in pH of the entire system for optimum activity presumably reflects the magnitude of the ΔpH. The pH optima of sorbed and nonsorbed cells or enzymes may, therefore, not be different.

This interpretation, however, does not explain the apparent ability of some microbial processes to occur at lower pH values in the presence than in the absence of charged particles. For example, nitrification can occur in soils having a measured acidity close to pH 4, whereas no nitrification occurs in liquid media, even in the presence of glass beads, below pH 6;[173] and bacteria grow at lower pH values in the presence than in the absence of certain clay minerals,[109] although this effect may result from the exchange of basic cations from the exchange complex with H^+ in the solution, where presumably growth is occurring.[108] It has been suggested that the relative concentration of cations and anions other than H^+ and OH^- within the double-layer might be important in these phenomena.[168] This suggestion, however, does not appear to parallel current concepts of protonation.[160,162,167]

Although pH appears to have a marked effect on microbes in soil, both on a gross level and at the level of the microhabitat, considerably more definitive data are required to understand the mechanisms whereby this important and fluctuating environmental factor influences the activity, ecology, and population dynamics of microbes in soil.

K. Oxidation-Reduction Potential

Perhaps the least understood environmental factor that presumably influences microbes in soil is the oxidation-reduction potential. Although it has been suggested that life can be defined as a continuous oxidation-reduction reaction,[174] the terminology, the methods of measurement, and — most important — the interpretation of measured potentials have not been resolved, either in pure culture or in natural microbial habitats.

The oxidation-reduction potential is essentially a measure of the tendency of a substance to lose or accept electrons; i.e., it is a quantitative measure of the free energy involved in electron transfer. Although both oxidation and reduction occur together, especially in biological systems, the redox reaction is composed of separate half reactions (half-cells or ionic couples), one of which involves a loss of electrons (oxidation) and the other a gain in electrons (reduction). The overall reaction is defined by the Peters-Nernst — commonly called simply Nernst — equation:

$$E_h = E_o + \frac{RT}{nF} \ln \frac{[\text{oxidized species}]}{[\text{reduced species}]}$$

where E_h = measured electrode potential (in V or mV), in reference to the standard hydrogen electrode; E_o = standard potential of the system (in V or mV), in reference to the standard hydrogen electrode; R = universal gas constant (8.315 volt-joules Coulombs); T = temperature (°K); F = Faraday (96,500 Coulombs); n = number of electrons participating in the potential system; ln = natural logarithm; and [oxidized-/reduced-species] = activity of the respective components, although concentration (in mol/l.) is used more commonly than activity. When a system is at midpoint (i.e., 50% reduction or oxidation = E_m), E_h = E_o. Most tables list electrode potentials of reactions of biological interest as the potential of a single ionic couple as E_o, which is equal to E_m at a given pH value, usually pH 7.[175]

From inspection of the Nernst equation, it is apparent that the direction in which a reaction proceeds is dependent on the number of free electrons in the system: if the number of electrons is increased, more reductant will be produced; if the number is decreased, more oxidant will be produced. Furthermore, the more highly oxidizing that a substance is, the more positive will be its electrical potential, and vice versa. In biological systems, the higher the proportion of oxidized substances present, the higher will be the E_h, and vice versa.

It is also apparent that E_h is affected by pH, temperature, pressure, and atmospheric composition. The pH of the environment controls the availability of protons and, therefore, the reversibility of the oxidation-reduction system, as well as the nature of the electroactive species (e.g., whether a weak acid will exist primarily in the dissociated or undissociated form). The relationship between E_h and pH, as derived from the Nernst equation, is, for example, E_h = -59 pH at 25°C. Consequently, if the ratio of oxidized to reduced components is kept constant (e.g., at 10:1), the E_h at pH 2 will be 51.2 mV and, at pH 8, 17.8 mV, demonstrating the importance of coupling E_h with pH measurements.

In addition to its direct thermodynamic effect, temperature also influences the rate of metabolism, thereby altering the rates at which both O_2 is consumed and reduced metabolites are produced. The level of available substrates also influences the rate of O_2 utilization and, therefore, the E_h. In flooded soils, an increase in readily available carbonaceous substrates increases the rate at which the E_h is reduced, and the E_h is inversely related to the rate of replenishment of O_2.[176-178]

Although the measurements of E_h are relatively simple procedures (e.g., potentiometrically, using noble metal and calomel electrodes) and relatively reproducible, the interpretation and significance of the measurements are often obscure. In most natural microbial environments, the measured E_h is the result of numerous mixed oxidation-reduction potentials derived from a series of reactions that may occur simultaneously but at different rates and which may not even belong to the same redox couple. The measured potentials are dependent on electrochemical reactions occurring at the electrode surfaces, and, as these surfaces are easily contaminated by trace quantities of components in the environment (e.g., metal ions, organic matter) which decrease the effective electrode surface, the kinetics of the electrode system is reduced and the measured E_h is unstable and drifts.[179-182] Many natural systems (e.g., nonflooded soil) lack sufficient poise (e.g., a buffering capacity to resist rapid changes in E_h as oxidizing or reducing agents are produced), and, therefore, marked and rapid fluctuations in E_h occur, which make precise measurements difficult.[178,179,182,183] Furthermore, many methods for measuring E_h in situ do not compensate for changes in temperature, pressure, ionic strength, volume, and O_2 tension. An apparatus for measuring E_h in waterlogged soils enables control of most of these environmental variables.[184]

Consequently, the measured E_h of biological systems, with the possible exception of simple, well-defined ones (e.g., electron transfer from one cytochrome to the next), is only a gross value reflecting the overall oxidation-reduction status of the systems. This inexactness may explain why there have been relatively few studies on E_h in microbial systems, either in pure culture[174,175,180] or in soil.[176-178,183,185]

Most studies on E_h in soil have been conducted with flooded soils, primarily because the problems associated with reduced conditions are most pronounced.[176-178,183,185] Although more studies on E_h have been conducted in water, muds,

sewage treatment processes, and sludge than in soil, the transformations of C, N, O, S. Fe, and Mn — the predominant chemical participants — are similar. In all these environments, bacteria are assumed to be primarily responsible for changes in E_h, although other microbial groups are also involved.

Among the microbial activities directly attributable to E_h in soil are mineral transformations, successions in microbial species from aerobes to anaerobes, alterations in organic products, changes in pH, and some plant diseases.

The transformation of minerals, as influenced by E_h, involves both the direct and indirect intervention of microbes.[186] Direct intervention includes the oxidative activities of chemoautotrophs and heterotrophs. Indirect interventions, which are dependent on the products of microbial metabolism, include corrosion, precipitation, adsorption, and chelation. The E_h, at pH 7, for the sequential reduction of various biologically important compounds in soil is approximately +224 mV for NO_3^-, +220 mV for Mn^{+4}, +122 mV for Fe^{+++}, and -150 mV for $SO_4^=$.[178,184,187-189]

The succession of microbial species is a function of both the E_h and pH of the environment. For example, sulfate reducers require a minimum E_h of approximately -200 mV at pH 7 for initiation of growth, but, at pH 4.15, they can function at +115 mV, and, at pH 9.92, at -450 mV;[189,190] the thiobacilli function best at potentials below -75 mV at pH 7, but they are active at +855 mV at pH 1.0 and at -190 mV at pH 9.2; denitrifiers can function from +665 mV at pH 6.2 to -205 mV at pH 10.2.[189]

In submerged paddy soils, the succession of bacteria, enumerated by standard dilution techniques, was from aerobic and facultative to total anaerobic to sulfate-reducing bacteria and was correlated with decreases in E_h, NO_3^-, and O_2 and with increases in CO_2, NH_4^+, Mn^{++}, Fe^{++}, short-chain fatty acids, sulfides, H_2, and CH_4.[176] These changes apparently occurred in two distinct stages. In the first stage, aerobic and facultative bacteria decomposed soil organic matter with the production of CO_2 and NH_4^+; with reductions in NO_3^-, Mn^{+4}, and Fe^{+3}; with no significant accumulations of organic acids, H_2, or CH_4; and with decreases in E_h from approximately +600 to +300 mV. In the second stage, anaerobes initially fermented soil organic matter to produce organic acids, which were utilized by sulfate reducers and methane producers to form sulfides, H_2, and CH_4, and the E_h dropped from approximately +200 to −150 mV.

In some submerged soils, the E_h may be as low as −300 mV, and the entire range can be as great as 1000 mV (from +700 mV when first flooded to −300 mV). The E_h of aerated soils usually ranges from approximately +400 to +700 mV, although reproducibility is poor, presumably because of insufficient poising.[178,183] The E_h of soil maintained at the 0.3 bar water tension was not significantly affected until the O_2 content of the soil was decreased below 4%.[191]

During flooding, 15 to 20% of the total soil nitrogen may be lost as N_2, although repeated cycles of submergence and drying decreased the amount of N_2 lost, as well as the rate at which E_h decreased.[178,192,193] Moreover, as the E_h of the soils decreased, extractable $PO_4^=$, Fe^{++}, Mn^{++}, and sulfide increased, as did pH. This increase in pH, however, changed with the duration of submergence; e.g., immediately after flooding, the pH increased 1 unit/61.8 mV drop in E_h; after 30 days of submergence, the ratio decreased to 1 unit/36 mV drop.

This change in pH with time further demonstrates that many interacting environmental events influence the gross E_h of soil. For example, at low E_h, levels of extractable Fe^{++} should be considerably higher, based on the amounts of Fe^{+++} initially present in soil, than those found. At low E_h, however, sulfate-reducing bacteria become active, and the resultant sulfide precipitates Fe^{++} as FeS. Furthermore, the production of metabolites toxic to other species, the capability and efficiency of species to use various available substrates as energy sources, and the ability of species to metabolize in the absence of O_2 are indirect influences of reduced E_h on microbial activity and interactions.

Although tolerances of microorganisms to E_h (at various pH values) have been established for numerous microbial groups,[189,190,194] and there are probably few habitats where all microbes are excluded primarily by the prevailing E_h,[189,194] nothing is known about the E_h within soil microhabitats. The growth of aerobes and anaerobes in the same soil having the same gross E_h (and pH) would indicate that individual microhabitats within a soil sample vary greatly in E_h (perhaps as much as 1,000 mV). Furthermore, nothing is known about the effects of E_h on

various physiological processes of microbes in soil (e.g., enzyme activity, spore germination of anaerobes, release of reducing metabolites by dying cells, toxin production, nutrient uptake), although some of these effects have been studied in pure culture.[174,175,179]

Consequently, although E_h apparently has a gross effect on microbial events in soil and, therefore, may have some predictive value for general microbial transformation, the definition of its effects on the activity, ecology, and population dynamics of microbes will depend on the capability to measure this environmental factor at levels approaching the microhabitat. Until then, caution should be used in invoking this factor to explain microbial events in soil, even though simple gross measurements, and mathematical treatment of them, are available.

L. Particulates

The predominance of a solid phase is one of the major characteristics that distinguishes soil from most other microbial habitats. This phase is composed of inorganic and organic particulates, as well as of the microbes themselves. Inorganic particles occupy more than 50% of the volume of most surface soils. Although microbes are aquatic creatures and even in highly structured environments, such as soil, may be physiologically restricted to the liquid phase, they are influenced by the physicochemical properties of the solid phase, especially by clay minerals and particulate organic matter. Sand- and silt-size particles, because of their relatively large size and inert surfaces, do not long retain water films, and thereby probably do not maintain permanent microbial populations, nor are they effective concentrating surfaces for nutrients. For example, clay-sized particles with an average diameter of 2 μm have 50 to 100 times more surface area than an equivalent amount of silt or sand. Most clay minerals also have charged surfaces which increase their reactivity, and some (e.g., montmorillonite, vermiculite) also have available intermicellar surface.

Although organic particulates may have larger specific surfaces and exchange capacities than clay minerals (e.g., the cation-exchange capacity of "humus" can be as high as 200 mEq/100 g as compared to 100 to 150 mEq/100 g for the most reactive clays),[195] the permanence of organic particles is dependent on their rates of degradation. These rates, as well as the effects of microbial alterations on the surface area and exchange capacity of soil organic particulates, are not well defined. Furthermore, the complexing of organic with inorganic particles may reduce their exchange capacity while increasing the resistance of the organic particulates to degradation.

Most microbial activity in soil is probably associated with the clay mineral fraction, which occurs primarily in aggregates or as coatings on larger, usually inert, particles.[1] Clay minerals vary greatly in their physicochemical properties (e.g., structure, specific surface, cation-exchange capacity, surface charge density, water retention, swelling capacity), which result primarily from their chemical and mineralogical composition.[77,196,197] These variations in properties are probably responsible for both the differential effects of clays on microorganisms and the many contradictory reports on these effects.[292]

Particulates, especially clay minerals, directly and indirectly influence the activity, ecology, and population dynamics of microbes in soil by their effects on the structure and pH of soil, on the availability and concentration of nutrients and water, on the adsorption and possible inactivation of inhibitors, and probably by altering the metabolic activity of cells. The literature on these and other aspects of the effects of clays on microbes in soil has been reviewed elsewhere[198,292,304] (see also Part IV), and, consequently, only a brief discussion is presented here.

The structure of soil influences its porosity, which, in turn, determines the relative amounts and rates of movement of water and gases. The structure is determined by the size and shape of the soil aggregates, and these depend, to a large extent, on the types and concentrations of clay minerals present and on the mechanisms — both physical and biological — binding the aggregates together.[199-201,384,386]

Particulates, especially those having ion-exchange capacities, can serve as concentrating surfaces for nutrients, both inorganic and organic. Mineral nutrients, especially cations, are retained by ion exchange, which reduces their rate of removal from the biologically active zone of soil by percolating water. These inorganic nutrients are probably utilized by microbes by an ion-exchange mechanism, whereby H^+ (and probably anions) produced during metabolism exchange various minerals from the exchange complex, thereby providing both nutrition and some buffering of pH

within microhabitats.[108,109,202] The magnitude of these nutritional and buffering capacities depends on the type of particulates present and, for clay minerals, increases as the cation-exchange capacity increases. Although the importance of diffusion rates of ions in microhabitats on microbial events has not been studied, the rates also appear to depend on the types of clay present; for example, the order in which Cu^{++}, Mn^{++}, and Zn^{++} diffused in the presence of various clay minerals was kaolinite $>$ illite $>$ montmorillonite $>$ vermiculite.[203]

Numerous studies have been conducted on the adsorption of organic materials, both substrates and inhibitors, by soil particulates, especially clay minerals, and on their relative availability to and activity against microbes. The results of these studies have been conflicting, with some studies showing increased utilization of presumably sorbed substrates and others indicating that sorption protects substrates from microbial or enzymic degradation.[72,165,198,292,309,389] There are similar conflicting reports on the antimicrobial activity of sorbed antibiotics[204,205,257,258,292] and pesticides.[206,207] Despite these conflicting reports, however, there is little doubt that differential sorption and availability of organics, whether potential nutrients or inhibitors, by different types of particulates affect microbial ecology in soil.

Particulates have also been shown either to increase, decrease, or not influence multiplication, growth, and various biochemical activities of microbes, although most of these studies have been conducted with pure cultures.[292] Soil particulates also appear to protect microbes from various environmental stresses, such as desiccation and elevated temperatures,[208] x-rays,[209] and hypertonic osmotic pressures.[108,391] Soil-borne viruses may be protected from degradation in soil by clay minerals,[210-213] and the sorption of bacteriophages and actinophages by clays may protect their respective hosts from lysis (unpublished).[214,215] The types of particulates present also appear to influence the establishment, survival, and spread of plant and animal pathogens in soil.[216-220,370,391] In most of these situations, clay minerals appear to be the primary particulates involved, and the effects are usually greatest when cation-exchange capacity and specific surface are high, as with montmorillonite. Unfortunately, not enough is known about the effects of organic particulates, primarily because they are difficult to isolate from soil and to purify without altering their physicochemical characteristics excessively.[221]

Despite the many studies on the effects of soil particulates, especially clay minerals, on a variety of microbial activities, most studies have raised more questions than they have answered. Many studies have been conducted only in artificial model systems, with little or no attempt to prove in soil the occurrence of the phenomena observed, or, when soil was used, the involvement of particulates as a determining environmental factor was only presumed but seldom proved. Furthermore, too few experiments have adequately shown that the variables under investigation were actually being evaluated. For example, few of the many studies on the availability to microbes or enzymes of adsorbed organics have examined the possibility that some of the organics were desorbed during the experiment and that only the utilization of the desorbed portion — not the adsorbed material — was really being measured. Conversely, precise physicochemical studies on adsorption of organics by soil particulates have seldom been concerned with the microbiological and ecological aspects of the resultant complexes. In other words, studies on particulates and their effects on microbial events have been approached from either physicochemical or microbiological viewpoints but infrequently from both views simultaneously. If the effects of particulates on the activity, ecology, and population dynamics of microbes in soil are to be clarified, critical experiments and definition of all the variables involved are necessary (e.g., characteristics of the particulates used, such as type, particle size, specific surface, exchange capacity, cation saturation; pH and ionic status of the experimental systems; mechanisms and sites of sorption and orientation of sorbed materials; methods for measuring sorption and desorption; stability and treatment of particulate-organic complexes before and during exposure to microbial or enzymic degradation; chemical, physical, and mineralogical characteristics of soils from which pathogens or other indicator organisms have been isolated).

Although many reports on the effects of particulates on microbes are conflicting, there is little doubt that the solid phase of soil, especially the clay mineral component, is an important determinant of the activity, ecology, and popula-tion dynamics of microbes in soil. Some examples of the effects of this determinant are presented in Part IV.

M. Spatial Relations

Research subsequent to that of Bail[222] and other early investigators has cast doubt on the concepts of the M-concentration and of space as a limiting factor for microbial development.[223-229] Although these concepts are still controversial, even in pure culture systems, they have not been studied adequately in natural habitats, especially in soil. There is obviously a great difference between the continuous space in a container of liquid medium or on an agar plate and the discontinuous space in a structured environment such as soil.

Microbial cells usually constitute less than 1% of soil. In a sand dune soil, only 0.02% of the available surface area of sand grains was colonized by bacteria, and 60% of the total bacteria were on organic particles which constituted 15% of the colonizable surface.[298] On the basis of such values, it is difficult to assume that space is a limiting factor to microbial proliferation in soil. However, microbes do not grow everywhere in soil but are probably restricted to areas that provide the necessary ingredients for metabolism. Consequently, it is the "active" or biological space, rather than the total or physical space, that is important in microbial ecology in soil.

Inasmuch as microbes require water, nutrients, and a favorable physicochemical environment for growth, the critical question is how many microhabitats in soil provide the necessary combination of these requisites for growth and survival of the various species comprising the soil microbiota. Considered from this perspective, few habitats in soil are probably endowed, for long, with an environment conducive to many soil microbes.

Even within microhabitats initially suited for microbial development, it is doubtful that actual space for proliferation — i.e., "Lebensraum" — becomes limiting before either some other essential factor (e.g., substrate, O_2) is reduced to limiting concentrations or the environment is altered to the detriment of the resident populations (e.g., reductions in pH or E_h, production of inhibitors).

Conversely, in microhabitats wherein conditions for growth are maintained for long periods, proliferation may continue until all biological space (probably water films) is filled with living and dead cells. Further proliferation in such habitats would be a function of the decomposition rate of the cells. Inasmuch as microbial cells and some of their products appear to be decomposed relatively slowly in soil,[22,230-232] growth within such microhabitats may be limited by available space. Microhabitats wherein such conditions prevail may be in surface layers of soils containing a continual supply of organic materials and in which the water content is optimal, so that gas exchange is not restricted and water bridges between microhabitats prevent both accumulation of toxic products and ensure a steady supply of soluble nutrients. Such microhabitats probably also contain particulates having a high exchange capacity (e.g., montmorillonite), which would buffer the environment against rapid decreases in pH, concentrate nutrients from the dilute soil solution, and adsorb and inactivate inhibitors.[233,292] Similar situations may also exist in localized areas where nutrition is adequate, e.g., near plant roots and pieces of fresh organic matter, where the structure and water content of the surrounding soil prevent other environmental factors from becoming limiting before space does.

It is difficult to demonstrate unequivocally such limitations in active space. However, some studies, both in soil[29,31] and in pure culture,[222,224] have indicated that, under conditions where other environmental factors have presumably been prevented from becoming limiting, available space can restrict further microbial development. Studies with model systems (e.g., when glass beads are used as the solid phase) have also suggested that there may be a need for "appropriate" space for various microbiological functions. For example, Curvularia sp. and Cochliobolus sp. produced conidia only in air-filled pores, whereas those of Fusarium culmorum were produced primarily within water films adjacent to beads. The hyphae of F. culmorum also penetrated further into water-filled areas than did those of the other two fungi.[234]

The particle size distribution of the particulates comprising the microhabitat, as well as the types of clay minerals and organic matter present, influences the ratio of active to total space. Most microbial activity in soil is probably restricted to water films, contact rings, and capillaries between soil particulates, and sand- and silt-size particles are not as effective in retaining water as are smaller particulates. Both organic and inorganic colloids possess a large surface area in relation to the volume that they occupy. The presence of these particulates, which in many cases are highly charged, may increase the active space within microhabitats by providing concentrating surfaces for organic and inorganic nutrients and by influencing other physicochemical factors.[233,292] These materials may act similarly to the solid phases used in laboratory biphasic systems to increase bacterial numbers above the M-concentration.[223,229] In soil and in these laboratory systems, the microbes are probably growing in the solution phase, with the solid phase being both a source of nutrients and/or O_2 and a repository for toxic materials.[229]

The dense growth of a microbial population in a restricted area probably also inhibits colonization of that microhabitat by other populations. For example, the mycelial development of Sclerotinia fructrigena and Cordyceps militaris may be so intense that the penetration of hyphae of other species is prevented.[235] Whether this prevention is caused solely by physical constraints on the available space or by alterations in the environment that are inimicable to the invaders requires clarification.

The dense growth of microbes in a physically restricted space can also confer both positive and negative advantages to the resident populations. For example, the lag phase of growth is decreased in many pure cultures when the number of inoculated cells is increased, either because of larger pools of precursors, adsorption of toxic constituents of the medium, or other mechanisms.[104] In dense populations, whether in soil or in pure culture, syntrophism is enhanced, the probability for genetic recombination and plasmid transfer is higher, and the inner cells of a colony may be protected against freezing, predation, and other environmental stresses.

Conversely, the movement of nutrients and O_2 toward, and of CO_2 and toxic metabolites from, cells is retarded in dense populations, especially with respect to inner cells.[229] In addition to such retardation, which in soil is probably greatest in water films forming "necks" between particles, hysteresis may affect the rate of movement of gases and dissolved materials. Some microbial species can adapt to environmental limitations induced by dense populations developing in limited active space by altering their morphology (e.g., increasing their surface to volume ratio; producing slime layers, fimbriae, and protoplasmic extrusions).[137] The occurrence of and the importance to microbes of such morphological changes in soil are not known (see Section N).

In addition to the possibility that limitations in available space influence the activity, ecology, and population dynamics of microbes in soil, spatial relations between microbes, substrates, and exoenzymes could also influence these phenomena. Because soil is a structured and discontinuous microbial environment, spatial accessibility to, rather than the total content of, nutrients is critical. Nutrients may sometimes be unavailable to microbes or their exoenzymes by being occluded within soil aggregates or narrow capillaries[236,237] or by being adsorbed to soil particulates, especially clay minerals. Exoenzymes may not reach their target substrates, or, if they do, the hydrolyzed subunits may not reach the cells elaborating the enzymes, as the enzymes or products may be intercepted by either adsorption or other microbes which degrade them. Similarly, interactions between and among species (e.g., growth factor exchange, alterations in atmospheric composition, genetic recombination, amensalism) will depend on the spatial proximity of the participating populations and the continuity and volume of water films joining their microhabitats.

Spatial relations assume major importance when survival and growth of a species are dependent on a single substrate, which, in turn, is elaborated by another species. Nitrification is an example of such a sequential dependency, especially as the populations involved in each step appear to be vertically distributed and the distributions depend on the concentration and location of their respective substrates. The kinetics and geography of nitrification are being evaluated by vector biochemical analyses.[238,239]

Although convincing evidence for both limitations in available space for microbial proliferation and the need for adequate spatial relations is lacking, both in soil and in other habitats including pure cultures, there can be little doubt that these factors influence microorganisms in soil. The effects of spatial relations are probably expressed directly as well as indirectly through their influence on other environmental factors (e.g., atmospheric composition, accumulation of inhibitory substances, selective depletion of nutrients, pH, E_h, microbial interactions).

N. Characteristics of the Microorganisms

In the final analysis, the success of an organism in any habitat depends on the extent and rapidity of its physiological responses to the prevailing environmental conditions. These responses, in turn, are constrained by the genetic capabilities of the organism. Although considerable information has been obtained in vitro on phenotypic responses of microorganisms to changes in various physicochemical properties, definitive evidence for similar responses in soil is lacking. This lack is the result primarily of the absence of adequate methods for studying microbes directly in the soil environment.

For example, although the generation time of individual species and even of simple mixed populations is easy to measure in the laboratory, essentially no data are available on the generation time of species growing in soil. The difficulties inherent in determining generation times in natural habitats, however, are not restricted to soil.[240] There is little doubt that the generation times of microbes in soil are considerably longer than those in pure culture, inasmuch as the doubling times for different microorganisms in various habitats, as determined by a variety of techniques, greatly exceed those normally encountered for similar microbes under laboratory conditions (e.g., 2 to 200 hr in lake water; 75 to 130 hr in sea water; 8 to 10 hr in the spleen; 20 hr in the intestinal tract; 12.5 hr in the rumen; 29 hr for spirochetes in syphilitic lesions).[240,241]

Although doubling times for microbes in soil have not been directly measured, indirect methods have indicated that similar relationships occur. For example, generation times for *Nitrosomonas* sp. in soil ranged, depending on the soil pH, from 38 to 100 hr and for *Nitrobacter* sp. from 21 to 58 hr, as calculated by oxidation rates of NH_4^+ and NO_2^-, respectively, in perfusion experiments.[242] Using fluorescent antibody techniques, the doubling times of *Rhizobium* sp. were calculated to be 241 to 361 days in nonsterile soil and approximately 14 hr in sterile soil (personal communication, E. L. Schmidt). Other values, obtained by coupling empirical calculations with either direct counting of bacterial cells[243] or respiratory measurements,[241] range from approximately 1,200 hr to 28.5 hr, respectively. The reasons for these extended generation times, which may, in fact, be conservative estimates, have not been defined, but, as various physicochemical and biological factors greatly influence multiplication even in pure culture, these factors (e.g., temperature, osmotic pressure, pH, atmospheric composition, nutrition, inhibitors, negative interactions between microbes) undoubtedly have a differential influence within soil microhabitats.

Morphology is another phenotypic expression that is relatively well defined in pure culture and in some natural microbial habitats other than soil. The size and shape of cells reflect, in part, the prevailing environmental conditions.[137] Under conditions of reduced nutrition, the surface to volume ratio increases and cells assume an elongated configuration, sometimes even forming filaments. Under conditions of desiccation, the cells decrease their surface to volume ratio and become more spherical.[104] The importance, or even occurrence, of these morphological changes in soil has not been adequately studied. Based on a variety of direct microscopic observations, the dormant bacterial population of an unamended soil appears to be composed primarily of coccoid and coccoid-rod cells, ranging in diameter from 0.5 to 0.8 μm and covered with a capsule-like material.[53] Conversely, many of the cells observed in soil by electron microscopy[47,54,55] and with the pedascope[56] are thin and have an elongated configuration.

Cellular morphology can also be altered in vitro by changes in pH, E_h, temperature, atmospheric composition, and the concentration of toxic materials and other solutes. The interference by some metabolites in cell wall synthesis is relatively well defined in pure culture and in some in vivo habitats (e.g., in the nephrons), as is the requirement of high solute concentrations to maintain these involution forms. The relative resistance of these forms to some antibiotics and to attack by bacteriophages capable of attaching to the same species when containing an intact cell wall has been described in pure culture. It is not known, however, whether metabolites capable of interfering with cell wall synthesis are produced, or persist, in soil; whether the solute concentrations in the microhabitats are high enough to prevent plasmoptysis of the resultant cells; and whether the formation of involution forms is a defense mechanism whereby some species are protected against lysis and predation and, thereby, persist in soil. Similarly, it has not been determined whether the thicker cell wall of gram-positive bacteria, which would tend to offer more protection against

desiccation, is responsible, at least in part, for the apparent higher concentration of these bacteria in soil. Until the various physicochemical characteristics of the microhabitats can be critically defined, it will not be possible to correlate cell morphology with the prevailing environmental conditions.

The role of various cellular surface structures in the survival of microbes in soil also requires investigation. For example, do fimbriae function as hold-fasts, thereby enabling some bacteria to adhere to particulates in the microhabitats? Many of the forms observed by direct electron microscopy of soil are densely covered by pili.[47] These forms are seldom isolated by standard soil dilution plate techniques, in part, perhaps, because they are not released from adherence to inanimate particulates.

Many soil bacteria produce slime layers in culture, and the occurrence of capsule-like structures on bacteria has been observed directly in soil.[53] These structures, however, are apparently not similar biochemically to capsules produced in laboratory cultures. Mucoid surface layers may also impart adhesive properties to cells in soil, which would enhance attachment to soil particulates and the formation of cell aggregates. In addition, slime layers might increase resistance to drying, decrease susceptibility to ingestion by predators, provide a source of substrates during periods of low ambient nutrition, and, because of the relative hydrophobicity of some capsular materials, may orient some cells at the water-air interfaces in the microhabitats, thereby increasing their exposure to O_2. All these phenomena impart survival value to cells and should, therefore, be studied and defined in the soil environment.

The importance of flagella to cells in soil also requires clarification. Although it is known, from pure culture studies, that synthesis of flagella is dependent on environmental conditions — primarily nutrition, pH, and temperature — nothing is known about their formation in soil. Even if flagella are formed in soil, they may not be as important to the motility of cells in soil as in pure culture or in nonstructured natural environments. If the bulk of the soil microbiota is, in fact, restricted to the water films of the microhabitats, the cells may not produce sufficient energy for significant flagellar movement within these structured films. Many soil bacteria do, however, produce flagella in culture, and, with the present state of knowledge, it may be assumed that they function as agents of motility in soil. If motility occurs, then those species producing flagella probably have some survival advantage, as they could respond more rapidly to changes in their immediate environment (e.g., by escaping predators, exhibiting various taxes).

The occurrence of cytoplasmic inclusions (e.g., volutin, lipids, polymeric carbohydrates) in microbes in soil also requires investigation. Although many microbes growing in the luxuriant nutrient conditions of most laboratory media accumulate such inclusions, presumably as reserve food materials, it has not been established that microbes in soil, where the nutrient status is assumed to be considerably poorer, do so. Obviously, the ability to accumulate reserve food materials in times of nutritional plenty for use in times of nutritional poverty would be of ecological advantage to microbes in soil.

The ability to form resting or resistant structures also confers survival value on some microbes. Much is known about the formation, resistance, persistence, and germination of such structures, but, again, most of this information has been derived from pure culture studies.[244] For example, both the formation and germination of bacterial endospores require a pH, pO_2, a_w, and other environmental conditions that are not too different from those required for optimal growth, although the limits for sporogenesis and germination are usually narrower than for vegetative growth. The germination of spores in culture can be induced by a variety of methods, including mechanical disruption, heat activation, and presence of available carbon substrates, but the exact environmental factors that induce sporogenesis are not well defined in pure culture and not at all in soil.

Soils contain spore-forming bacteria, and the concentration of endospores appears to be higher in "dormant" soils (e.g., dried, stored, nutrient and O_2-deficient soils).[245] Although this suggests that spores are important in the resistance of some species to adverse environmental conditions, the differential effects of various environmental factors on production and germination of endospores in soil, especially in the microhabitats, are not known.

Even less is known about the formation, persistence, and reversal to vegetative growth in soil of resistant structures of other microbes (e.g.,

cysts of protozoa, algae, *Azotobacter*, slime molds, and myxobacteria; sexual and asexual spores and sclerotia of fungi). The longevity of various resting structures of different organisms has been compiled,[246] and some structures retain viability after storage for more than 60 years. This longevity, however, usually occurs under unnatural conditions of storage, and no information is available on the persistence of resting structures in natural soil. Inasmuch as conditions conducive to germination and vegetative growth probably exist intermittently in soil in situ, it is difficult to assume that many resting propagules persist unchanged for extended periods. Only within undisturbed soil aggregates, where entrance of water, nutrients, and other factors required for germination is reduced, can the persistence of such structures be assumed.

Although numerous soil organisms show various rhythmic physiological processes in pure culture, the occurrence of such rhythms, and their entrainment by prevailing environmental conditions in nature has not been adequately defined. The ability of an organism to regulate its functions to operate most actively at the most propitious time would enhance its adaptability and, therefore, its survival in the soil microhabitats.

All of these physiological and morphological adaptations to the prevailing soil environment are dependent upon the genetic capabilities of the organisms. Although there is considerable information about genetic recombination, mutation, mechanisms of enzyme induction and repression, etc. in microbes in pure culture, essentially nothing is known about whether and how such phenomena operate in soil.

Genetic recombination would greatly enhance the adaptability and survival value of microbes in soil, but little evidence has been provided that the various types of recombination, especially among procaryotes, do, in fact, occur in soil. Among eucaryotic microbes, genetic recombination undoubtedly occurs, as indicated, for example, by the presence of sexual fungal spores in soil and by the relatively rapid changes in some plant pathogenic imperfect fungi capable of undergoing heterocaryosis. In the procaryotes, only conjugation between bacteria (*E. coli*) has been demonstrated in soil, and then only in sterile soil.[23]

The occurrence of transformation and transduction in bacteria in soil has apparently not been investigated. If these phenomena do occur, they probably are restricted to extremely localized sites within the microhabitats. Bacteriophages or pieces of naked DNA probably do not persist long in soil, as both materials are excellent substrates and would be rapidly degraded. Although DNA and phages may be adsorbed by clay minerals and soil organic matter, their protection from microbial degradation and subsequent release in a form infective to an appropriate host has yet to be determined. Similarly, the transfer of various plasmids that might confer survival value to bacteria requires investigation in soil. Until such phenomena are conclusively demonstrated in natural soil, caution must be exercised in extrapolating laboratory observations to in vivo conditions and invoking the concept that genetic recombination is a mechanism whereby microbes, especially procaryotic ones, in soil adapt to prevailing environmental conditions.

Similar caution must be used in speculating about mutation rates in soil. Although mutagenic agents are undoubtedly present, their location relative to metabolizing microbes, their concentration, and their persistence must be evaluated before it can be assumed that mutation rates in soil are other than those occurring by chance alone.

Another fertile area for investigation is the induction and repression of enzyme synthesis in microbes in soil. Although many environmental factors (e.g., specific substrates, pH) have been shown to be involved in enzyme induction in pure cultures, the effect of these factors on induction in soil has not been elucidated. Induction undoubtedly occurs, and, from the anthropocentric viewpoint of adaptability, organisms containing few constitutive enzymes but capable of rapidly synthesizing required enzymes would have superior survival potentials. Although many substrate enrichment studies would suggest the induction of enzymes capable of utilizing specific substrates, it is not always clear whether induction has occurred or whether the added substrates have merely enriched for species already containing the necessary enzymes. Similarly, changes in soil atmosphere (e.g., decreases in O_2 and increases in CO_2) may not induce physiological adaptations but may result only in the enrichment of facultative and microaerophilic species already present.[128]

The occurrence in soil of enzyme repression, especially by feedback inhibition, also needs clarification. Because individual microhabitats are probably populated by several species, the metabolic products of one species may be utilized rapidly by others, thereby precluding their accumulation in concentrations sufficient to cause feedback inhibition. Conversely, the metabolic products of one species could result in sequential enzyme induction in a neighboring species.

The heterogeneous microbial composition of soil greatly enhances the biochemical potentialities of the microhabitats. Consequently, it is questionable whether compounds that, in pure culture, influence mutagenesis, genetic recombination, enzyme induction and repression, etc. actually accumulate in sufficient quantities and for a long enough time to influence markedly these phenomena in soil.

Although knowledge of the interactions between microbial characteristics and environmental factors is imperfect, any study of microbial ecology in soil must consider the genotypic capabilities and phenotypic expressions of the resident microbes. In soil, perhaps more than in any other environment, microbes are constantly confronted with either dying or adapting rapidly to constantly changing environmental conditions. The characteristics and capabilities of the microbes, therefore, are ultimately the deciding factors in their ecology and population dynamics in soil.

O. Interactions Between Microorganisms

One assumption inherent in the concept of the existence of microhabitats in soil is that each microhabitat, at one time or another, contains cells of different species and that these species interact. The results of such interactions can be either detrimental, beneficial, or neutral to one or all species involved, and, therefore, these interactions influence the relative success and ecology of the component species.

Much has been written about interactions between microbes. Nevertheless, the delineations between interactions, as well as the terminologies employed, are many and confusing. Perhaps the most widely used scheme for classifying interspecific interactions is that proposed by Odum.[247] This scheme, a modification of which is presented in Table 1, was developed as a model for interactions between any organisms, not only between microorganisms. However, the scheme is based only on interactions between two species, whereas, in natural habitats, especially in microbial habitats, several species in close juxtaposition probably interact both simultaneously and

TABLE 1

Some Interactions Between Species that Influence the Activity, Ecology, and Population Dynamics of Microbes in Soil[1]

Interaction		When not interacting A	B	When interacting A	B
Positive	Neutralism	0	0	0	0
	Commensalism	–	0	+	0
	Protocooperation	0	0	+	+
	Mutualism	–	–	+	+
Negative	Competition	0	0	–	–
	Amensalism	0	0	–	0
	Parasitism[3]	0	–	–	+
	Predation[3]	0	–	–	+

[1] Based on Odum.[2][4][7]
[2] 0 = no effect; + = positive effect; – = negative effect.
[3] Species A is the host and Species B is the parasite or predator.

sequentially. For example, the development of organism A may be stimulated by organism B in a commensal relation and the enhanced growth of A could result in the accumulation of sufficient amounts of an inhibitor against organism C (amensalism), which previously competed with and prevented the development of organism D; the enhanced growth of organism A may result in an increase in the predatory activity of species E, which may now decrease the density of A, and, as a result of prey depletion, eventually also of E. Such multiple interactions are probably more common in soil microhabitats than the model two-species interactions. In addition, the delineation of interactions is not discrete but probably can be better described as a continuum or spectrum merging positive and negative interactions (e.g., symbiosis can be considered an extreme form of mutualism or as an extreme form of balanced or controlled parasitism). For this reason many more permutations of positive, negative, and neutral interactions than normally described probably occur in soil.

Another factor that influences interactions in soil is the spatial distribution of the participants. For example, if organism A is dependent on a growth factor produced by organism B, the organisms must be close enough together so that A can exploit the growth factor before it is degraded by another species.

The environmental factors prevailing in the microhabitat will also influence the type and degree of interaction between resident species by altering the phenotypic expression of the genetic capabilities of the organisms. For example, lowering the pH by the production of acidic intermediary metabolites by organism A may alter the metabolism of organism B, which, at higher pH values, would elaborate enzymes capable of lysing organism C.

Despite the wealth of literature on microbial interactions,[5,6,12,16] relatively little is known about interactions between microbes in soil in situ. Most of the information has been derived from laboratory studies on simple mixed populations, which precludes an evaluation of higher order interactions between multiple microbial species and also between the microbiota and the various physicochemical environmental factors. However, because of the apparent importance of interactions between microbes on their ecology and population dynamics, a brief discussion of various types of interactions is presented. To simplify this discussion, interactions will be broadly divided into positive and negative interactions.

The absence of any interaction (i.e., neutralism) probably does not occur to any extent in soil. Because many microhabitats are in contact, via water bridges, the products of metabolism, either stimulatory or inhibitory, are eventually distributed throughout adjacent habitats. It is conceivable that neutralism occurs when two or more populations are sufficiently separated that their activities do not influence each other.

However, such monocultures are probably the exception in soil, and, where they do occur, they are probably the result of the ability of the organisms to prevent the establishment of other species, which undoubtedly are periodically introduced into the microhabitat by movement of gravitational water, hyphae, protozoa, metazoa, etc. It is also difficult to assume that two or more species will neutrally coexist in the same microhabitat if their nutritional and other environmental requirements are so different that competition does not occur. Many environmental requirements are similar for most microbial life, and the very fact that several species are growing together in a limited area must result in some competition for space and commonly required nutrients. Furthermore, the production of diverse, or even similar, metabolites must rapidly alter the microenvironment to the detriment of at least one of the inhabitants. Perhaps neutralism occurs in soil microhabitats only when the population densities of resident species are so low that competition is not a limiting factor and the deleterious effects of growth are buffered by the environment. Unfortunately, there has been little investigation of neutralistic interactions in soil.

In a commensal relationship, the growth of one species is enhanced by the presence of another, which presumably is not affected by the interaction. The association is not necessarily specialized, and different species provide the commensal (i.e., the species benefited) with the requisite factors. For example, many species of plants can provide the nutrients necessary for the growth of nutritionally fastidious microbes in the rhizosphere. Similarly, many microbial species probably synthesize growth factors required by a wide spectrum of auxotrophs, of either the same or different species. The conversion of a substrate (e.g., hydrolysis of a polymer into simple subunits; oxidation or reduction of aldehydes) into a form utilizable by a commensal that lacks the enzyme systems necessary to attack the original substrate can be accomplished by a variety of species. Conversely, some biogeochemical transformations (e.g., those of nitrogen and sulfur) are restricted to one or few species, but the commensal beneficiaries represent a wide spectrum of species.

In addition to nutritional relationships, commensal interactions may take the form of one or more species detoxifying the environment and, thereby, enabling a commensal sensitive to the toxic conditions to proliferate. This detoxification may involve degradation of an organic inhibitor, oxidation or reduction of inorganic ones, production of compounds that chelate toxic ions, lowering the pH to prevent dissociation and, thereby, penetration of toxic fatty acids, etc. Alteration in the physicochemical properties of the microhabitat (e.g., changes in pH, E_h, osmotic pressure, ionic strength, atmospheric composition) during growth of one or more unrelated species can enhance the development of other species. The casual association of a small organism with a larger one can provide the smaller with protection against predation and adverse environmental stresses, and the larger organism may not only provide a more rapid means of dispersal and introduction of the commensal into new microhabitats but also some nutrition and water films during the process. Many of the bacteria present in various "spheres" (e.g., rhizosphere, phyllosphere, hyphosphere, spermosphere, nematosphere) are probably commensals of this type.

The benefits derived by the commensal may be expressed by a change in its morphology, which may confer survival value to it under different environmental conditions. For example, endospore formation in some species of bacteria is enhanced by the metabolites of other bacterial species,[248] and growth, sporogenesis, spore germination, and other morphogenetic changes in fungi are induced by metabolites, both volatile and soluble, from the same or different species of fungi as well as from a variety of bacteria.[145,146]

The transfer of plasmids and genetic recombination in bacteria can be considered a type of commensalism, as the recipient may derive an ecological advantage, such as resistance to inhibitors and greater nutritional capabilities. Although such transfers have not been demonstrated unequivocally in soil, this form of commensalism would require close proximity between the commensal and the donor. In many other forms of commensalism (e.g., nutritional, detoxification, changes in environment), the actions of the independent species probably influence commensals in relatively distant microhabitats, especially if the factors involved are volatile.

Although these and other forms of commensalism are easily demonstrated in two-member cultures in the laboratory, not many have been conclusively demonstrated in soil. There is little doubt, however, that most occur in soil in situ, inasmuch as many processes in soil are probably dependent upon commensalistic relationships. For example, many soil organisms do not have the enzymic capabilities to degrade crude plant and animal residues and must wait for organisms containing lytic enzymes to provide them with the resultant monomers as substrates; the wide occurrence of microbes requiring growth factors indicates that these are, in part, provided in situ by other species; the presence of anaerobes, even in well-aerated soils, indicates that, in the microhabitats, O_2 is removed by neighboring aerobes; the production of substrates for chemo-autotrophs (e.g., NH_4^+, NO_2^-, H_2S, S) requires the activities of other organisms that may not be directly benefited by the subsequent action of the autotrophs.

In some commensal interactions, however, the host probably also benefits eventually by the action of the commensal. This type of "supra-commensalism" would be described, according to Odum's scheme, by $-O/++$ (not interacting/interacting). For example, a temperate phage requires its host bacterium for survival, but, even though the host is capable of existing without the phage, the bacterium may benefit by being protected against further infection by phage, and, by transduction, its nutritional and toxin-producing capabilities may be enhanced. The metagon system in some paramecia[249] is another example of this supracommensalism, inasmuch as the metagon particles (probably bacterial in nature) require the paramecia for their survival. The paramecia, which can function without the presence of the commensal, benefit by being able to kill other paramecia while, simultaneously, becoming immune to attack by other "killer paramecia." Although such interactions can be considered as a form of symbiosis, the relationship is obligatory for the survival of only one member. The importance and, even, the existence of supra-commensalism in soil await clarification.

Another relationship, which is casual and not obligatory to either member but which may benefit both members, is protocooperation. Most of the reported occurrences of protocooperation, which again are based primarily on two-member laboratory studies, involve some type of nutritional or detoxifying interaction. For example, *Azotobacter* sp. or other free-living nitrogen-fixing organisms may enrich a nitrogen-poor microhabitat, which, in turn, enhances the growth of species incapable of using atmospheric nitrogen but which provide the nitrogen-fixer with a source of readily utilizable carbon substrate; reciprocal growth factor exchange may permit both fastidious species to develop; degradation of some polymeric substrates (e.g., cellulose, chitin) may require the simultaneous contribution of enzymes from several species, none of which has the complete enzyme complement needed; the metabolites of one species, which may be auto-inhibitory or cause enzyme repression by feedback inhibition, can serve as substrates for a comparison species, thereby enabling the continued growth of both species.

Many of these protocooperative interactions have been designated as synergistic; i.e., two or more species growing together are capable of some performance that neither is capable of doing by itself. Although there are many examples of this phenomenon, in addition to those mentioned above (e.g., production of a metabolite toxic to a third species; formation of gas — the Castellani phenomenon; enhanced pathogenesis; some sequential biogeochemical transformations), their occurrence and their importance in the survival and ecology of microbes in soil have not been adequately documented. The results of synergism are often of value to man (e.g., in some industrial fermentations), and it is anthropocentric and teleological to ascribe ecological benefits to the organisms involved without sufficient evidence.

Furthermore, many of the protocooperative interactions require that the cooperators be in relatively close association. Even though common water films probably exist between many micro-habitats in soil, cooperative species — that are either exchanging growth factors, pooling enzymes for a common cause, providing substrates for a nitrogen-fixer, etc. — must be close enough together so that an interceding species does not degrade the compounds involved before the co-operators can make use of them. Although this interaction is not considered to be species specific, it is probably seldom that adequate cooperators are in a favorable juxtaposition in soil.

Protocooperative interactions may be intraspecific as well as interspecific. Studies on bacterial conjugation in soil have indicated that auxotrophic strains of the same species exhibit syntrophism (i.e., cross-feeding) in soil.[23] Because strains of the same species or closely related species may be more adapted to the same microhabitat, intraspecific and intrageneric protoco-operation may occur more frequently than inter-specific and intergeneric protocooperation.

A modification of protocooperation is where the growth of only one organism is stimulated (i.e., OO/+O, not interacting/interacting). This interaction differs from commensalism in that the stimulated organism is capable of developing independently but performs better when in the presence of another organism. The release of growth factors or precursors for some essential metabolite by plant roots or by neighboring microbes could stimulate the growth of a hetero-troph capable of synthesizing the necessary materials but which, by utilizing those present in its milieu, is able to redirect its metabolism and energy into other pathways. Because this type of protocooperation does not depend upon reciprocal benefits, it may be a more important interaction in soil than the more conventional type of proto-cooperation.

Perhaps the most widely studied and docu-mented interaction is mutualism, wherein the interaction is usually obligatory and specific for both partners; i.e., two (or more) species live together for mutual benefit, and neither species can live normally without the other. Many of these symbiotic relationships are of economic value to man (e.g., nitrogen-fixation, mycorrhizal associa-tions, lichens, in the rumen and cecum, primary productivity in waters) and, therefore, have been studied extensively. Furthermore, most symbioses involve a microbe and a larger organism, such as a plant or animal, and relatively little is known either about or the extent of microbe-microbe symbioses. This is especially true in the soil environment, where most attention has been directed toward symbiotic nitrogen-fixation and mycorrhizal associations.

Despite the lack of information about the extent of obligatory microbe-microbe mutualism in soil, there is little doubt that this interaction provides an ecological advantage to the symbionts. Among the benefits derived are more stable sources of nutrients; removal or neutralization of toxic materials; protection against predators, parasites, and environmental stresses; and, in general, a favorable and, in some instances, an isolated environment for maximum development.

Regardless of the type of positive interaction between microbes in soil, this confluence of enzyme systems is undoubtedly important in the sequential degradation and production of sub-strates, removal of inhibitors, production of growth factors, maintenance of favorable environ-mental conditions, etc. — phenomena that markedly influence the succession and survival, and, therefore, the ecology and population dynamics of the participating microorganisms.

Negative interactions between microbes are probably more common in soil and in other natural habitats than are positive interactions, which may account, in part, for their better definition. The relatively rapid growth rate of microorganisms and the intimate association between cells and their inanimate environment have resulted in the evolution of a variety of mechanisms whereby microbes exert their in-fluence to the detriment of members of the same or other species.

Perhaps the most common type of negative interaction is competition for one or more of the essentials necessary for microbial survival. Com-petition for carbon substrates, especially those that are readily utilized by a wide spectrum of species, is probably predominant. However, mineral nutrients, growth factors, O_2, water, and possibly space can also become limiting factors for which species within the same microhabitat compete. Although a specific factor may not be limiting in a microhabitat under a given set of conditions, alteration of those conditions may rapidly bring that factor into competition. For example, the concentration of mineral nutrients may be adequate to support several potentially competitive species with the prevailing amount of carbon substrates present, but the introduction of available carbon sources into the microhabitat, even though releasing the environment from one limiting constraint (i.e., carbonaceous substrate), can rapidly result in the depletion of the mineral nutrients to limiting concentrations. This deple-tion is usually sequential and depends on the amount of essential minerals originally present, on their relative requirement by the species during carbon utilization, and on their turn-over or cycling rates. In the example cited, once the carbon limitation has been removed, the order of depletion is nitrogen, phosphorus, and sulfur. The addition of phosphorus and sulfur without nitro-gen does not relieve the competition nor increase the rate of utilization of the carbon sub-strate.[29-31]

In microhabitats adequately endowed with

organic and inorganic nutrients, the proliferation of cells may exceed the rate at which O_2 is replenished, and competition occurs for the diminishing O_2. This depletion of O_2 not only excludes many species incapable of growing under microaerophilic or anaerobic conditions, but the concomitant increase in CO_2 inhibits species sensitive to elevated concentrations of this gas.[127,128]

Limitations in available space for proliferation are difficult to demonstrate in soil as well as in other habitats. However, competition for attachment sites and for space within the water films of microhabitats may occur under favorable conditions of growth.[31]

The extent of competition probably varies between microhabitats, depending on their structure, nutritional composition, and other chemical, physical, and mineralogical properties. Many environmental factors (e.g., pH, E_h, temperature) probably alter the competitive ability of species, primarily by influencing their metabolic rates and pathways.

The genetic capabilities and the phenotypic response rates of organisms are probably also major factors in competition. For example, species having a rapid growth rate on simple substrates may be more capable of exploiting these nutrients when they are present in short supply, whereas organisms with a slower growth rate, but with a more extensive enzyme complement (e.g., extracellular hydrolytic enzymes), may be more competitive in an environment containing complex substrates; asporogenous bacteria probably initiate growth more rapidly after introduction of substrates than species present as spores; fungal species having rapidly germinating spores would exploit an environment more rapidly than other species.

The phenotypic expressions of species are probably also modified by prevailing environmental conditions. For example, release from a specific limiting factor could alter the generation time of a species, thereby enhancing its competitive capabilities. Consequently, a slight alteration in only a single environmental factor could result in a minor component of a mixed population eventually becoming the dominant species within a microhabitat. Similarly, a slight genetic alteration could result in intense intraspecific competition, with the mutant becoming dominant if the genetic change conferred a competitive advantage.

Although there is little doubt that competition is a major phenomenon in soil, conclusive evidence for most of the mechanisms, in situ, is not available. Many of the data have been derived from simple laboratory studies, wherein most environmental factors have been kept constant. Even in such studies, it is often difficult to distinguish between competition and amensalism. For example, a reduction in fungal growth elicited by bacteria in dual cultures in soil could be the result of either substrate or mineral nutrient depletion, of the release of inhibitors, or of changes in abiotic environmental factors by the bacteria.[250-253]

Any of these phenomena could result in zones of inhibition. In some dual cultures, a zone develops between species from which neither fungi nor bacteria can be isolated, suggesting that amensalism, rather than competition, is involved (unpublished).

Furthermore, there is no evidence to indicate that the results of competition in soil are always detrimental to all species involved, as suggested by Odum's scheme. A more usual result would presumably be the exclusion of one species, with the other species being either unaffected (i.e.,–0, when interacting) or eventually benefited (i.e., –+, when interacting).

In amensalism, one species gains the ascendency in a microhabitat by its ability to excrete products that adversely affect the development of other organisms of the same or different species. The effect can be direct, as in the production of inhibitors, or indirect, as in the production of metabolites that alter one or more environmental factors.

Many types of metabolites produced by one species may be inhibitory or lethal to others. These metabolites may be organic or inorganic, soluble or volatile, and effective in minute amounts or only in relatively high concentrations. The inorganic inhibitors include NH_4^+, NO_2^-, NO_3^-, $SO_4^=$, CN^-, H_2S, H_2O_2, CO, and CO_2. Some of these materials act directly on the cell, others alter the environment, and some do both. For example, high concentrations of NH_4^+ may directly poison some species and inhibit those incapable of developing at high pH values; $SO_4^=$ probably inhibits primarily by reducing the pH; CO_2, at elevated concentrations, differentially inhibits many species as well as inducing morphological changes in others.[5,148]

Among organic compounds inhibitory only at relatively high concentrations are short-chain fatty acids (e.g., formic, acetic, propionic, butyric, lactic), some alcohols, aldehydes, and other metabolic products. The activity of some of these materials, especially of the acids, is pH-dependent; e.g., organic acids are more toxic at a pH near or below their pK values, as the undissociated acids apparently penetrate cells better. The production of these metabolites is also dependent on environmental conditions; e.g., alcohols and other products of glycolysis (e.g., heterofermentative products) will be produced in greater quantities in microhabitats low in O_2.

Perhaps the most often ascribed form of amensalism in soil is that of antibiotics (i.e., substances produced by microorganisms that are inhibitory, in extremely low concentrations, to other organisms). Because of the importance and success of antibiotics in medicine, and because many antibiotic-producing microorganisms have been isolated from soil, the success of one microbial species over another in soil is often attributed to antibiotic production. There is, however, little direct evidence for significant antibiotic production in soil in situ. Many species that produce antibiotics in culture do not appear to be dominant species in soil, and many of the species commonly isolated from soil in large numbers do not produce antibiotics in culture. In commercial antibiotic production, optimal nutritional and environmental conditions (e.g., of pO_2, pH, E_h) are required, and many antibiotics are produced only late in the growth cycle. Comparable environmental and nutritional conditions probably seldom occur in soil in situ, and the amensalistic value of an inhibitor produced so late after colonization is speculative. Furthermore, there is little evidence for the isolation of known antibiotics from soil, except in a few instances when soils are heavily inoculated with known antibiotic producers or in carbon-rich loci.[254,255]

This lack of direct evidence, however, does not preclude the possibility that antibiotics are sometimes formed in some microhabitats wherein the prevailing environmental conditions are conducive.[254,256] For example, after the introduction of available carbon, the population density rapidly increases, and any antibiotics formed would have a profound, albeit spatially restricted, effect on the population diversity. This local production of antibiotics would preclude their detection by extraction of the soil as a whole, especially if the antibiotics were bound to clay minerals and organic matter. Furthermore, the difference in environmental conditions between soil in situ and the laboratory may differentially induce antibiotic formation. For example, preliminary experiments have indicated that species that produce antibiotics against numerous microorganisms on agar plates or in liquid media do not inhibit the same microbes in soil plates, whereas some species that have no inhibitory effect in artificial laboratory media show marked apparent amensalism in soil plates (unpublished). Whether this amensalism results from antibiotic production or from competition or other environmental changes in the soil plates is not yet known.

Antibiotics produced in the soil microhabitats may also either be rapidly degraded[257] by species not sensitive to them or adsorbed, especially at pH values below their isoelectric point. Such adsorption may inactivate some antibiotics but not others, although much more investigation is needed on this aspect.[258] Most studies on the effects of antibiotics have evaluated only one antibiotic at a time, whereas in the in vivo situation, several antibiotics may be elaborated and function synergistically.

Such a confluence of inhibitors may be responsible for the widespread mycostasis observed in soil, even though no specific inhibitors have yet been isolated.[259,260] Although this mycostasis may be the result of substrate depletion (inasmuch as it can be reversed by carbonaceous amendments), of lysis, or of unfavorable environmental conditions, it does appear to be of biological origin and, consequently, could be the result of toxic inhibitors, either antibiotics or other organics present in soluble or volatile form.

Although species resistant to purified antibiotics do not appear to be more numerous in soil than in other microbial habitats, as indicated by tests on isolated microorganisms, some species may be more resistant to some antibiotics in some soil microhabitats. Although neither the widespread existence of such resistant forms nor the mechanisms of resistance (e.g., transfer of plasmids for drug resistance, mutations, morphological adaptations) have been adequately defined in soil, such resistance would provide an ecological advantage to some species.

In addition to inhibitors active against other species, some microorganisms produce substances that are intraspecifically toxic. Many bacteria, in culture and in some habitats, produce bacteriocins, i.e., proteinaceous materials inhibitory against strains of the same species or against closely related strains. These materials, the synthesis of which appears to be governed by autonomous episomal plasmids, are produced by both gram-positive and gram-negative species, many of which have been isolated from soil.[261-263a]

Many fungal species produce autoinhibitors that prevent either the germination of their spores, initiation of vegetative growth from resting propagules, or mycelial extension.[145,146, 264,265] In some species, these inhibitory materials accumulate within the structures, and extensive leaching is required before their concentration is reduced to noninhibitory levels.[264] Conversely, *Saccharomyces cerevisiae* liberates a toxic protein which kills sensitive strains in culture, similar to the action of some bacteriocins.[266]

Among the protozoa, several intranuclear and intracytoplasmic particles have been described, many of which confer an intraspecific amensalistic advantage to the host. Perhaps the best documented are the particles of some species of *Paramecium*.[249,267,268] These particles are probably bacterial endosymbionts which synthesize lethal toxins, some of which are specific for only one exconjugant of a mating pair. The presence of these intracellular particles also protects the host against the toxic agents liberated.

Although these mechanisms of intraspecific amensalism could have ecological consequences, there is no evidence that they occur in soil. Intraspecific inhibition would result in the elimination of organisms that, because of similar requirements, probably offer the greatest competition in a microhabitat, and the survival of the best suited strain would improve the competitive prospects of the species. It would seem, therefore, that this aspect of amensalism in soil should be studied extensively.

Regardless of the mechanisms involved, amensalism is undoubtedly an important factor in the survival and ecology of microbes in soil. Not only can such an interaction influence the success of both the amensal and inhibitor species involved, but the entire succession in a microhabitat may be redirected. For example, an inhibitor produced by species A may eliminate species B, thereby permitting species C to develop. If the inhibitor persists and establishes a zone of inhibition in the microhabitat, the recolonization by species B and the establishment of other populations detrimental to either species A or C may be prevented. Obviously, the unlimited ecological permutations that are possible indicate that methods must be developed for studying amensalism in soil in situ.

The categorization of lysis is difficult: some lytic organisms are saprobes, and their ability to lyse cells is a form of amensalism; others, however, can be best described as obligate exploiters. Regardless of rigid classification, however, there is little doubt that lysis is an important negative intramicrobial interaction in soil. Many organisms capable of lysing living or dead cells of other species have been isolated from soil, and, in some instances, the lytic enzymes responsible have been purified and identified. Furthermore, dead cells do not accumulate in soil, indicating that active lysis occurs. A wide spectrum of bacteria, actinomycetes, fungi, slime molds, and myxobacteria is capable of lysing a wide assortment of microorganisms.

The lysis may take the form of autolysis, wherein a cell enzymically destroys its own cellular membranes, thereby causing plasmoptysis or other disruption of the cell. Although all cells apparently contain autolytic enzymes, these are normally under cellular control. This control, however, can be disrupted by environmental changes brought about by a neighboring species (e.g., elucidation of antibiotics or other inhibitors; alteration of ambient pH; depletion of nutrients and O_2).

In heterolysis, lytic enzymes are released by one species and act on a neighboring one. These enzymes, which attack primarily cell wall components (e.g., chitinase, glucanases, lysozyme), and their mode of action on a variety of susceptible and non-susceptible species have been studied extensively in pure culture.[5,22,269-272] The synthesis of such lytic enzymes, as that of other cell metabolites, is dependent upon environmental conditions. Consequently, even though a species has the capability to synthesize such enzymes in culture, synthesis may not occur in the microhabitats under the prevailing environmental conditions. Similarly, the resistance of some species to lysis — whether the result of the formation of endospores, chlamydospores, sclerotia, cysts, pigments, or other mechanisms — is probably also mediated by the environmental conditions within the microhabitats. Furthermore, the persistence of lytic enzymes in an active state in soil is not known, and they may be rapidly degraded by insensitive species or be adsorbed to clay minerals and other particulates and, thereby, rendered ineffective.

Consequently, even though lysis is undoubtedly an important ecological factor in soil, much more must be learned about its dependence on the prevailing environment before it can be invoked as a primary factor influencing the ecology and population dynamics of species that, in the laboratory, either produce lytic enzymes or are sensitive or insensitive to lysis.

Another negative interaction is direct exploitation, wherein one organism becomes the primary food source for another. In parasitism, the exploiter is usually smaller than the host, and it is to the nutritional advantage of the parasite to keep the host alive. In predation, the prey is usually, but not always (e.g., *Bdellovibrio* sp., nematode-trapping fungi), smaller than the predator, and the prey is usually consumed and destroyed during feeding. As with other microbial interactions, not too much is known about the importance of exploitation in the ecology of microbes in soil.

The presence in soil of large numbers of organisms which, when isolated into gnotobiotic culture, consume other organisms indicates that predation undoubtedly occurs in soil. Protozoa, nematodes, myxobacteria, myxomycetes, and cellular slime molds are prodigious feeders on bacteria, and many also consume fungi and algae.[5,6,273,274] Most predators show specificity in their prey selection, thereby indirectly conferring an ecological advantage to those species not consumed.

Because of the voracious appetite of most predators, large quantities of a prey species may be consumed within a microhabitat, thereby rapidly altering its microbial ecology. Although data from two-member studies have indicated that fluctuations in population densities of predators are directly related to population densities of prey species, this relationship has not been conclusively demonstrated in soil. A predator could decimate a prey species within a microhabitat, but many predators are highly motile and probably range widely in soil in search of additional prey. Once the predator has left a microhabitat, the prey survivors probably proliferate rapidly, inasmuch as

the environment is now safe and relatively free of competition for nutrients, space, etc. Because predators persist in soil, decimation of prey species is apparently not complete. The structured nature of soil probably provides the prey with many sites in which it is inaccessible to the larger predator. Furthermore, a predator feeding on microorganisms may, in turn, be preyed upon by another, possibly larger, predator, thereby maintaining a balance between microbial preys and their predators.

An interesting type of predation, wherein the predator is smaller than the prey, is illustrated by the predaceous fungi. These fungi are saprobes, rather than obligate predators, that have the capacity to develop morphological adaptations for the capture of animals of microscopic dimensions. Although much is known about prey specificity, mechanisms of capture, induction of capture structures, geographic distributions, etc.,[274-276] there is little agreement about the importance of predaceous fungi in significantly affecting the ecology and population dynamics of their prey in soil.[277]

Predation undoubtedly plays an important ecological role in soil, but its importance, both within individual microhabitats and in the larger soil mass, is not well defined. Even less defined are the effects of environmental factors on this relationship. These factors could influence the development of either or both the prey and predator, as well as the prey-predator relationship. For example, under certain conditions, the prey, but not the predator, might proliferate, thereby eliminating predation as a negative factor impinging on the prey. The nutritional status of the microhabitat may be conducive to the production of protective mechanisms (e.g., slime layers, specific toxins) by the prey, thereby protecting it in this microhabitat but not in an adjacent one poorer in nutrition. The presence of certain types of clay minerals may adsorb the chemical compounds responsible for attracting the predator or for inducing morphological changes in predaceous fungi, myxobacteria, myxomycetes, and cellular slime molds. The structure and mineralogy of the microhabitat may also determine the degree of refuge and escape available to the prey. Temperature and moisture may influence the relative motility of macropredators or the faunal prey of predaceous fungi. Until the environmental effects are clarified in situ, predation, as an ecological determinant and as a means of biological control of plant and animal pathogens, will remain essentially an interesting laboratory phenomenon.

The extent and importance of parasitism in the soil environment has also not been sufficiently plumbed. The isolation of organisms from soil capable of parasitizing other organisms from soil provides indirect evidence that parasitism does occur in soil in situ. The delineation between parasitism and predation, however, is not always easy, even in pure culture; e.g., some predaceous fungi do not immediately kill their prey, but instead grow and sporulate, either endo- or ectoplasmically, at the expense of the living host.

Some forms of parasitism are also difficult to distinguish from symbiosis. For example, some mycoparasites appear to obtain nutrients from living cells, often other fungi, with little or no apparent damage to the host, regardless whether the parasitic fungus ramifies within the host or remains external and feeds either by means of penetrating haustoria or by altering the permeability of the host cell membranes.[278] The intracellular particles of various protozoa[249,267,268] can also be considered an example of such balanced parasitism, a term that attempts to bridge parasitism and symbiosis.

Although lytic bacteriophages can be defined as parasites, temperate phages in the prophage stage can, perhaps, be better classified as symbionts, especially if they confer some additional capabilities to the host. The importance of bacteriophages in the ecology and population dynamics of microbes in soil has not been clarified.[279] Because phages can be isolated from soil and their persistence in the absence of host cells has not been adequately defined, it can be assumed that they are parasitizing cells in soil. Although their distribution, persistence, etc. in soil need further study, a virulent phage could result in replacement of a sensitive species by a resistant mutant having altered genetic capabilities or, in the absence of such resistant mutants, could cause the elimination of a species from a microhabitat. Conversely, mutation to phage resistance may decrease the biochemical potential of the host, as indicated by the inability of phage-resistant strains of *Rhizobium trifolii* to fix N_2 and, in some cases, to form root nodules.[279,280] A temperate phage would not only confer protection on a host against further infection by a lytic form of the phage, but,

by transduction, could increase the survival potential of the host. The lysogenic bacteria probably provide protection for the phage, and such cells may be the primary source of vegetative phages in soil.

Although little is known about the ecological impact of bacteriophages in soil, even less is known about the occurrence and importance of viruses attacking fungi,[281] blue-green algae,[282-284] protozoa,[285] or other members of the soil microbiota.

Despite the large amount of literature on *Bdellovibrio* species,[286] a bacterium parasitic on other bacteria, its effect on the ecology and population dynamics of its hosts in soil is not clear. These parasites (although they may also be described as predators) have been shown to be present in soil and to attack numerous bacterial species common in soil.[287-290] Although originally thought to be obligately parasitic, some strains are apparently capable of growing saprophytically,[286] an ability which would enhance their persistence in soil in the absence of immediate hosts.

Despite the wealth of information on the types and mechanisms of intramicrobial parasitism, presumably occurring in soil, essentially no data are available on the effects of environmental factors on this interaction. Because attachment of phages to host cells, even in pure cultures, is dependent on pH, E_h, ionic strength, and other environmental variables,[67] these and other factors undoubtedly affect virus-host interactions in soil. Similarly, the motility of bdellovibrios, but not of their hosts, is inhibited in culture by low pH, anaerobic conditions, and elevated concentrations of NaCl; attachment to the host is mediated by prevailing physicochemical conditions; and the burst size and duration of the latent period of the bdellovibrios are affected by temperature.[286,291] The adsorption of viruses and bdellovibrios by soil particulates has not been adequately studied, nor has the interference that a structured environment may pose to a parasite in finding or fortuitously coming in contact with its specific host. The role of environment in the development of structures or compounds that confer on a potential host resistance to parasitism or, in the case of bacteriaphage interactions, the formation of protoplasts to which phages cannot attack, must be clarified.

Although there is considerable overlap in the classification and modes of action between and among positive and negative interactions, there can be little doubt that such interactions play a major role in the ecology and population dynamics of microbes in soil. What is not adequately defined, however, is the role that various environmental factors, alone or in different combinations, play in these interactions. Until such knowledge is available, man will not be able to utilize these interactions for his benefit. For example, why is it usually not possible to enrich soil, even by heavy exogenous inoculation, with organisms that, in pure culture, produce compounds active against either undesirable microbial pathogens or various insect pests? Why do some soils, but not others, permit the establishment and proliferation of some exogenous organisms (e.g., some fungi and bacteria pathogenic to man and plants)? Why do enteric bacteria disappear more rapidly in some soils than in others? Undoubtedly, the success or failure of an organism to establish itself in a soil micro-habitat depends, to a large extent, on its ability to interact with its neighbors, but these interactions, in turn, depend on the prevailing environmental conditions.[5,292] Consequently, the demonstration that microbes capable of or requiring positive or negative interactions are present in soil is not sufficient evidence that these interactions occur commonly in soil. If the environmental conditions in the microhabitats are not suitable, then the specific interactions observed in the laboratory may not be functioning, and the success or failure of a species must be attributed to some phenomenon other than intramicrobial relations. The continued study of microbial interactions without concomitant evaluation of the effects of environment on these interactions will, therefore, probably not result in much new information about microbial ecology in soil.

III. METHODS OF STUDYING MICROBIAL ECOLOGY IN SOIL

The variety of methods that have been used to study the activity, ecology, and population dynamics of microbes in soil reflects the complexity of this microbial habitat. Despite the multitude of methods, few, if any, have completely achieved their purpose, i.e., to study microbes *in soil*. Most methods either remove microbes from soil and study them in the laboratory or move soil to the laboratory and there study it for some function. No methods currently available begin to approach the level of the microhabitat, and, therefore, they only provide data which are some mathematical function of enumerable individual microhabitats. This is stated not to denigrate soil microbiologists but to emphasize the difficulties inherent in studying an environment as complex as soil and to underscore the ingenuity that soil microbiologists have demonstrated in their attempts to resolve these difficulties.

Perhaps the greatest dereliction on the part of many investigators has been the tendency to extrapolate from relatively simple experiments — even though these may be well designed and employ extremely sophisticated instrumentation and techniques — to the in situ situation without verifying that the phenomena observed in vitro do occur, in essentially the same manner in vivo. Because the real purpose of microbial ecology is to determine what is occurring in the environment of interest — in this instance, soil — and not in the test tube, the investigator must shuttle continuously back and forth between model experimental designs and the in situ condition. Consequently, the soil microbial ecologist, perhaps more than any other microbial ecologist, must operate simultaneously on many levels of experimental complexity, ranging from natural field conditions to artificial pure culture systems and, where necessary, to the molecular level. In this multipronged approach, imagination, ingenuity, and the asking of critical questions are the key ingredients, and, although modern instrumentation can be an asset in many experiments, much can yet be learned by applying standard and relatively inexpensive methods to well-designed experiments.

Most methods for studying microbes in soil are either of the indirect or of the more indirect type. One of the prerequisites for ecological studies, especially those that go beyond the purely descriptive stage, is that changes in the composition and activities of populations be followed over a period of time in, ideally, an undisturbed environment. This can be achieved in many studies of animal and plant ecology, as the observer and his sampling and measurement activities are usually small in relation to the ecosystem under investigation. In some microbial habitats (e.g., waters), the investigator can arrange his experiments so that continuous or intermittent observations can be made over an extended period without greatly disturbing the integrity of the system. The discontinuity and opacity of soil, however, preclude such measurements, inasmuch as direct observations are limited and any additions to and removals from soil disturb its spatial relationships.

Direct microscopic observations of microbes in soil are restricted by the opacity of soil; by the similarity in size, color, and shape of many microorganisms and soil particulates; by limitations in the resolving power, depth of focus, and working distance at the magnifications required; and by the common retention by cells and inanimate particles of many nonspecific dyes. Although modifications of conventional microscopy have been applied to soil microbial ecology, most microscopic techniques are still plagued by the problem of having to remove or disturb the soil in some manner.

Fluorescent antibody techniques[293-298] provide greater specificity to staining methods, but only one or few marker species can be studied at one time, the specificity of the staining reaction is not absolute, the spatial integrity of the soil is usually disturbed, and continuous observations of the same soil site are usually not possible.

Incident light microscopy, using water-immersion objective lenses or "dipping cones," overcomes some of the difficulties in opacity and in the working distance between the soil surface and the objective. However, the limited depth of focus, the necessity for immersion oil or similar fluids at adequate magnifications, and the need for fluorescent staining or other techniques[243,298-300] to aid in distinguishing between soil particles and microbial cells limit the adaptability of this technique to observations over a significant time span. The development of a method that diffracts the reflected incident light in a manner that living cells are differentiated from dead and dying cells and from soil particulates is encouraging,[53,301,302] as is the use of infrared photography to visualize bacteria in the presence of soil particles.[303] Similarly, the use of crossed nicols, in conjunction with bright field and phase contrast illumination, should be investigated further.[304] If these and related techniques can be applied to undisturbed soil for extended times, many of the problems of direct microscopy may be resolved.

The increased construction of high-voltage electron microscopes may overcome some of the problems (e.g., removal of soil samples for examination, artifacts produced during preparation and examination) currently associated with both transmission and scanning electron microscopy. Although electron microscopy has provided much information on the morphology of soil microbes[47,54,55] and some on particulate-microbe[304-308] and clay-protein complexes,[309] this technique should probably be used presently only as a confirmatory method in conjunction with a variety of indirect techniques. With sufficiently high energy electron sources, it might be possible to view microbes in thin slices of soil — that have been allowed to come to some biological and physical equilibrium after preparation and before examination — without evacuation, shadowing, and other manipulations that can result in misleading artifacts. By these means, the location of microbial cells relative to inanimate particulates and water films might be determined, thereby providing some insights into the actual structure of microhabitats.

The spatial distribution of microbes in natural soil can be demonstrated, in part, by soil sectioning techniques.[310-314] However, because the soil is removed and usually hardened with an impregnating resin, water films are replaced and small, but possibly significant, movements of cells may occur. Furthermore, it is not possible to observe the growth and distribution of organisms in the same microhabitat over a period of time nor to study the structure of cells, soil particles, and their complexes. Similarly, dissection of soil clods,[315] while permitting examination of microbes in situ, is not adapted to continual observations. In addition, these techniques are relatively tedious, and the value of the information obtained must be balanced against the time and effort expended.

The numerous modifications of immersion methods[315-323] provide some information about the type of organisms present, their relative location on a gross scale, their rate and nature of colonization of the immersed object, and, if carefully designed, on their interactions. In these methods, some object (e.g., slides, meshes, media-filled tubes, baits and traps), designed to attract or entrap microbes, is placed into soil, removed after a given time, and the organisms present are evaluated by either direct microscopic observations[316,319,320] or on isolations made from the subsequently removed immersed object.[315,317,318,322,323] These methods, however, are highly selective, and, unless the purpose of their use is restrictive (e.g., to determine the presence and colonization ability of a specific plant pathogen), the data obtained can be misleading. For example, the moisture regime around a glass slide or coverslip will be different, as a result of condensation, from that of the soil microhabitat, thereby encouraging rapidly growing organisms that require a high a_w. Furthermore, primarily sessile organisms will attach, and the concentration of populations on relatively large and artificial (i.e., in relation to soil microhabitats) surfaces may encourage grazing and predation, which may be so selective that, when the object is removed, the actual dominant populations may be absent and only the remainder unpalatable to the predators is erroneously considered the dominant population.

Similarly, immersion tubes containing selective and abnormally high supplies of substrate may be rapidly overgrown by precocious species having either a shorter generation time or a constitutive enzyme complement capable of utilizing these media, thereby excluding the colonization by slower growing species. In addition, gas relations on or within the immersed object may differ from those in the in situ microhabitats; spatial relations between soil particles, water films, substrates, and microbial cells are altered by placement and removal of the immersed object; because the object must be removed for analysis, only single observations on single locations are possible and, therefore, time-course studies are precluded; techniques of placement, removal, isolation, and analysis are tedious and time-consuming; and reproducibility, interpretation, and statistical evaluation of the observations are difficult.

A current modification of the immersion method that has elicited much interest among soil microbial ecologists is the peloscope/pedoscope developed by Perfil'ev and Gabe[324,325] and used extensively by them, Aristovskaya, and others.[52, 326,327] In this approach, rectangular capillaries, with small (e.g., 3 to 25 μm) and variously shaped cross sections, are inserted for various periods of time in soils, muds, and waters and then examined microscopically. Because the capillaries, which become filled with the ambient solution, presumably simulate the surrounding nutritional environment, the organisms which develop therein presumably reflect the dominant forms present in that environment. Although numerous new and bizarre-looking microorganisms (some of which appear to be primary agents in the precipitation and accumulation of iron and manganese, espe-

cially in aquatic sediments, and others which form special appendages and are predators on other microbes) have been discovered by these techniques and the vertical distribution of these organisms over very small distances has been demonstrated, some of the same questions raised about other immersion techniques must be applied. In addition to the disturbance of the soil by the insertion and removal of the capillaries, the physicochemical environment within the capillaries probably differs from that of the ambient soil environment. Because of their small inside diameter, the interfacial energies between the glass and the solution are probably high, resulting in an increase in the quasicrystalline structure of the water, which, in turn, would influence, among other factors, rates of gas exchange, movement of nutrients and waste products, and possibly directly the metabolism, cell wall synthesis, and cell division of organisms that are either drawn into or have the energy to enter the capillaries. All these factors would tend, as in other immersion methods, to select certain species, thereby not providing an accurate picture of the microbial ecology of soil in situ.

A first step in the study of the ecology of any environment is an enumeration of the species present, regardless of their spatial distribution. Such a compendium is not too difficult to compile for animals and plants and even for microbes in some homogeneous habitats. In soil, however, the types and numbers of microorganisms present have not yet been definitively assessed, despite attempts for more than 100 years. Although the literature is replete with numbers of bacteria, fungi, and other members of the microbiota in various soils, at various times of the year, under various plant covers, etc., the variations in numbers are great, even for repeated analysis of the same location. Although the reasons for these fluctuations are not exactly known, these numbers have been derived from both dilution techniques and microscopic counts of soil smears, immersed contact slides, soil thin sections, etc.

Dilution plating techniques normally yield about 1 to 10% of the number of cells determined by direct microscopic counts (ca. 10^8 to 10^{10} cells/g oven-dry soil). Some workers, notably Nikitin,[47,328] Perfil'ev and Gabe,[324,325] and Aristovskaya,[52,326,327] have suggested that dilution techniques yield even lower percentages, approaching 0.01 to 0.001% of the total viable cell population as determined by either electron microscopy or the pedoscope. The differences in cell numbers obtained between dilution and microscopic methods are usually ascribed primarily to the inability to distinguish accurately between living and dead cells and to the failure of many soil microbes to grow on laboratory media. The judicious application of fluorescent microscopy, using either nonspecific stains, such as acridine orange,[299,300] or specific fluorescent-antibody conjugates,[293-297] as well as infrared and light-diffraction techniques,[53,301,302] should help in distinguishing living from dead or dying cells.

The main limitation in dilution plating techniques is probably that they permit the development of only some of the microbes present in soil. The failure of many soil microbes to develop into visible colonies on laboratory media has been ascribed to nutritional fastidiousness, dormancy, inhibition by neighboring cells and colonies, and differences in physicochemical properties between laboratory and soil environments. For example, *Fusarium oxysporum* f. *cubense* is seldom isolated on dilution plates prepared from soils in which it is present, because, at the dilutions necessary for its detection (i.e., 1:10 to 1:100), the fungus is rapidly overgrown by more rapidly germinating and growing species. Inhibition of these fungal species by incubation of the dilution plates in an atmosphere of high CO_2 permits the development of colonies of the fusaria.[127] Many fungi are also inhibited from growing and sporulating when cultured in the presence of some bacterial species, even when these are not in physical contact with the fungi.[145,146]

In an attempt to circumvent many of these problems, routine use is made of nutritionally selective media, media with different pH values, incorporation of broad-spectrum or specific antibiotics and other inhibitors, incubation under altered environmental conditions of temperature, aeration, etc., and numerous other variations whose primary purpose is to eliminate or reduce the growth of some species while encouraging that of others. However, regardless of the manipulations used, it is obvious that dilution plating techniques yield only a partial picture of the types and numbers of microbes in soil and essentially no information on their relative location and distribution.

The restrictiveness of dilution plating is, perhaps, greatest with fungi, inasmuch as the majority of species that develop colonies on plates are those that produce abundant spores.[321] Soil washing techniques,[329-331] the purpose of which is to remove spores by water rinses before dilution plating, are not entirely effective, as many of the fungal colonies that subsequently develop on agar plates are derived from retained spores as well as from vegetative propagules. Nonsporulating or sparsely sporulating fungi can be isolated by incubating soil crumbs directly with a nutrient medium or by isolating — with the aid of microscopes and micromanipulators — and culturing hyphae directly from soil.[321] These methods, however, are not only tedious, but they also provide little information on ecological relations among fungi and between fungi and other microbial groups.

Although the limitations in both microscopic and dilution plating techniques have long been recognized, most soil microbiologists have tacitly assumed that either method, or both together, provides a reasonable picture of the composition of the soil microbiota. The continual demonstrations that soil harbors, both as resident and transient forms, not only a vaster microbiota, but also one that contains morphological, physiological, and pathological forms that differ greatly from those usually observed on dilution plates or by direct microscopy, have in recent years reversed this view. In addition, these demonstrations may provide some explanations for the long-observed discrepancies between the activities of microbes in soil and the numbers observed or isolated.[29,241,298,332]

One milestone in this reversal was the discovery of *Bdellovibrio bacteriovorus*,[286,333] a bacterium parasitic on other bacteria, which, among other scientific considerations, appears to have stimulated a renewed interest in soil inhabitants (e.g., myxobacteria) capable of lysing other forms.[22] Other recent advances include the isolation of a catalase-negative, microaerophilic, coccoid microorganism, which may have taxonomic relations with the Actinomycetaceae and Mycobacteriaceae, from soil at dilutions of 10^9.[334] This organism, which may comprise the majority of the coccoid forms observed in soil and may be present in greater numbers than the entire soil microbiota normally isolated, is probably excluded from soil dilution plates by less fastidious and faster growing species.[334] Another example of the current interest in defining the composition and distribution of the soil microbiota is the increased activity of medical mycologists in directly isolating human pathogens from soil and various dungs.[335]

One of the most exciting recent advances has been the observation and — unfortunately in only a few cases — the isolation of "unusual" soil microbes. Bystricky and co-workers[54,55] found unusual forms in soils incubated with a carbon-free medium. One of these forms — which is of the general size of bacteria, but strongly contoured with longitudinal rows of spherical subunits, either entirely or over just part of its surface — has been observed in soils collected from various parts of the world which differ markedly in physico-chemical characteristics.[55] The morphology of this organism, tentatively named "helicoidal poly-spheroid," does not appear to be an artifact of the electron microscopy procedures used nor of the medium, as identical handling of "normal" (i.e., taxonomically known) soil bacteria has not resulted in similar shapes.

Unusual microbes in soil have also been observed in the laboratories of Nikitin, Perfil'ev and Gabe, and Aristovskaya. The latter two laboratories[56,324-327] have observed these forms primarily in capillaries inserted into muds (peloscope) and soils (pedoscope). Nikitin, by examining soil suspensions directly with transmission electron microscopy, has described a wide spectrum of new organisms, some that are similar in size to common soil bacteria and some that are as small as 0.1 μm, but most of which have morphologies unlike those of microbes usually isolated from soil.[47] These forms, which do not grow on most laboratory media, have been classified as bacteria-like (forms with unusual appendages; thread-like with stalks); protozoa-like (star-like; organisms with pores; rod-shaped with spherical convex protrusions); algae-like (small diatoms and algae-like); and virus-like. Many of these forms have long appendages, some the size of pili but some considerably thicker, which suggests that they may act as holdfasts to secure these organisms to soil particles and, therefore, prevent their release and isolation by conventional soil dilution plating methods.

Nikitin has suggested that these forms comprise 80 to 85% of the microbiota of podzolic soils and may exceed 2 x 10^9 viable cells/g soil,[47,328] a figure that is greater, by one to three magnitudes, than the numbers of viable cells normally enumerated by dilution techniques. The number of these unusual forms is positively correlated with concentration of the "fulvic acid" fraction, as extracted by Nikitin et al.,[336] and is inversely related to the number of usual forms. This fulvic acid fraction, as well as agarized soil, serves as a medium for the laboratory growth of some of these unusual forms. One of the organisms observed by Nikitin and isolated in pure culture by both Nikitin and Bystricky appears to be identical in morphology to "helicoidal polyspheroid" (personal communications).

Even without the use of the electron microscope, new and different forms of microbes have been reported from soil. For example, a gram-negative *Bacillus* sp. isolated from soil irradiated with 5 megarads derived from ^{60}Co appears on some synthetic media as a long, slender, tightly coiled cell bearing no morphological relation to known bacteria but, on an agarized-soil medium, exhibits a palisade arrangement of cells with well-developed endospores similar to *Bacillus circulans*.[337] This pleomorphism suggests that care must be exercised in describing new forms isolated from soil, as they may be just morphological variants of known species, perhaps induced by the physicochemical status of the soil at the time of observation. This polymorphic potential, of course, has been long appreciated by medical and other mycologists.

Another impetus for the study of novel microbial forms in soil has come from paleontology, through the description of presumed fossil microbial forms in ancient geological deposits.[338-346] One of these forms, presumably a relict of *Kakabekia umbellata* Barghoorn, the 1900±200 x 10^6-year-old Precambrian microfossil from the Gunflint formation,[338] has been reported to have been isolated from soil under ammonia-rich atmospheres.[118,143,144] However, in view of the apparent ubiquity and variety of the many novel and pleomorphic microbial forms now being observed in soil, the assumption that these are prehistoric living organisms, primarily because their morphology bears no resemblance to hitherto known species, must be critically evaluated.

Despite the progress that has been made in the isolation, enumeration, and identification of soil microorganisms, it is doubtful whether enough is currently known to be as important an aid to understanding the ecology and population dynamics of microbes in soil as compendia of plants and animals are to the ecologist of "macro-organisms."

Another requirement in studying the ecology and population dynamics of an ecosystem is measurement of the rates of growth and distribution of individual species comprising the heterogeneous population. These measurements are also difficult to perform in soil, for many of the reasons stated above, and few successful techniques have been developed. A successful technique should permit recovery of marker species over a given time span with minimum disturbance of the soil, in order to prevent movement of cells, substrates, and soil particles; simulate conditions that exist in soil in situ; provide precise information on the rates of growth and microgeographic distribution of individual species, even in a mixed population; be rapid and reproducible; and enable both alteration of the environment and statistical analysis of the results. No single method meets all these requirements, but the soil replica plating method,[252] coupled with other techniques, approaches the requirements.

Briefly, this technique involves the precise placement of inocula and amendments at specific loci in soil contained in petri dishes and the measurement of growth of individual species by periodically replicating from the "soil plates" to selective agar media. The replicator is constructed of acrylic plastic, through which thin (e.g., 20 gauge) stainless steel nails have been inserted. The pattern of placement and the number of nails is dependent primarily on the number of organisms studied simultaneously and on the information desired. Precise placement of inoculum is achieved by means of a template. The replicator needles and the template are sterilized by flaming with alcohol. A line is inscribed on the tops of the template and replicator and on the sides of the soil plates and the petri dishes to which replication is made. By aligning all lines during inoculation and replication, the geometric integrity of the system is maintained, and time vs. spread curves can be constructed for each species analyzed.

The soil plates are incubated at constant temperature and in a saturated atmosphere, which is required to maintain the soil water content (usually adjusted to the 0.3 bar tension) throughout an extended incubation period. A saturated atmosphere is attained in a conventional incubator by inserting a pan of water, the temperature of which is maintained several degrees above ambient with an immersion heater. An alternate method is to place the soil plates on a rack over a layer of water in a glass chromatography jar covered with a desiccator top, the opening of which is plugged with cotton, and permit the temperature to fluctuate several degrees. With either method, losses in soil water are usually no greater than 2 to 4%.

When single organisms, or a few organisms sufficiently separated and easily distinguishable, are inoculated into sterilized soil, replications can be made to a nonselective medium. When combinations of organisms are inoculated, replications are made to differentially selective media, not only to facilitate reading of the replica plates but also to prevent confusing interactions between organisms on the replica plates with those occurring in the soil plates. For example, when a bacterium and a fungus are growing in the same soil plate, the distribution of the bacterium can be measured by replicating to a medium containing rose bengal and streptomycin to inhibit the bacterium, and the distribution of the bacterium can be measured by replicating to a medium containing cycloheximide to inhibit the fungus. In systems containing many organisms, the rate of spread of individual species can be measured by coupling differential media with other distinguishing characteristics of the organisms (e.g., pigmentation, colony morphology, substrate specificity, and lysis) and by taking advantage of any unique environmental requirements (e.g., temperature, anaerobiosis) of the species. The rate of spread of individual species is recorded on "maps" maintained for each soil plate, and from these maps growth curves for each species are constructed.

This technique has been used successfully in measuring rates of growth of different microorganisms in soils differing in physicochemical composition (Tables 3, 4, and 5), in studies on interactions between microbes and the influence that soil characteristics have on these interactions,[16,252] and in studies on genetic recombination of bacteria in soil.[23]

The utility of the replica plating technique can be extended by coupling it with fluorescent staining procedures (e.g., identification of marker species in mixed population by staining colonies isolated on the replica plates with conjugated fluorescent antibodies), with autoradiological techniques (e.g., using ^{14}C-labeled cells as inoculum), and with specific genetic markers (e.g., auxotrophic or lysogenic strains).

Although the technique is best suited to and has been used primarily in studies conducted with inoculated sterile soils, it can be and has been used with natural soils.[252] By sequentially replicating to a variety of selective media and with the incorporation of some of the other modifications mentioned above, a pattern of the density and changes in distribution of microbes in natural soils can be obtained.

The replica soil plate technique permits the construction of complex and controlled ecosystems in a single petri dish (e.g., substrates, mineral nutrients, pH, osmotic pressure, inhibitors, etc. can be altered in different sections of the soil), and interactions between microorganisms and the influence of environmental variables on these interactions can be studied directly in soil over prolonged periods (e.g., some soil plates have been studied continuously for more than two months). Because soil itself is used, the influence of the physical and chemical characteristics of soil on microbial events can be directly evaluated, thereby eliminating extrapolations to soil of observations obtained from non-soil systems. The technique is rapid and nontedious, and many variables, with a statistically sufficient replication of these, can be evaluated simultaneously (e.g., data can be put on punched cards and multiple regression analyses between growth rates and the characteristics of the soils and organisms can be made by computer).

The generation time of microbes in soil might also be computed from growth data obtained by this technique. Brock,[240] in a recent discussion that emphasized the lack of information on the doubling time of microbes in natural habitats, suggested a number of techniques (e.g., isotopic labeling, genetic markers) that might be applicable to determining the generation time of microbes in several habitats, all of which, however, are less complex than soil. One of the problems that might arise in the application of these techniques to soil is the recovery of marker species without unduly altering the structural and spatial integrity of the soil during repeated samplings. The use of the replicator in conjunction with these techniques might minimize this problem and provide much needed information on the generation time of microbes in soil.[241]

The most widely employed methods for studying microbes in soil are based on their metabolic activities, with respiratory methods being perhaps the most used. Respiratory measurements are usually well correlated with other indices of microbial activity, such as transformations of carbon, nitrogen, phosphorus, and sulfur, accumulation of metabolic intermediates, pH changes, organic matter content, etc.[332] Maximum rates of respiration, however, usually precede by several days to weeks the maximum numbers of microbes isolated, suggesting that respiratory rates reflect the metabolic activity rather than the numbers of at least that portion of the soil microbiota enumerated by conventional dilution plating methods.

Either CO_2 evolution or O_2 uptake can be measured. The former, however, has wider applications, as CO_2 is evolved from aerobic as well as anaerobic environments, and ^{14}C-labeled substrates or cells can be used.

Although respiration is considered indicative of a metabolizing microbiota, some of the CO_2 evolved from soil may have nonbiological origins; e.g., CO_2 may be produced by chemical decarboxylation, by cell-free, heat-stable enzymes, and by the action on carbonates of added amendments or organic acids produced during metabolism.[347,348] Oxygen may be consumed during chemical oxidation,[347] and both CO_2 and O_2 may be adsorbed on soil-water interfaces.[349] Even soil presumably sterilized by ionizing radiation still respires,[350] although the absence of growth on dilution plates does not constitute unequivocal proof that the soil is devoid of living organisms.

The many methods of measuring soil respiration, ranging from relatively simple (e.g., manometry, titrimetry) to more sophisticated techniques (e.g., infrared, gas chromatography, polarography), have been described with their advantages and disadvantages.[332,351] Although many respiratory methods are simple, inexpensive, and relatively reproducible, as well as permitting the simultaneous study of many variables, it is important that the size of the soil sample used be large enough to be representative of the many microhabitats comprising the soil under investigation.[124,332]

Regardless of which techniques are used, the respiratory activity is, under most circumstances, only a measurement of the overall metabolic activity of microbes in soil and, therefore, does not provide much information about which members of the heterogeneous microbiota are responsible for the CO_2 evolved or O_2 consumed nor about the biochemical pathways involved. Despite these limitations, however, the technique has applicability in studies on the microbial ecology of soil, as it does provide an indication of the metabolic potential of the soils under investigation. Furthermore, if the questions posed are specific and the experiments judiciously designed, valuable information can be derived.

For example, if a nonspecific substrate, such as glucose, is added to a soil, the gross metabolic activity of that soil can be determined, but little is learned about the kinetics of individual populations comprising the microbiota of that soil. If, however, a substrate is added that can be utilized only by some species (e.g., certain pesticides), then respiratory measurements can provide much information about the development of those species and the effect on them of the environmental characteristics under study. An example of this is shown in Figures 6 and 8: aldehyde oxidation in soil appears to be restricted to only some species, whereas the homologous acid shows no such specificity; consequently, considerable information on the influence of clay minerals on aldehyde utilizers, but little on acid utilizers, can be obtained from these respiratory studies. For more definitive answers (e.g., whether there is an increase in the number of aldehyde utilizers or an induction in existing populations of enzymes capable of oxidizing or reducing the aldehyde moiety), it would be necessary to couple these respiratory measurements with other techniques (e.g., enumeration and identification of aldehyde utilizers).

Microbial activity in soil can also be evaluated by following the disappearance of substrates; appearance of intermediary metabolites; changes in pH; mineralization, immobilization, or volatilization of various nutrients; changes in numbers of specific microbial groups; heat production; etc.[352]

Many of these changes can be conveniently studied by soil perfusion techniques, either those that recirculate the perfusate[353-357] or are based on the chemostat principle.[358] As with respiratory methods, the activities studied must be restricted to few species if perfusion techniques are to provide information on specific population dynamics rather than on the gross metabolic activity of the soil. In addition, perfusion techniques are highly artificial, as there is a continual reorientation of cells, substrates, and products throughout the soil column, as well as a water status seldom encountered in soils in situ. Nevertheless, this method can provide much information on microbial transformations, changes in specific microbial populations, induction of specific enzyme systems, and the effects of selected environmental parameters on these and other events. An example of the type of information that can be derived from this technique is shown in Figure 4. In addition to being a simple and inexpensive method, the perfusion technique is readily coupled with other methods, especially respiratory ones.

Cell-free enzymes have been used as indicators of soil microorganisms,[359-363] as enzyme activity in soil appears to be correlated with the concurrent or immediate past activity of intact cells. The presence of cell-free enzymes is usually deduced from the conversion of simple substrates in sterile soil (i.e., after microbes capable of developing on various media have been killed by chemicals or ionizing radiation). Few enzymes, however, have actually been isolated from soil and purified. Many enzymes have been detected, but, with the exception of some transaminases, each is restricted to a single step catalysis involving only a single substrate. This apparent restriction can probably be explained by the necessity for sequential enzymic activity of a complex of enzymes, with accessories, in a spatial arrangement that enables the transfer of metabolic intermediates and energy from one enzyme system to another. Consequently, the activity of cell-free transaminases in soil is surprising, as these require concomitant spatial accessibility to two substrates. However, as enzyme activity in even sterilized soil may occur within recently killed cells and cell-free enzymes can be adsorbed on clays and organic matter, multienzyme activity in soil is possible.

Despite the many contradictions that exist in the literature relating to the sorption of enzymes, the activity of sorbed enzymes, the persistence of enzymes in soil, etc., several aspects recommend enzymic assays for the study of microbial ecology of soil; e.g., assays for specific enzymes are highly sensitive and most assays are relatively easy to conduct. However, as with respiratory and perfusion techniques, the enzymes to be studied must be unique to only a few species if enzymic data are to provide information on the ecology and population dynamics of component species rather than be just another indicator of gross metabolic activity.

Similar considerations should be given to the use of the luciferin-luciferase assay for ATP. Although increased use is being made of this technique for the determination of total biomass in soil and in other microbial habitats,[364] several problems have not yet been resolved (e.g., quantitative recovery of ATP from soils; accumulation of ATP by adsorption to soil organic matter and clays; importance of potential contamination from a single leaf or large soil microfauna). Until these and other problems are resolved, it is dubious whether the application of this technique to the study of microbial ecology of soil provides data that could not be obtained by cheaper and more specific methods.

Little application to soil microbiological studies has been made of bacteriophages. Because of their high host-specificity, bacteriophages should be good indicators of changes in host populations. Reasonable success has recently been obtained in following fluctuations of various coliforms (i.e., IMViC types) in river water by changes in coliphage titers (unpublished). If potential problems, such as the quantitative recovery of phages from and selective adsorption of phages by soil, can be resolved, modifications of this technique might provide valuable information on the population dynamics of microbes in soil.

Similarly, the use of lysogenic phages, abortively transduced gene loci, plasmids, and other genetic markers[240] may provide increased specificity for studying the ecology and population dynamics of individual species within the heterogeneous soil microbiota.

Regardless of the methods employed, the soil microbial ecologist must exercise considerable ingenuity both in designing and in interpreting his experiments. In many instances, the questions posed are more critical than the methods used to obtain the answers. Seldom will one type of experiment provide the definitive answers to events occurring in the complex environment that is soil, and the investigator may have to ask — and answer — his questions on several levels of experimentation. Furthermore, the complexity of soil predicts that the variability in data will, in many instances, be great, especially as the level of the microhabitat is approached. Consequently, application of the principles of experimental design and statistical analysis of data should be foremost in the mind of the soil microbial ecologist.

IV. CLAY MINERALS AND MICROBIAL ECOLOGY

A. Introduction

The primary purposes of this section are (1) to demonstrate briefly how the activity, ecology, and population dynamics of microbes in soil can be influenced, both directly and indirectly, by a single environmental factor expressing its influence alone and in concert with other factors and (2) to emphasize that, because the study of microbes in soil is complex, an integrated method of investigation must be employed. This complexity requires that all components of the ecosystem (e.g., microorganisms, clay minerals, intact soil) be studied individually, in various combinations, and in their totality, and that these studies be conducted essentially concomitantly on various levels of experimental complexity, ranging from natural field conditions to artificial model systems. Too many studies of ecological phenomena in soil are limited either to recording empirical observations of events in situ or, at the other extreme, studying the activities and interactions of a few species derived from soil under laboratory conditions and extrapolating back to the complex natural system. The integrated and multipronged approach being used in the study described herein enables a more complete evaluation of the relevant events and mechanisms, inasmuch as it provides for maximum use of personnel and facilitates the rapid investigation of observations made, for example, in pure culture studies in systems of increasing complexity and vice versa.

A schematic representation of studies on the influence of clay minerals on microbial ecology in soil at various levels of experimental and system complexity is shown in Figure 2. In both this schema and in the discussion, the chronology of experimentation is more apparent than real and has been organized as such solely to facilitate presentation. In actuality, most of the levels of complexity have been and are under simultaneous investigation.

For data supporting most of this discussion, reference is made to published papers and only unpublished data are presented here.

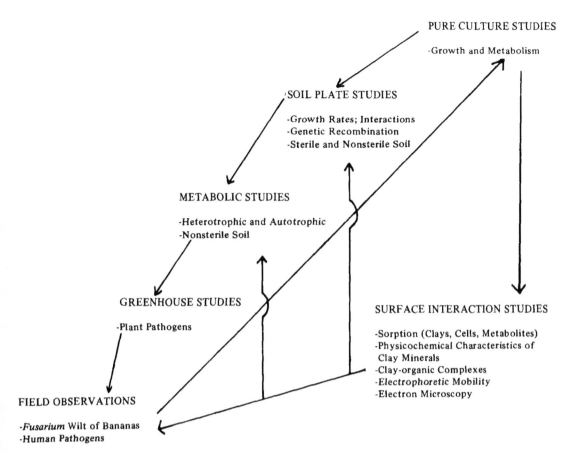

PURE CULTURE STUDIES

-Growth and Metabolism

SOIL PLATE STUDIES

-Growth Rates; Interactions
-Genetic Recombination
-Sterile and Nonsterile Soil

METABOLIC STUDIES

-Heterotrophic and Autotrophic
-Nonsterile Soil

GREENHOUSE STUDIES

-Plant Pathogens

SURFACE INTERACTION STUDIES

-Sorption (Clays, Cells, Metabolites)
-Physicochemical Characteristics of
 Clay Minerals
-Clay-organic Complexes
-Electrophoretic Mobility
-Electron Microscopy

FIELD OBSERVATIONS

-*Fusarium* Wilt of Bananas
-Human Pathogens

FIGURE 2. Flow diagram of studies being conducted at various levels of experimental and system complexity on the effects of clay minerals on the activity, ecology, and population dynamics of microbes in soil.

B. Field Studies

The initial indication that clay mineral composition influences the ecology of microbes in soil resulted from studies to explain the differential rate of spread of *Fusarium* wilt of banana. The causal fungus, *Fusarium oxysporum* f. *cubense* (E. F. S.) Snyd. and Hans., is soil-borne, invades the plant through the roots, and becomes distributed in the vascular system, resulting eventually in wilting and death. The rate of disease spread, expressed as the effective banana-producing life of a plantation, differs considerably from one soil to another: wilt-susceptible bananas can generally be produced on "short-life" soils for 3 to 10 years and on "long-life" soils for more than 20 years, with some soils being in continuous banana cultivation for more than 70 years.

Analyses of 143 soils from banana plantations in Honduras, Guatemala, Costa Rica, Panama, Colombia, Ecuador, and the Cameroons showed no correlations between the effective banana-producing life and 17 chemical and physical properties of the soils.[365-368] A high correlation, however, was apparent between the effective banana-producing life and the clay mineralogy of the soils: a particular species of clay mineral was present in almost all long-life soils, but was absent from almost all short-life soils. The distinguishing clay was a crystalline, three-layer, hydrous alumino-silicate, which, when homoionic to K^+, air-dried, and then solvated with glycerol, expanded to give either a distinct x-ray diffraction peak at a *d* value greater than 14.5 A or a peak between 14 and 14.5 A, which was skewed toward higher spacings.[77] When homoionic to Mg^{++} or Ca^{++}, air-dried, and solvated with glycerol, all clays from long-life soils and some from short-life soils gave an 18 A reflection.[369] Although a species name was deliberately omitted, the distinguishing clay mineral tended toward montmorillonite in the montmorillonite-vermiculite sequence. All soils, regardless of their disease history, contained various combinations of other clay minerals (e.g., vermiculite, illite, mica, chlorite, kaolinite/halloysite), but there was no correlation between their presence or absence and the rate of spread of *Fusarium* wilt.

To determine whether this apparent correlation between the rate of spread of *Fusarium* wilt of banana and clay mineralogy is unique to this plant disease or is indicative of a basic ecological phenomenon, attempts are being made to find other plant diseases which show distinct geographic distributions that could be used as indicators for further in vivo correlations. To date, no comparable plant disease systems have been found, primarily because most soil-borne plant diseases occur on annual crops, which are usually rotated over a period of several years, and, with the relatively few perennial crops as well as with most nonrotated annual crops, the farmer does not usually keep the detailed disease records necessary for correlation studies. Consequently, although the search for additional indicator plant diseases continues, attention is being directed to other soil-borne indicator organisms.

Numerous fungi pathogenic to humans are widely present in soil and several show apparent discrete geographic distributions. *Histoplasma capsulatum*, on the basis of its reported isolation from soil and the prevalence of histoplasmosis and histoplasmin sensitivity, appears to exhibit such distributions.[370,371] The results obtained to date with 100 soils, representing 34 geographic locations in 8 countries, from which *H. capsulatum* had been isolated were similar to those obtained with *Fusarium* wilt of banana, in that the distinguishing clay mineral was absent from all soils, except for some from three locations (Table 2).[372] Neither the presence nor absence of other clay minerals, nor the pH, the only other soil characteristic determined and which ranged from 5.4 to 8.0, was related to the distribution of the fungus.

Although there were no "controls" in this correlation comparable to the long-life banana soils and there are undoubtedly numerous areas having soils which contain neither *H. capsulatum* nor the distinguishing clay mineral, the absence of this clay mineral in so many soils from which the pathogen has been isolated strongly indicates a causal relationship. Furthermore, differences were apparent in the clay mineral composition of the soil samples from which the pathogen was isolated and of the soils predominantly present in the general area from which the samples were derived: i.e., in many locations, the distinguishing clay was present in the general area, whereas it was absent in the samples containing the pathogen.[372]

These studies indicate that the clay mineral composition of soils is related to the ecology of at least two fungal species. Preliminary investigations with other soil-borne pathogens indicate that the clay composition of soils is not implicated in the geographic distribution of *Coccidioides immitis*, but possibly in that of *Cryptococcus neoformans*, *Blastomyces dermatitidis*, and enzootic leptospirosis (unpublished).

With the former two fungi, moreover, the distribution and physical introduction of their propagules into soil do not appear to be limiting factors. The rate of spread of *Fusarium* wilt of banana does not appear to be related to the frequency or intensity of natural floodings, to the use of surplus irrigation waters from fields with high disease incidence on low incidence fields, nor to the use of diseased planting stock. In numerous instances, both long- and short-life plantations were simultaneously planted with rhizomes obtained from the same established plantations which sometimes had previously been abandoned to the disease.

Although *H. capsulatum* appears to be associated with animal droppings, primarily those of birds and bats, the essentially unrestricted geographic distribution of animal manures and the high saprophytic ability of the fungus suggest that type and availability of energy sources are no primary limiting factors in its ecology.[370, 373,374] Morevoer, the widespread distribution of avian and chiropteran habitats indicates further that the potential exists for the dispersal and introduction of propagules of *H. capsulatum* into almost any soils.

The apparent differences between the potential and actual distributions of these two fungi therefore, appear to be related to factors that influence their establishment and subsequent development in, rather than to their introduction into, soil. These factors probably involve competitive forces imposed by the indigenous soil microbiota. Because clay minerals appear to affect differentially the growth and activity of this microbiota, the types of clays present would

TABLE 2

Clay Mineralogy of Soils From Which *Histoplasma capsulatum* Has Been Isolated

Location	Samples	M	V	Mi	I	K
Augusta, Ga.	1			+		+
Tennessee	1			+		+
Sturgis, Miss.	1			none		
Selma, Ala.	1			none		
Laurens Co., S.C.	2		+	+	+	+
Timmonsville, S.C.	1		+			+
Mississippi Co., Ark.	4				+	+
Washington, D.C.	4		+	+	+	+
Clarksburg, Md.	9		+	+		+
Libertytown, Md.	2		+	+		+
Walkersville, Md.	1			+	+	+
Washington Co., Md.	11		+	+	+	+
Conesus Lake, N.Y.	1			+	+	+
Alexandria Bay, N.Y.	1			none		
Milan, Mich.	2		+	+	+	+
Kankakee, Ill.	3			+	+	+
Lima, Ohio	2			+	+	+
DeSoto, Kans.	2			+	+	+
Lexington, Ky.	6		+	+	+	+
Bryantsville, Ky.	1			+	+	+
Mexico, Mo.	3			+	+	+
Kansas City, Mo.	1			+	+	+
Cass Co., Mo.	1	+		+	+	+
Cass Co., Mo.	1			+	+	+
Audrain Co., Mo.	1			+	+	
Bellevue, Iowa	1			+	+	+
Mason City, Iowa	3			+	+	+
Grand Island, Nebr.	1			+	+	
Transvaal, So. Africa	9			+		+
Mbujimayi Area, Congo	12			+	+	+
Tingo Maria, Peru	1		+	+		+
Lara, Venezuela	1			+		
Oropuche, Trinidad	1			+		
Madden Lake Cave, Panama	6	+				+
Emilia-Romagna Reg., Italy	2	+	+	+	+	+
Total number of soils	100					
Total number of locations	34					

V = Vermiculite; Mi = Mica; I = Illite; K = Kaolinite; M = "Montmorillonite". pH range: 5.4–8.0.
(Stotzky, unpublished)

influence microbial interactions and, thereby, the ability of these fungi to become established.

C. Pure Culture Studies

To determine whether and how clay minerals influence microorganisms, the effect of various clay minerals, differing in physicochemical characteristics, on growth and activity of microorganisms in pure culture was studied. From some 100 samples of clay minerals and various particles which possess some of the characteristics of clays, essentially only samples of montmorillonite stimulated the respiration of bacteria, primarily by maintaining the pH of the environment suitable for sustained growth.[108,109] This mechanism was confirmed with more than 20 species differing in morphology, motility, Gram reaction, stages of development, etc. Additional mechanisms, not yet defined, were apparently also involved, as montmorillonite stimulated respiration even when the pH of the systems was buffered.

The stimulation of bacterial respiration was related to the cation exchange capacity, possibly the specific surface, but not to the particle size distribution of the clays.[375] The effectiveness of maintaining a suitable pH was related to the relative basicity of the hydroxides of the cations on the clay exchange complex, which determined the extent to which metabolically produced H^+ were removed from the solution phase wherein the bacteria were presumably metabolizing.[109] Because clays apparently exchange H^+ produced during metabolism with basic cations from their exchange complex, clays with a high cation exchange capacity (e.g., montmorillonite) are capable of exchanging more H^+ and, therefore, of maintaining longer the pH of the ambient solution at a level suitable for bacterial metabolism. The lack of correlation between surface charge density and respiration suggests that the total cation exchange capacity of, rather than its distribution within, the microhabitats determines the extent to which metabolically produced H^+ are removed from the solution phase.

In soil in situ, the accumulated H^+ on the exchange complexes within the microhabitats are probably displaced in turn by basic cations dissolved in the surrounding soil solution. The rapidity of this displacement, however, is dependent on many factors (e.g., source of cations, continuity of water films between microhabitats, concentration gradients), suggesting that it is not an instantaneous nor continuous process. Consequently, maintenance of a favorable pH within the microhabitats in the interims between "recharging" of the exchange complexes will depend on the capacity of the clays to exchange basic cations for H^+. Expandable clays with a high cation exchange capacity (e.g., montmorillonite) apparently do this best.

The respiration of mycelial homogenates of 27 fungal species representing four classes was generally not affected by any clays at concentrations below 4% (w/v), regardless of the initial pH of the systems or of the media used.[376] The buffering ability of clay minerals was apparently not a decisive factor in the growth of fungi, which are generally more tolerant to low pH than are bacteria.

With higher concentrations of clay, however, the metabolic activity of most fungal species was markedly inhibited. This inhibition occurred with montmorillonite at concentrations of 4% and above, but with kaolinite (used as an internal control) essentially only at concentrations above 40%, and the inhibition was greater as the metabolic activity increased.[376] Comparable concentrations of other clay minerals and similar particles resulted in degrees of inhibition intermediate between those caused by montmorillonite and kaolinite (unpublished).

The inhibition did not result from limitations in carbon substrates, but was related to the viscosity of the systems, which apparently influenced the rate at which O_2 reached the sites of metabolism. The viscosity was increased more, per unit weight of clay, in montmorillonite than in kaolinite systems, reflecting the structure and physicochemical properties of these clay minerals: kaolinite is a non-swelling 1:1 clay, whereas montmorillonite, a 2:1 clay, swells enormously by trapping water between its lattices. As gas exchange occurs more slowly in a gel structure than in a particulate suspension, respiration was inhibited more by montmorillonite than by kaolinite at comparable or even vastly greater concentrations. Although differences in water structure at the clay surfaces may have been involved, this aspect was probably of minor importance, as the systems were continuously shaken and the water structure thereby continuously disrupted.[292,376]

The rate of radial colonial growth of fungi on the surface of agars supplemented with various types of clays was also reduced by increasing concentrations of some clays, especially montmorillonite (unpublished). Because aeration was not an apparent limiting factor in these experiments, additional mechanisms whereby certain clays inhibit fungal development are indicated.

The effects of clay minerals on germination of fungal spores were less pronounced than on mycelial respiration. However, increasing concentrations of montmorillonite, but not of kaolinite or other particles, reduced the rate of germination of most species whether determined with a particle analyzer[377] or by respirometry (unpublished).

The inhibitory effects of montmorillonite became more pronounced after the spores had germinated, presumably because impairment of O_2 replenishment was more limiting to mycelial metabolism than to spore germination.

The effects of clay minerals on the respiration of actinomycetes appear to be different and more complex than their effects on bacteria and fungi. The respiration of some actinomycetes was stimulated by low levels of montmorillonite, apparently also because of pH-mediation, and inhibited by higher concentrations (unpublished).

Montmorillonite also appears to "protect" microorganisms against hypertonic osmotic pressures. The metabolic activity of microbes decreases as the solute concentration of the medium increases, but this decrease was less in the presence of montmorillonite than in its absence or in the presence of kaolinite.[108] When the solute concentration was maintained at a constant high level, respiration increased in proportion to the amount of montmorillonite present; comparable quantities of kaolinite had no such effect. These studies, which have been conducted with a variety of microorganisms and solutes (both inorganic and organic), indicate that the effect is greater with bacteria than with fungi (unpublished).

Although the mechanisms involved have not been defined, the ability of montmorillonite to enable microbial growth in the presence of hypertonic osmotic pressures provides another means whereby certain clays may differentially influence microbial ecology in soil. Although the solute concentration of normal soil solutions is usually not high enough to suggest osmotic inhibition of the soil microbiota, the concentrations may increase to inhibitory levels during periods of reduced moisture and increased nutrition, and the concentration in the microhabitats is undoubtedly higher than that measured in the soil solution. For example, the protection of bacteria by montmorillonite against hypertonic osmotic pressures may be important in explaining the apparent restricted distribution of *H. capsulatum*, as bird droppings contain a high salt content and the fungus grows and sporulates best in media containing approximately 3% NaCl.[378] Inasmuch as this concentration of salt is inhibitory to many terrestrial bacteria, the apparent correlation between bird droppings and the distribution of *H. capsulatum* may reflect an osmotic inhibition of bacteria in some soils, which would enhance establishment of the fungus. This inhibitory effect of avian manure on bacteria may be minimized in soils containing montmorillonite, as this clay would protect the bacteria against the increased osmotic pressure and thereby limit the establishment of the fungus.

From the field correlations between the presence or absence of the distinguishing clay mineral and the rate of spread of *Fusarium* wilt of banana and the isolation of *H. capsulatum* and from the results of the pure culture studies, working hypotheses have been developed of the mechanisms whereby clays influence the activity, ecology, and population dynamics of microorganisms in soil. In extrapolating the results of pure culture studies to natural, heterogeneous, microbial populations, certain assumptions, however, must be made. One assumption is that individual natural microhabitats may not be as complex as the sum of their components would suggest. In soil, most microbial activity is probably restricted to water films encompassing soil particulates and adjacent to clay surfaces, as these sites best provide the factors requisite for microbial development (e.g., greater permanence of water films, which are more tenaciously held by the activity of the clays than the rest of the soil water; concentration of both carbon and mineral substrates; differential sorption and inactivation of inhibitors). The microbial, chemical, physical, and mineralogical composition of each microhabitat undoubtedly differs, and the ability of an organism to become established and proliferate within a particular microhabitat, and then extend into adjacent ones, will depend on the characteristics of the organism and the habitats.

Another assumption is that *H. capsulatum* and *F. oxysporum* f. *cubense*, based on their apparent restricted geographic distributions, are not, except in soils where they have become established, "normal" soil inhabitants and should be considered as soil invaders.[379] A corollary question to this assumption is how these fungi become established as inhabitants in some soils but not in others. There are two main mechanisms by which an invader can become an inhabitant. One is for the organism to adapt, either physiologically or genetically, to the environment. Although most microbes are capable of some degree of adaptation, this process is relatively slow and haphazard, even in eucaryotic organisms, and is basically dependent on the genetic plasticity of the organism, with the environment's role being primarily, and perhaps only, to select certain adaptations. Consequently, assuming that distribution and introduction of these fungi into soil are not limiting factors, adaptation to all environments should occur, and the fungi should be much more widely present in soil. As they are apparently not universally present, despite their probable wide dispersal by avian, chiropteran, and other animal vectors, by water, wind, etc., the importance of adaptation in their establishment as soil inhabitants is probably minor.

The second mechanism by which an organism can become established in a specific habitat is for the habitat to change so that the antagonistic capabilities of the indigenous population are reduced sufficiently to enable the invader to become so entrenched that it can withstand the onslaught of the indigenous population when the latter regains its full capabilities. Inherent in this mechanism is the assumption that distribution and physical introduction of the organism into the habitat has occurred and that the organism has sufficient biochemical capabilities to grow in and exploit its new environment.

Based on the premise that the critical prerequisite for growth and development of an organism in a heterogeneous population is its establishment as a member of the population and that mere distribution and physical introduction of an organism into a habitat is insufficient, a sequence of ecological events in two microhabitats differing in clay mineral composition can be hypothesized. These sequences are schematically represented in Figure 3.

When substrates are sporadically introduced

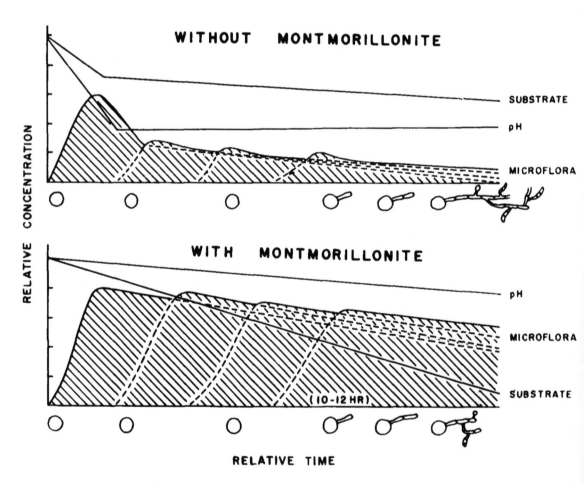

FIGURE 3. Schematic representation of events presumed to occur in the microhabitats of soil with and without montmorillonite. The assumption is made that substrate is introduced at the origin when fungi are present predominantly as spores. The hatched lines indicate sequential development of primarily bacterial populations. See text for further details. (Stotzky, unpublished)

into the microhabitats, the first utilizers are probably certain asporogenous bacteria, followed sequentially by populations of other bacteria, actinomycetes, and then fungi — the latter populations have a slower growth rate, must first germinate, require growth factors provided by previous populations, etc. In the absence of montmorillonitic clays, the dearth of exchangeable basic cations to buffer the environment against the accumulation of acidic metabolites restricts the development of acid-sensitive populations. When fungal spores present in the microhabitat eventually germinate, the environment is favorable for fungal development: the level of competitive populations is low and some substrate remains, thereby permitting the fungi, with their higher tolerances to low pH values, to proliferate within the microhabitat and to use it as a food base for growth into adjacent ones. Consequently, the fungi become established before the pH again becomes favorable for resumption of growth by the inhibited competitive populations, and the fungi eventually ramify throughout the soil.

In the presence of montmorillonitic clays, the primary and secondary populations develop to much higher levels, as the pH is buffered by exchange of basic cations from the clays for H^+ produced during metabolism. When fungal spores germinate (assuming here no production of germination inhibitors by prior populations), the environment is unfavorable for the development of the fungi, as they must compete with established and active populations for O_2, substrates, "Lebensraum", etc., which by then have been reduced to low levels. In addition, the prior populations have probably produced mycostatic materials and increased the CO_2 tension. The combination of these and, perhaps, other factors reduce fungal proliferation, especially extension to adjacent microhabitats. This reduction in spread from one microhabitat to another is reflected in situ, in, for example, the difference in the rate of spread of *Fusarium* wilt of banana: the disease eventually spreads in both long- and short-life soils (significantly in a radial pattern), but the rate differs.[365,366]

This simple model invokes only one direct (i.e., pH mediation) and several indirect mechanisms (e.g., substrate depletion, changes in atmospheric composition) by which clay minerals indirectly influence the spread of fungi in soil. Other mechanisms, acting both directly and indirectly

(e.g., differential sorption of substrates and inhibitors, differential protection against hypertonic osmotic pressures, genetic recombination), are undoubtedly also operative and mediated by clays.

D. Soil Replica Plate Studies

The above hypotheses and models were derived from in situ observations on the one hand and from highly simplified and artificial pure culture systems on the other. To evaluate the validity of these hypotheses at a level of experimental complexity intermediate between these two extremes, the growth and interaction of microorganisms in intact soils differing in chemical, physical, and mineralogical composition are being studied. Because of the opaqueness of soil, continuous direct microscopic observations are not possible, and, to circumvent this and other problems, the soil replica plating technique was used.[252]

The results of studies[253] on the spread of bacteria, fungi, and actinomycetes in different soils agree with the results obtained with the same organisms in pure culture: the spread of bacteria and actinomycetes was almost twice as rapid through soils containing the distinguishing clay mineral usually absent from short-life banana soils and from soils whence *H. capsulatum* has been isolated, whereas the spread of fungi was almost twice as rapid in soils not containing such clay (Tables 3, 4, and 5) (unpublished). The differential growth of microorganisms did not appear to be primarily correlated with any of 18 identified soil characteristics other than the absence or presence of the distinguishing clay mineral, although, within each clay group, a secondary relationship between pH and growth of bacteria and actinomycetes are apparent: the higher the pH, the better the growth. The addition of montmorillonite to soils not naturally containing this clay species enhanced the spread of bacteria and actinomycetes and retarded that of fungi (Table 6) (unpublished).

The faster spread of bacteria and actinomycetes through soils containing montmorillonite suggested that pH-mediation was involved. The slower spread of fungi, however, was probably not related to the increased viscosity and reduced O_2 replenishment elicited by high levels of montmorillonite in pure culture.[376] The soils were maintained at or near the 0.3 bar water content, where the thickness of water films associated with the clays and the replenishment of O_2 were not limiting factors. The mechanisms whereby

TABLE 3

Linear Extension of Bacteria in Soils With (A) and Without (B) the Distinguishing Clay Mineral (30-day Growth Period)

Bacterium	Soils A	Soils B
	mm/day	
Azotobacter chroococcum	0.24	0.12
Proteus vulgaris	0.28	0.18
Escherichia coli	0.30	0.13
Agrobacterium radiobacter	0.43	0.21
Flavobacterium sp.	0.52	0.29
Pseudomonas aeruginosa	0.64	0.19
Achromobacter eurydice	0.86	0.36
Mycobacterium smegmatis	0.13	0.06
Sarcina lutea	0.24	0.16
Cornynebacterium sp.	0.30	0.10
Bacillus cereus	0.27	0.19
Bacillus megaterium	0.50	0.22
Micrococcus rhodocrous	0.19	0.08
Arthrobacter globiformis	0.51	0.16
Mean	0.39	0.18

Average of 8 A-Soils and 10 B-Soils
(Stotzky, unpublished)

TABLE 4

Linear Extension of Actinomycetes in Soils With (A) and Without (B) the Distinguishing Clay Mineral (30-day Growth Period)

Actinomycete	Soils A	Soils B
	mm/day	
Streptomyces flavovirens	0.34	0.25
Streptomyces sp.	0.37	0.23
Streptosporangium sp.	0.36	0.19
Micromonospora fusca	0.37	0.22
Micromonospora chalcea	0.38	0.19
Nocardia paraffinae	0.41	0.28
Nocardia corallina	0.51	0.34
Mean	0.39	0.24

Average of 8 A-Soils and 10 B-Soils
(Stotzky, unpublished)

TABLE 5

Linear Extension of Fungi in Soils With (A) and Without (B) the Distinguishing Clay Mineral (15-day Growth Period)

Fungus	Soils A	Soils B
	mm/day	
Absidia sp.	1.55	2.75
Zygorhynchus moelleri	1.20	3.25
Rhizopus stolonifer	1.30	3.15
Saccharomyces cerevisiae	0.25	0.25
Microascus cinereus	0.90	1.40
Chaetomium sp.	1.40	2.85
Sordaria fimicola	4.00	6.00
Armillaria mellea	0.16	0.23
Fomes annosus	0.23	0.26
Schizophyllum commune	1.50	2.70
Rhizoctonia sp.	1.60	2.70
Fusarium oxy. f. conglutinans	1.00	1.15
Fusarium oxy. f. cubense	1.20	2.30
Fusarium nivale	1.50	2.50
Penicillium egyptiacum	1.60	2.50
Penicillium jenseni	1.35	2.85
Aspergillus niger	1.25	2.70
Aspergillus terreus	1.40	2.80
Sphaerostilbe repens	1.70	2.80
Trichoderma viride	2.25	4.50
Mean	1.37	2.48

Average of 8 A-Soils and 10 B-Soils
(Stotzky, unpublished)

montmorillonite retards the growth of fungi in soil plates are not known at present. However, as these studies were conducted with autoclaved soils, microbial interactions were not involved, indicating that montmorillonite also has a direct inhibitory effect on fungi. (Autoclaving does not destroy the structure of clay minerals.[380])

Preliminary results with mixed populations of microbes inoculated into soil plates indicate that antagonistic effects of bacteria and actinomycetes against fungi are more pronounced in soils containing the distinguishing clay mineral.[252] The mechanisms involved in this antagonism have not been defined; however, indications are that some organisms that do not appear to produce diffusable inhibitory substances in culture do produce them in soils containing montmorillonite. Conversely, many organisms that are active producers of inhibitors on agar or in liquid media appear to be ineffective in soil. Clays do not appear to influence negative interactions between fungi, at least not with the species studied (unpublished).

Regardless of the mechanisms involved, the soil-plate studies are confirming, at another level of experimental complexity, the observations derived from field and pure culture studies that

TABLE 6

Linear Extension of Microorganisms in a Soil not Naturally Containing the Distinguishing Clay Mineral (Check) and Amended with 5, 10, and 20% Kaolinite or Montmorillonite.

Microbe	Check	Kaolinite			Montmorillonite		
	mm/day	5	10 mm/day	20	5	10 mm/day	20
Fungus							
Zygorhyncus moelleri	5.81	4.68	3.70	5.00	3.03	2.01	2.53
Absidia sp.	3.18	3.43	3.43	3.59	2.69	3.13	2.89
Fusarium nivale	3.31	3.31	3.06	3.13	2.15	2.26	1.93
Penicillium egyptiacum	3.38	3.13	3.44	3.59	3.03	2.15	2.47
Aspergillus terreus	5.25	3.75	4.25	4.13	2.75	2.22	2.81
Aspergillus niger	4.13	3.75	4.13	5.00	3.12	2.93	3.37
Rhizoctonia sp.	3.38	3.18	3.37	3.25	2.18	3.00	3.12
Bacterium							
Flavobacterium sp.	0.13	D	D	D	0.15	0.20	0.21
Achromobacter eurydice	0.08	D	D	D	0.23	0.23	0.35
Arthrobacter globiformis	D	D	D	D	0.07	0.18	0.23
Agrobacterium radiobacter	D	D	D	D	0.12	0.27	0.21
Actinomycete							
Streptomyces sp.	D	0.05	0.07	0.02	0.31	0.20	1.57
Nocardia paraffinae	D	D	D	D	0.15	0.29	0.31
Micromonospora fusca	D	D	D	D	D	0.08	0.18

D = growth disappears after a few days.
(Stotzky, unpublished)

clays differentially affect the growth of microorganisms in soil.

The soil replica plate technique is also being used to study the effects of clay minerals on genetic recombination of bacteria in soil. The addition of montmorillonite appears to enhance the rate of conjugation in *Escherichia coli*, in part by stimulating the growth of the bacteria and thereby increasing the probability for effective pairing and conjugation of donor and recipient cells.[23] These studies, which also indicate that montmorillonite enhances the growth of bacterial auxotrophs in soil — perhaps by acting as concentrating surfaces for nutrients — add another dimension to the differential role that clay minerals may play in the adaptation and ecology of microbes in soil.

E. Metabolic Studies

Because most of the soil replica plate studies have been conducted with sterile soils inoculated with only one or several microbial species, microbial activity in nonsterile soils is being studied to evaluate the effects of clay minerals at another level of experimental complexity. Soils not naturally containing the distinguishing clay were amended with approximately 5, 10, and 20% montmorillonite or kaolinite, either just before the experiments were conducted or more than four years earlier. The latter soil-clay mixtures were maintained in a greenhouse with alternate wetting and drying and periodic cropping.

When these mixtures were perfused with glycine, the heterotrophic conversion of the amino acid to NH_4^+ and the autotrophic conversion of NH_4^+ to NO_2^- and NO_3^- were enhanced by increasing increments of montmorillonite but not of kaolinite (Figure 4) (unpublished). All the mechanisms responsible for this stimulation are not yet defined, but maintenance of favorable conditions (e.g., pH) in the microhabits for the nitrifying bacteria appears to be involved. The stimulatory effects of montmorillonite persisted for more than one year of continuous perfusion — the duration of the experiment — with fresh glycine being added every 20 to 30 days.

The rate and extent of decomposition of various organic substrates in soils, either naturally containing different clay minerals or amended

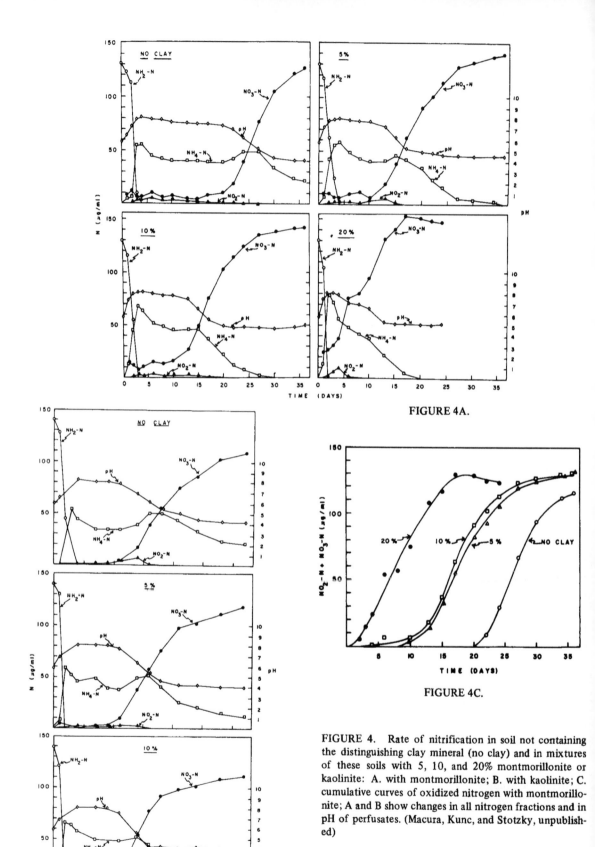

FIGURE 4A.

FIGURE 4C.

FIGURE 4. Rate of nitrification in soil not containing the distinguishing clay mineral (no clay) and in mixtures of these soils with 5, 10, and 20% montmorillonite or kaolinite: A. with montmorillonite; B. with kaolinite; C. cumulative curves of oxidized nitrogen with montmorillonite; A and B show changes in all nitrogen fractions and in pH of perfusates. (Macura, Kunc, and Stotzky, unpublished)

FIGURE 4B.

112

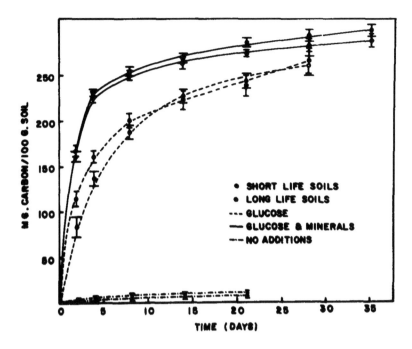

FIGURE 5. Average cumulative rate of C evolved as CO_2 from 84 soils with (long-life soils) or without (short-life soils) the distinguishing clay mineral; 1% glucose added where indicated. (Stotzky, unpublished)

with montmorillonite or kaolinite, were also influenced by the type and concentration of clay present. No significant differences in substrate utilization — as measured by CO_2 evolution[332] — were apparent when glucose was added, either alone or with mineral nutrients, to 84 soils, some containing the distinguishing clay mineral (i.e., long-life soils) and others not (i.e., short-life soils) (Figure 5) (unpublished). Similar results were obtained with several other substrates (e.g., starch, casein, soybean meal) added to soil-clay mixtures.[381] With some substrates, however, the presence of montmorillonite, but not of kaolinite, markedly stimulated both the rate and extent of decomposition (Figure 6). The decomposition of other substrates was depressed by the addition of montmorillonite but not of kaolinite (Figure 7) (unpublished).

The most dramatic effect of clay minerals on the decomposition of organics has been observed with aldehydes (e.g., vanillin, syringaldehyde, benzaldehyde, heptaldehyde, acetaldehyde), the decomposition of which was markedly stimulated by increasing increments of montmorillonite but not of kaolinite (Figure 8).[382] This effect appears

to be restricted to the aldehyde, as the decomposition of the corresponding acid was not stimulated to the same extent (Figure 8) (unpublished). Although the mechanisms involved in this specific stimulation are not yet known, the similarity between the curves for nitrification (Figure 4) and aldehyde utilization (Figures 6 and 8) (i.e., the decrease in the lag periods by the addition of montmorillonite) suggests that the methods by which montmorillonite influences these two processes may be similar, even though one process is autotrophic and the other heterotrophic. In both processes, however, specific groups of microorganisms are involved, i.e., nitrifiers and aldehyde oxidizers,[383] and the clay may affect these two divergent processes indirectly by establishing a favorable environment for the specific organisms rather than directly affecting their metabolism.

These studies with nonsterile soils also demonstrate, at another level of experimental complexity, the differential effects of clay minerals on microbial activities and indicate additional mechanisms whereby clays may influence the ecology and population dynamics of microbes in soil.

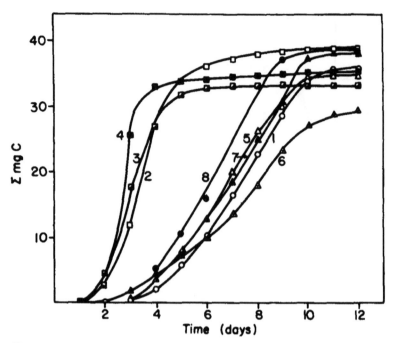

FIGURE 6. Cumulative rate of C evolved as CO_2 from 100 g soil-clay mixtures enriched with syringaldehyde at a concentration equivalent to 0.05% C. CO_2 evolved from mixtures not enriched with substrate (i.e., endogenous respiration) was subtracted. Curve 1 = no clays added; Curves 2, 3, and 4 = 5, 10, and 20% montmorillonite added, respectively; Curves 5, 6, and 7 = 5, 10, and 20% kaolinite added, respectively; Curve 8 = 5% mica-vermiculite (Zonolite) added. (Kunc and Stotzky, unpublished)

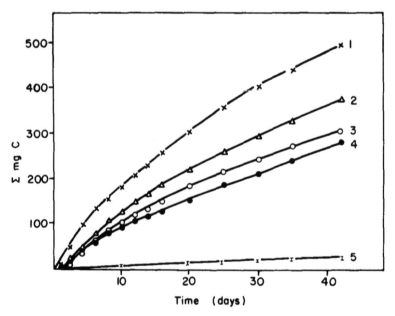

FIGURE 7. Cumulative rate of C evolved as CO_2 from 100 g soil amended with montmorillonite and enriched with Na-oleate at a concentration equivalent to 1% C. Curve 1 = no clay added; Curves 2, 3, and 4 = 5, 10, and 20% montmorillonite added, respectively; Curve 5 = all soil-clay mixtures without Na-oleate added (i.e., endogenous resipiration). (Kunc and Stotzky, unpublished)

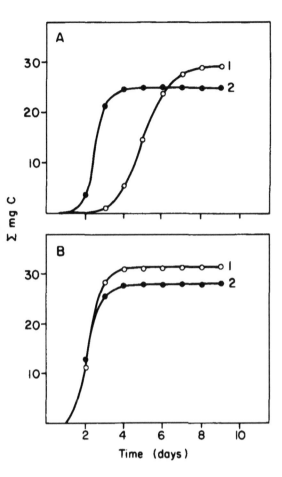

FIGURE 8. Cumulative rate of C evolved as CO_2 from 100 g soil enriched with vanillin (Curve 1) or vanillic acid (Curve 2) at concentrations equivalent to 0.05% C in absence (A) or presence (B) of 20% montmorillonite. CO_2 evolved from soil-clay mixtures not enriched with substrates (i.e., endogenous respiration) was subtracted. (Kunc and Stotzky, unpublished)

F. Greenhouse Studies

To bridge further the field and pure culture studies, nonsterile soils, with and without the distinguishing clay mineral, and soils amended with montmorillonite or kaolinite have been planted in the greenhouse with various species susceptible to soil-borne root-infecting pathogens, inoculated with the appropriate pathogen, and rates of disease spread measured. The rates were significantly slower in soils either naturally containing the distinguishing clay or amended with montmorillonite than in soils not containing this clay (unpublished). For example, in soils planted with cabbage and centrally inoculated with *F. oxysporum* f. *conglutinans*, the rate of disease

spread was slower in soil containing the distinguishing clay mineral and decreased in soils not containing this clay when they were mixed with the former soil (Figure 9). Because all soils were partially sterilized by fumigation or steam and inoculated with a suspension of the nonsterile soil containing the distinguishing clay prior to planting and introduction of the pathogen, the results were probably not related to differences in the original indigenous microbiota but to differences in the clay mineral composition of the soils, thereby simulating the field observations with *Fusarium* wilt of banana. The reasons for the increase in the rates of disease spread in mixtures of soils not containing the distinguishing clay are not known, but this increase indicates further that the decreases obtained with the other mixtures were not the result of the physical disturbance of the soils but of the addition of the distinguishing clay mineral.

These studies have not been as exhaustive as would be desirable, primarily because of the difficulty in distinguishing symptoms by the inoculated pathogens from those caused by others (e.g., viruses, insects) and by the ambient environment (e.g., temperature, moisture) and because of

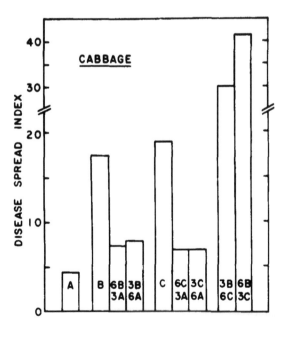

FIGURE 9. Rate of spread of *Fusarium* wilt of cabbage in various soil mixtures. Soil A contains the distinguishing clay mineral; soils B and C do not. Soils mixed in ratios of 6 to 3 and 3 to 6, as indicated. (Stotzky, unpublished)

the time and space requirements. Nevertheless, these greenhouse studies have provided valuable information on another level of experimental complexity of the effects of clay minerals on microbial ecology in soil.

G. Surface Interaction Studies

Because the physiology of sorbed and non-sorbed cells may differ, and the adsorption of substrates and metabolites by clays and other soil constituents might affect their availability and activity — both phenomena which could profoundly influence the ecology of microbes in soil — surface interactions between clay minerals and microorganisms and their metabolites are being studied. Both microbial cells and clay minerals normally carry a surface charge: on clay minerals, especially on 2:1 minerals, the charge results primarily from isomorphous substitution within the crystal, and, even though positive sites are present, the net electrical charge is negative;[77] on microbial cells, the charge results from ionic changes in surface components and is determined by the isoelectric point (IP) or dissociation constant (pK) of the components; most organics involved with microbial life have either no charge or one also dependent on the ambient pH. The pH of microhabitats in soil is presumed to be usually above the IP or pK of most components, and, therefore, the net charge on cell surfaces would be expected to be negative and that of organics to be either negative or neutral. Consequently, it is necessary — even before attempting to compare the activity of sorbed and nonsorbed cells or organics — to explain whether and how, in view of their high electrokinetic potentials and same charge, sorption occurs between populations of clay, cells, and organics. Regardless of the type of bonding (whether coulombic, London-van der Waals, hydrogen, coordination, etc.), there must first be a reduction in the electrokinetic potentials for the populations to come close enough together for secondary bonding to occur.

Sorption, as determined by measuring changes in particle size distributions with a Coulter Counter[384,385] and by electron microscopy,[308] between a variety of bacterial species and clay was evident in vitro only at pH values below the IP of the cells (i.e., when the cells were positively charged, and, as the clays remained negatively charged, coulombic interaction apparently occurred) or when the electrokinetic potentials were decreased by polyvalent (especially tri- and tetravalent) cations. Some charge reversal also occurred with some cations (e.g., Al^{+++}, Fe^{+++}), presumably because their hydrous oxides coated the particles. Sorption increased as the valency of the cations present on the clay and in the suspending electrolyte increased (Table 7).

Neither the size, motility, nor Gram reaction of bacteria nor the size of the clay appeared to influence sorption. Sorption occurred between bacteria and montmorillonite homoionic to monovalent cations only at pH 2 to 4, but did not occur at any pH with cells of *Saccharomyces* sp., (The surface of yeast cells has also been shown to be different from bacterial cells in electrophoretic mobility studies.) Heat-killed yeast cells lost their ability to sorb, whereas cells killed with benzalkonium chloride did not.

The flocculation of clays by microbial metabolites (e.g., filtrates from cultures of bacteria, fungi, and actinomycetes; purified rhizobial polysaccharides) was influenced by the source of the metabolite and by the types of cations present, and, as with intact cells, increased with the valency of the cations.[386]

These studies indicated that sorption between clays and microbial cells occurs essentially only at pH values below the IP of cells or in the presence of polyvalent cations. Under natural soil conditions, however, sorption appears to occur at pH values above the apparent IP of cells and in the absence of large quantities of tri- or tetravalent cations (e.g., the dominant cations in the soil solution are usually K^+, Na^+, Ca^{++}, Mg^{++}, H^+).

The contradiction between apparent sorption of microbial cells in vivo (e.g., removal of cells from waters in percolation beds, the failure to wash substantial numbers of microbes from soil in perfusion experiments) and the absence of sorption in vitro, except under these rather extreme conditions, can perhaps be reconciled by at least two possibilities. One is that microbes are retained in soil by entrapment in narrow channels between particles and/or by surface tension rather than by direct physicochemical attraction.

The other possibility is based on observations that the IP of some bacteria changes with the type of cations present in the ambient environment.[387,388] For example, the IP of several bacteria in the presence of low concentrations (ionic strength = 3×10^{-4}) of mono- or divalent

TABLE 7

Apparent Relative Sorption Between Microorganisms and Montmorillonite Homoionic to Different Cations as Determined by Particle Size Distribution Analysis (CC) and Electron Microscopy (EM). (Stotzky and Bystricky, unpublished)

Cation	pH[1]	Alcaligenes faecalis CC	Alcaligenes faecalis EM	Agrobacterium radiobacter CC	Agrobacterium radiobacter EM	Chromobacterium orangum CC	Chromobacterium orangum EM	Bacillus megaterium CC	Bacillus megaterium EM	Bacillus subtilis CC	Bacillus subtilis EM	Escherichia coli CC	Escherichia coli EM	Enterobacter aerogenes CC	Enterobacter aerogenes EM	Serratia marcescens CC	Serratia marcescens EM	Saccharomyces cerevisiae CC	Saccharomyces cerevisiae EM
Na	6.8	1[2]	±	0	0	1	±	0	0	-	0	0	0	-	0	-	0	0	0
H	3.0	-	1	-	1	-	2	-	1	-	1	-	1	-	1	-	1	0	2
Ca	6.9	3	2	0	1	3	3	1	2	-	3	-	3	-	3	-	3	0	2
Mg	6.8	-	3	-	1	-	3	-	-	-	3	-	2	-	2	-	3	0	2
La	6.3	4	2	1	3	4	3	4	3	-	3	-	3	-	2	-	3	4	3
Al	4.8	4	3	1	3	4	4	4	4	-	4	-	4	-	4	-	4	4	4
Th	3.8	-	4	-	4	-	4	-	3	-	3	-	4	-	4	-	4	-	2

[1] pH of 3.3% clay suspension (0.2 - 2µ).
[2] Arbitrary scale from 0 to 4; - = not determined; ± = questionable.

117

chloride salts was at pH 2.5 to 3.5, whereas, in the presence of LaCl$_3$ or CrCl$_3$, it was at approximately pH 5.0, and in the presence of FeCl$_3$ or AlCl$_3$, at pH 7.0 (Figure 10). The IP was also shifted to higher pH values, and the electrophoretic mobility decreased at all pH values as the cation concentrations were increased. These effects were also observed in dilute soil extracts, sea water, and various laboratory media. Clays, however, remained negatively charged at all pH values in solutions of low ionic strengths, and only in the presence of higher concentrations of tri- and tetravalent cations did their charge reverse. Microbial metabolites either increased or decreased the electrophoretic mobility of clay minerals depending on the microbial species.

These observations may explain how sorption between clays and microbes occurs in nature. As low concentrations of polyvalent cations and macromolecules are usually present in soil microhabitats, microbes may be positively charged, or, at least, have a lower net negative charge at normal pH values in vivo than at comparable pH values in vitro, which would facilitate sorption with negatively charged clays. The possible importance of the actual pH at charged surfaces (i.e., the pH$_s$) must also be considered.

Preliminary studies on the adsorption of organics by clays indicate that montmorillonite homoionic to various mono-, di-, tri-, and tetravalent cations has no affinity for glycine and only a slight affinity for dimethylamine. Montmorillonite homoionic to H$^+$ exhibited a strong affinity for dimethylamine and a limited affinity for glycine. The glycine, however, was easily removed by repeated washings with water. The retention of water by montmorillonite was also influenced by the type of cation saturation (e.g., H$^+$ > other mono- > di- > tri- > tetravalent cations) (Figure 11), and dimethylamine but not glycine altered the relative retention (unpublished).

To study the adsorption of larger molecules, which serve both as substrates and as examples of biologically-active materials, various proteins (e.g.,

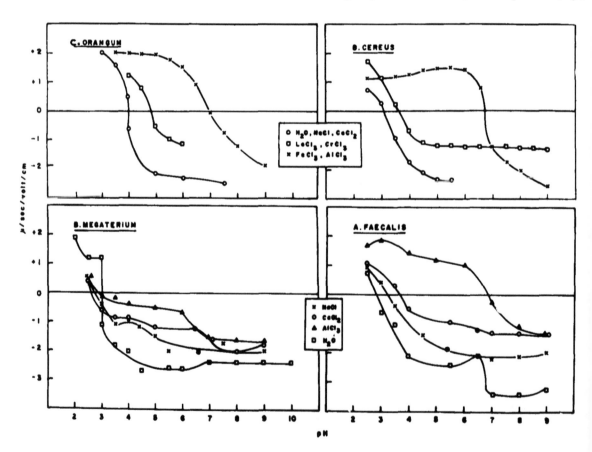

FIGURE 10. Electrophoretic mobility (in μ/sec/V/cm) of four bacteria in distilled water and in different electrolytes (ionic strength = 3 x 10^{-4}) at different pH values. Note shift in isoelectric points. (Santoro and Stotzky, unpublished)

118

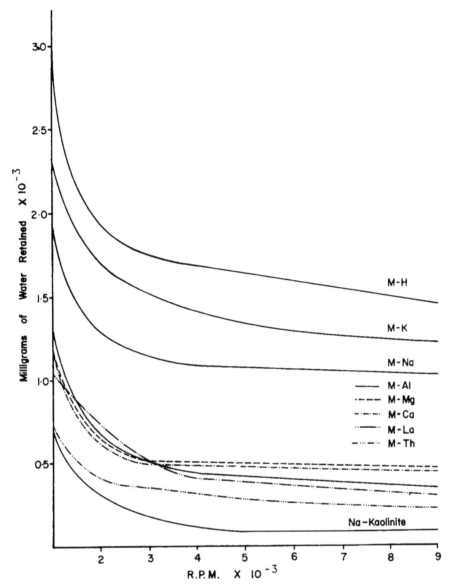

FIGURE 11. The retention of water by montmorillonite homoionic to different cations and by Na-kaolinite. (Colom and Stotzky, unpublished)

casein, catalase, chymotrypsin, edestin, egg albumin, invertase, lactoglobulin, lysozyme, ovomucoid), ranging in molecular weight from approximately 14,000 to 310,000 and in IP from approximately pH 1 to 11, were adsorbed onto montmorillonite homoionic to H^+, Na^+, Ca^{++}, La^{+++}, Al^{+++}, or Th^{++++}. Adsorption was observed in every combination of clay and protein, even when the ambient pH of the clay suspensions, which ranged from approximately pH 3 to 7, was several pH units above the IP of the proteins.[389] The amount of protein adsorbed and the shapes of both the equilibrium adsorption and binding isotherms (i.e., protein retention after ultimate washings) were more related to the molecular weight of the proteins, to the number of active sites on the proteins available for reaction with the adsorption sites on the clays, and to the valency of the cation of the clay exchange complex than to the ambient pH and IP of the proteins. The mechanisms involved in this adsorption are not clearly defined, but pH may not be a critical factor in adsorption of large complex molecules such as proteins.

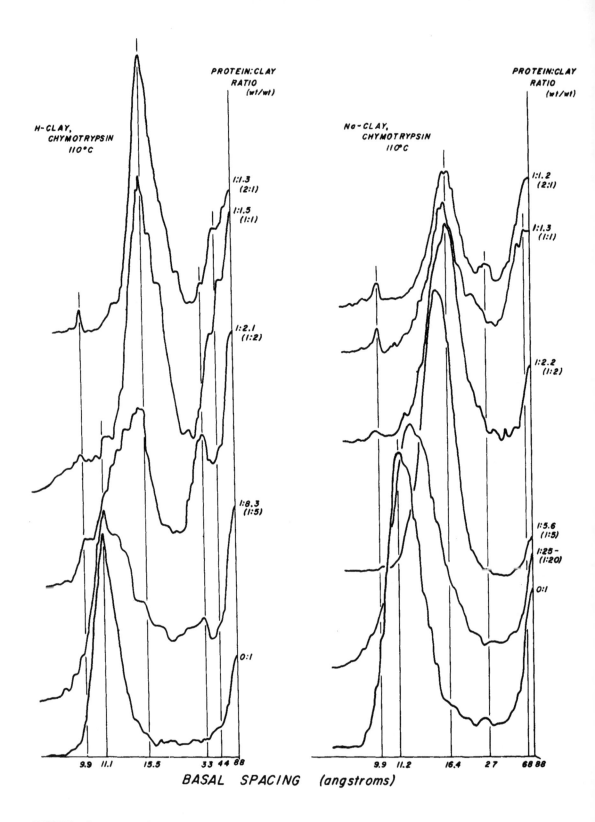

FIGURE 12. X-ray diffraction spectra of oven-dried (110°C) complexes formed by addition of chymotrypsin to montmorillonite homoionic to H[+] and Na[+]. Protein:clay ratios of final (and original) complexes are indicated. See Harter and Stotzky[389] for details of formation of clay-protein complexes. (Harter and Stotzky, unpublished)

120

Most of the proteins appeared to be adsorbed between the clay lattices, which they expanded in proportion to the amount of protein adsorbed (Figure 12). Some proteins (e.g., catalase) appeared to be only slightly adsorbed intermicellarly as essentially no expansion of the lattices was evident by x-ray diffraction analysis. Examination by electron microscopy indicated that these proteins were adsorbed on the flat surfaces of the clays.[309]

When individual clay-protein complexes were the sole source of carbon for bacteria in respirometric experiments, the availability of the complexed proteins, but not of comparable concentrations of free proteins, was dependent on the type of protein and on the cation saturating the clay exchange complex (unpublished). In general, casein, chymotrypsin, and lactoglobulin were utilized by the bacteria when complexed with montmorillonite, but to a considerably lesser extent than when not complexed (Figure 13). Catalase, invertase, and pepsin were not utilized at

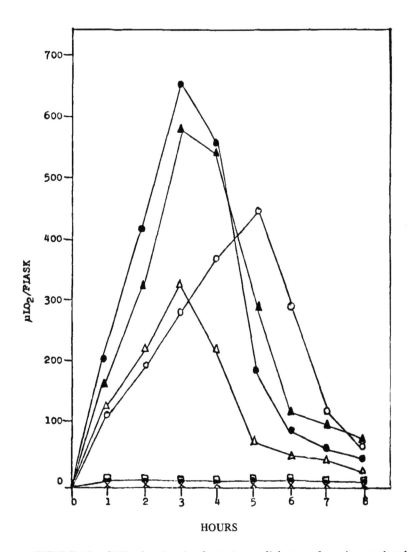

HOURS

FIGURE 13. Utilization by *Agrobacterium radiobacter* of casein complexed with montmorillonite homoionic to Ca. Δ = Ca-montmorillonite-casein complex and bacteria; ▲ = uncomplexed casein and bacteria; ○ = uncomplexed casein and Ca-montmorillonite and bacteria; ● = clay-casein complex and 50% uncomplexed casein and bacteria; □ = bacteria only; ■ = clay-casein complex only; x = Ca-montmorillonite only. (Friedlander and Stotzky, unpublished)

all when complexed, although they were rapidly utilized when not adsorbed (Figure 14).

Even though catalase, and, to a lesser degree, pepsin, were adsorbed primarily on the surface of montmorillonite, rather than between the lattices as were the other proteins, they were not susceptible to microbial degradation. Conversely, the enzymic activity of catalase appeared to be greater when adsorbed by montmorillonite than when in solution (unpublished).

These differential effects of adsorption of organics by clay minerals on their nutritional availability and enzymic activity suggest further mechanisms whereby clays may influence the activity, ecology, and population dynamics of microbes in soil. Additional studies on the adsorption and availability of other substrates, on the adsorption and reactivity of inhibitors, on the

physiology of sorbed and nonsorbed cells, etc. are obviously needed to resolve these and other mechanisms.

H. Summation

It is apparent, even from these incomplete studies, that a single environmental factor, such as the type of clay minerals present, will greatly influence the activity, ecology, and population dynamics of microbes in soil, and that this influence is manifested in a series of changes in other environmental factors (e.g., pH, nutrition, ionic environment). Moreover, it is also apparent that the definition of the influence of even a single factor in as complex an environment as soil requires studies on various levels of experimental complexity.

These investigations also suggest at least one

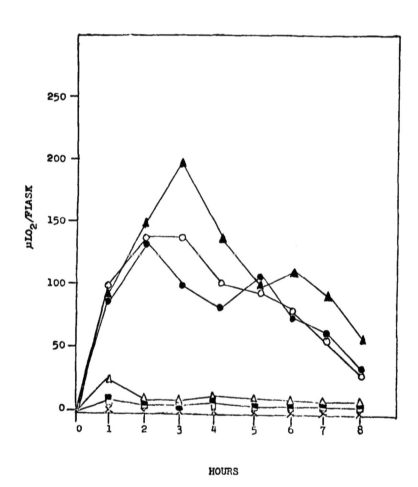

FIGURE 14. Utilization by *Agrobacterium radiobacter* of catalase complexed with montmorillonite homoionic to Al. See Figure 13 for details. (Friedlander and Stotzky, unpublished)

possible method to alter the soil environment for the purpose of eradicating or controlling undesirable microorganisms (e.g., *H. capsulatum*), i.e., the incorporation of montmorillonite. The greenhouse studies have indicated that new, stable soils can be established by amending natural soils with mined clay minerals: e.g., when soils were mixed with various concentrations of montmorillonite or kaolinite in a cement mixer, placed in flats, and maintained for at least four years in the greenhouse under a regime of intermittent cropping and fluctuating temperatures and moisture, the mixtures, especially those containing montmorillonite, behaved as natural soils in soil replica plate, perfusion, and substrate utilization studies. Various analyses (e.g., chemical, particle size distribution, aggregate stability) also indicated that stable soil-clay mixtures resulted. Consequently, the addition of montmorillonite to soils in the field should result in stable new soils from which clay will not be leached rapidly from the zone of maximum microbial activity, and, therefore, infrequent, if any, further additions of clay will be required.

The possible eradication of *H. capsulatum* and other undesirable fungi from soil by alteration of the soil environment by the incorporation of montmorillonite should be more successful and more lasting than the use of fungicides. Fungicides kill organisms already present in soil but do not alter the environment for long, and, consequently, pathogens could establish themselves again if reinoculated. The history of fungicide use against plant pathogens indicates that after an initial reduction in both the saprophytic and pathogenic soil microbiota most microorganisms reestablish themselves, sometimes at levels higher than those present prior to fumigation. The necessity for repeated fumigation against soil-borne plant pathogens attests to the short duration of control and suggests that for long-term control it is necessary not only to kill the pathogen but also to alter the environment and, thereby, the ecology and population dynamics of the nonpathogenic soil microbiota so that reestablishment of the pathogens is prevented. Such an approach to eradication of certain pathogens from soil would not only be cheaper and more permanent but would also reduce the introduction of high concentrations of chemicals into the biosphere.

Regardless of what methods are used to alter the ecology of microbes in soil, much more information than is presently available about the factors that influence their ecology and population dynamics in soil is necessary. For example, does the apparent restricted geographic distribution of *H. capsulatum*, as indicated by both isolation from soil and histoplasmin sensitivity, imply environmental control in areas where it is not present?[370] Why are large areas endemic to histoplasmosis, as indicated by skin testing, and yet the fungus has not been isolated from soils in these areas? Conversely, why can the pathogen be isolated from soils in some areas where the human population is predominantly histoplasmin negative?[390] Why is it impossible sometimes to isolate *H. capsulatum* and other pathogens from stored soils or field sites that were once positive, and why, in some areas, does there appear to be a seasonal incidence of isolation from soil?[370] Does this imply an influence of climate on the competitive soil microbiota, which in turn results in fluctuations in the population of the pathogen, or does it imply recurring alterations in the forms in which the pathogen exists in soil at different times? What are the propagules that the fungus produces to persist and proliferate in soil and what mechanisms (e.g., fungitoxins, lysis) does the soil microbiota use to control the fungus? Is the correlation between the presence of manures and the isolation of *H. capsulatum* more apparent than real, especially as the fungus is present in many areas devoid of dung?[374] What stimulators of the pathogen and/or inhibitors of the competitive soil microbiota are present in manures? Does the apparent correlation between manure and the isolation of *H. capsulatum* in some areas reflect only a secondary effect? Perhaps the clay mineralogy of the soils is the primary determinant through its influence on the antagonistic microbiota, and the droppings act either by stimulating the fungus or inhibiting the antagonists? When manure is present, the pathogen may grow better, but, if montmorillonite is also present, the stimulation of the pathogen is masked by the stimulation of the antagonists by the clay, with the net result that the fungus is prevented from becoming established. The stimulatory or inhibitory substances in manure may also be adsorbed and rendered ineffective by montmorillonite, or the clay may negate the effect of the high solute concentrations derived from the manure and thereby indirectly prevent the establishment of the fungus.

From even these few possibilities, it is apparent that too little is known about the chemical, physical, and biological factors that influence microbes in soil. Until these factors are better defined, ecological eradication of undesirable microorganisms will be a trial and error situation at best.[391]

V. CONCLUSIONS

The soil is a complex microbial habitat in which numerous physicochemical and biological factors interact. Consequently, the influence of any single environmental factor on the activity, ecology, and population dynamics of microbes in soil is difficult to evaluate, inasmuch as a slight change in one factor can trigger and magnify changes in other factors — changes that often are unanticipated and unappreciated. This cascading effect in most natural microbial habitats, but especially in soil, can be compared to poking a balloon filled with water: a change is caused not only at the point of indentation, but also throughout the balloon and along its entire perimeter.

To study the effects of these meshing environmental factors on the soil microbiota, the investigator must be a scientific assimilator and dilettante capable of applying, in both breadth and depth, concepts and techniques from all scientific disciplines. Experiments must be precisely designed to ensure that the factors under consideration are actually being studied, that others are being controlled and monitored, and that the results can be statistically evaluated. The investigator should also be prepared to move back and forth between in vitro model systems and soil in situ.

The difficulties inherent in studying microbes in soil should not be a deterrent but rather a goad to an investigator considering the study of microbial ecology. Although soil is, undoubtedly, the most complex of microbial habitats and techniques for its study are inadequate, life on Earth is largely dependent on the biogeochemical activities in soil. Consequently, the study of microbes in soil should be embraced by the most enthusiastic and competent scientists in order to utilize more effectively this most important natural resource.

The necessary progress in microbial ecology in soil — and in other natural habitats as well — will only be made through the development of new concepts and knowledge and the innovative application of old concepts and knowledge. In an era of rapidly increasing environmental pollution, this progress must be made now.

Acknowledgments
Some of the studies reported herein were supported over the years, in part, by Public Health Service Research Grants AI-05810, from the National Institute of Allergy and Infectious Diseases, AP-00440, from the National Center for Air Pollution Control, and FR-07062, a Biomedical Sciences Support Grant to New York University; by Grant 1 from the Forest Service, Department of Agriculture; and by Grant R-800671 from the Environmental Protection Agency. The outstanding contributions of many co-workers to these studies are also gratefully acknowledged, as are the typing, proof-reading, and editorial skills of N. Anderson, M. Bohren, F. Roales, K. Stotzky, and H. Babich.

REFERENCES

1. Brewer, R., *Fabric and Mineral Analysis of Soils*, John Wiley & Sons, New York, 1964.
2. Alexander, M., *Introduction to Soil Microbiology*, John Wiley & Sons, New York, 1961.
3. Alexander, M., Biochemical ecology of soil microorganisms, *Annu. Rev. Microbiol.*, 18, 217, 1964.
4. Alexander, M., Biochemical ecology of microorganisms, *Annu. Rev. Microbiol.*, 25, 361, 1971.
5. Alexander, M., *Microbial Ecology*, John Wiley & Sons, New York, 1971.
6. Brock, T. D., *Principles of Microbial Ecology*, Prentice-Hall, Englewood Cliffs, N.J., 1966.
7. Burges, A., *Micro-organisms in Soil*, Hutchinson, London, 1958.
8. Burges, A. and Raw, F., Eds., *Soil Biology*, Academic Press, New York, 1967.
9. Williams, R. E. O. and Spicer, C. C., Eds., *Microbial Ecology*, Symp. Soc. Gen. Microbiol., Vol. 7, Cambridge University Press, Cambridge, 1957. (N.B., many papers relevant to this topic can also be found in other volumes.)
10. Gray, T. R. G. and Parkinson, D., Eds., *The Ecology of Soil Bacteria*, Liverpool University Press, Liverpool, 1968.
11. Parkinson, D. and Waid, J. S., Eds., *The Ecology of Soil Fungi*, Liverpool University press, Liverpool, 1960. Press,
12. Krasil'nikov, N. A., *Soil Microorganisms and Higher Plants*, Akad. Nauk. SSSR, Moscow, 1958 (transl. by Israel Program Sci. Transl., Washington, D. C., 1961).
13. McLaren, A. D. and Peterson, G. H., Eds., *Soil Biochemistry*, Vol. I, Marcel Dekker, New York, 1967.
14. McLaren, A. D. and Skujins, J., Eds., *Soil Biochemistry*, Vol. II, Marcel Dekker, New York, 1971.
15. Quastel, J. H., Soil metabolism, *Annu. Rev. Plant Physiol.*, 16, 217, 1965.
16. Gray, T. R. G. and Williams, S. T., *Soil Micro-organisms*, Hafner, New York, 1971.
17. Baker, K. F. and Snyder, W. C., Eds., *Ecology of Soil-Borne Plant Pathogens*, University of California Press, Berkeley, 1965.
18. Doeksen, J. and van der Drift, J., Eds., *Soil Organisms*, North Holland Publishing Co., Amsterdam, 1963.
19. Russell, E. J., *Soil Conditions and Plant Growth*, 9th ed., rewritten by E. W. Russell, Longmans, London, 1961.
20. Macura, J. and Vancura, V., Eds., *Plant Microbes Relationships*, Czechoslovak Academy of Science, Prague, 1965.
21. Gilmour, C. M. and Allen, O. N., Eds., *Microbiology and Soil Fertility*, Oregon State University Press, Corvallis, 1965.
22. Webley, D. M. and Jones, D., Biological transformation of microbial residues in soil, in *Soil Biochemistry*, Vol. 2, McLaren, A. D. and Skujins, J., Eds., Marcel Dekker, New York, 1971, Chap. 15.
23. Weinberg, S. R. and Stotzky, G., Conjugation and genetic recombination of *Escherichia coli* in soil, *J. Soil Biol. Biochem.*, 4, 171, 1972.
24. Delwiche, C. C. and Finstein, M. S., Carbon and energy sources for the nitrifying autotroph, *Nitrobacter, J. Bacteriol.*, 90, 102, 1965.
25. Alexander, M., Persistence and biological reactions of pesticides in soils, *Soil Sci. Soc. Am. Proc.*, 29, 1, 1965.
26. Alexander, M., Biodegradation: problems of molecular recalcitrance and microbial fallibility, *Adv. Appl. Microbiol.*, 7, 35, 1965.
27. Bremner, J. M., Organic nitrogen in soils, in *Soil Nitrogen*, Bartholomew, W. V. and Clark, F. E., Eds., Am. Soc. Agron., Madison, 1965, Chap. 3, 93.
28. Nommik, H., Ammonium fixation and other reactions involving a nonenzymatic immobilization of mineral nitrogen in soil, in *Soil Nitrogen*, Bartholomew, W. V. and Clark, F. E., Eds., Am. Soc. Agron., Madison, 1965, Chap. 5, 198.
29. Stotzky, G. and Norman, A. G., Factors limiting microbial activities in soil: I. The level of substrate, nitrogen, and phosphorus, *Arch. Mikrobiol.*, 40, 341, 1961.
30. Stotzky, G. and Norman, A. G., Factors limiting microbial activities in soil: II. The effect of sulfur, *Arch. Mikrobiol.*, 40, 370, 1961.
31. Stotzky, G. and Norman, A. G., Factors limiting microbial activities in soil: III. Supplementary substrate additions, *Can. J. Microbiol.*, 10, 143, 1964.
32. Stotzky, G. and Mortensen, J. L., The effect of crop residue and nitrogen additions on the decomposition of an Ohio muck soil, *Soil Sci.*, 83, 165, 1957.
33. Zajic, J. E., *Microbial Biogeochemistry*, Academic Press, New York, 1965.
34. Erlich, H. L., Biogeochemistry of the minor elements in soil, in *Soil Biochemistry*, Vol. 2, McLaren, A. D. and Skujins, J., Eds., Marcel Dekker, New York, 1971, Chap. 12.
35. Keeler, R. F. and Varner, J. E., The metabolism of molybdate and tungstate in *Azotobacter*, in *Trace Elements*, Lamb, C. A. and Beattie, J. M., Eds., Academic Press, New York, 1958, 293.
36. Weinberg, E. D., Trace metal control of specific biosynthetic processes, *Perspect. Biol. Med.*, 5, 432, 1962.
37. Weinberg, E. D., Roles of metallic ions in host-parasite interactions, *Bacteriol. Rev.*, 30, 136, 1966.
38. Guirard, B. M. and Snell, E. E., Nutritional requirements of microorganisms, in *The Bacteria*, Vol. 4, Gunsalus, I. C. and Stanier, R. Y., Eds., Academic Press, New York, 1962, Chap. 2.
39. West, P. M. and Lochhead, A. G., The rhizosphere in relation to the nutrient requirement of soil bacteria, *Can. J. Res.*, 18, 129, 1940.
40. Lochhead, A. G. and Chase, F. E., Qualitative studies of soil microorganisms. V. Nutritional requirements of the predominant bacterial flora, *Soil Sci.*, 55, 185, 1943.
41. Wallace, R. H. and Lochhead, A. G., Amino acid requirements of rhizosphere bacteria, *Can. J. Res.*, 28, 1, 1950.

42. Lochhead, A. G. and Thexton, R. H., Bacteria requiring B$_{12}$ as a growth factor, *J. Bacteriol.*, 63, 219, 1952.
43. Lochhead, A. G. and Burton, M. O., Qualitative studies on soil microorganisms. XIV. Specific vitamin requirements of the predominant bacterial flora, *Can. J. Microbiol.*, 3, 35, 1957.
44. Lochhead, A. G., Qualitative studies of soil microorganisms. XV. Capability of the predominant bacterial flora for synthesis of various growth factors, *Soil Sci.*, 84, 395, 1957.
45. Lochhead, A. G. and Burton, M. O., An essential bacterial growth factor produced by microbial synthesis, *Can. J. Bot.*, 31, 7, 1953.
46. Koser, S., *Vitamin Requirements of Yeasts and Bacteria*, Charles C Thomas, Springfield, Ill., 1968.
47. Nikitin, D. I., Vasil'eva, L. V., and Lokhmacheva, R. A., *New and Rare Forms of Soil Microorganisms*, Nauka Publ., Moscow, 1966 (in Russian).
48. Moyed, H. S. and Umbarger, H. E., Inhibition in biosynthesis of amino acids, *Physiol. Rev.*, 42, 444, 1962.
49. Brown, A. D., Aspects of bacterial response to the ionic environment, *Bacteriol. Rev.*, 28, 296, 1964.
50. Gibbons, N. E. and Payne, J. I., Relation of temperature and sodium chloride concentration to growth and morphology of some halophilic bacteria, *Can. J. Microbiol.*, 7, 483, 1961.
51. Neidhart, F. C., Effects of the environment on the composition of bacterial cells, *Annu. Rev. Microbiol.*, 17, 61, 1963.
52. Markovitz, A. and Sylvan, S., Effect of sodium sulfate and magnesium sulfate on heteropolysaccharide synthesis in gram-negative soil bacteria, *J. Bacteriol.*, 83, 483, 1962.
53. Casida, L. E., Jr., Microorganisms in unamended soil as observed by various forms of microscopy and staining, *Appl. Microbiol.*, 21, 1040, 1971.
54. Nemec, P. and Bystricky, V., Peculiar morphology of some microorganisms accompanying Diatomaceae. Preliminary report, *J. Gen. Appl. Microbiol.*, 8, 121, 1962.
55. Orenski, S. W., Bystricky, V., and Maramorosch, K., The occurrence of microbial forms of unusual morphology in European and Asian soils, *Can. J. Microbiol.*, 12, 1291, 1966.
56. Aristovskaya, T. V., *Microorganisms in Soil*, Publishing House of Moscow University, Moscow, 1965 (in Russian).
57. Meyer E. A. and Moskowitz, M., The effect of calcium binding agents on the virulence of *Clostridium perfringens* for the white mouse, *J. Bacteriol.*, 69, 111, 1955.
58. Surgalla, M. J., Properties of virulent and avirulent strains of *P. pestis*, *Ann. N.Y. Acad. Sci.*, 88, 1136, 1960.
59. Jannof, A. and Zweifach, B. J., Inactivation of bacterial exotoxins and endotoxins by iron, *J. Exp. Med.*, 112, 23, 1960.
60. Shankar, K. and Bard, R. C., The effect of metallic ions on the growth and morphology of *Clostridium perfringens*, *J. Bacteriol.*, 63, 279, 1952.
61. Webb, M., Effects of magnesium on cellular division in bacteria, *Science*, 118, 607, 1953.
62. Asbell, M. A. and Eagon, R. G., Role of multivalent cations in the organization, structure, and assembly of the cell wall of *Pseudomonas aeruginosa*, *J. Bacteriol.*, 92, 380, 1966.
63. Humphrey, B. and Vincent, J. M., Calcium in cell walls of *Rhizobium trifolii*, *J. Gen. Microbiol.*, 29, 557, 1962.
64. Lilly, V. G., Barnett, H. L., and Krause, R. F., Effects of the alkali metal chlorides on spore germination, growth, and carotenogenesis of *Phycomyces blakesleeanus*, *Mycologia*, 54, 235, 1962.
65. Halmann, M. and Keynan, A., Stages in germination of spores of *Bacillus licheniformis*, *J. Bacteriol.*, 84, 1187, 1960.
66. Rees, M. K., Effect of chloride on oxidation of hydroxylamine by *Nitrosomonas europaea* cells, *J. Bacteriol.*, 95, 243, 1968.
67. Puck, T. T., Goren, A., and Cline, J., The mechanism of virus attachment to host cells: the role of ions in the primary reaction, *J. Exp. Med.*, 93, 65, 1951.
68. Luria, S. E. and Steiner, D. L., The role of calcium in the penetration of bacteriophage T5 into its host, *J. Bacteriol.*, 67, 635, 1954.
69. Cummings, D. J. and Kozloff, L. M., Biophysical properties of bacteriophage T2, *Biochim. Biophys. Acta*, 44, 436, 1960.
70. Kay, D., The effect of divalent metals on the multiplication of coli bacteriophage T5, *Br. J. Exp. Pathol.*, 33, 228, 1952.
71. Rose, A. H., *Chemical Microbiology*, Plenum Press, New York, 1965.
72. McLaren, A. D. and Skujins, J., The physical environment of microorganisms in soil, in *The Ecology of Soil Bacteria*, Gray, T. R. G. and Parkinson, D., Eds., Liverpool University Press, Liverpool, 1968, 3.
73. Cameron, R. E., Properties of desert soils, in *Biology and the Exploration of Mars*, Pittendrigh, C. S., Vishniac, W., and Pearman, J. P. T., Eds., Natl. Acad. Sci., Natl. Res. Council, Washington, D. C., 1966, Chap. 8.
74. Bouyoucos, G. J. and Cook, R. L., Measuring the relative humidity of soil at different moisture contents by the Gray Hydrocal hygrometer, *Soil Sci.*, 104, 297, 1967.
75. Low, P. F., Physical chemistry of clay-water interaction, *Adv. Agron.*, 13, 269, 1961.
76. Low, P. F., Effect of quasi-crystalline water on rate processes involved in plant nutrition, *Soil Sci.*, 93, 6, 1962.
77. Grim, R. E., *Clay Mineralogy*, McGraw-Hill, New York, 1953.
78. Quastel, J. H. and Scholefield, P. G., Biochemistry of nitrification in soil, *Bacteriol. Rev.*, 15, 1, 1951.
79. Zagallo, A. C. and Katznelson, H., Metabolic activity of bacterial isolates from wheat rhizosphere and control soil, *J. Bacteriol.*, 73, 760, 1957.

80. Birch, H. F., The effect of soil drying on humus decomposition and nitrogen availability, *Plant Soil,* 10, 9, 1958.
81. Gooding, T. H. and McCalla, T. M., Loss of carbon dioxide and ammonia from crop residues during decomposition, *Soil Sci. Soc. Am. Proc.,* 10, 185, 1946.
82. Mortensen, J. L. and Martin, W. P., Decomposition of the soil conditioning polyelectrolytes, HPAN and VAMA, in Ohio soils, *Soil Sci. Soc. Am. Proc.,* 18, 395, 1954.
83. Soulides, D. A. and Allison, F. E., Effect of drying and freezing soils on carbon dioxide production, available mineral nutrients, aggregation, and bacterial population, *Soil Sci.,* 91, 291, 1961.
84. Pauling, L., A molecular theory of general anesthesia, *Science,* 134, 15, 1961.
85. Davey, C. B. and Miller, R. J., Correlation of temperature dependent water properties and growth of bacteria, *Nature,* 209, 638, 1966.
86. Davey, C. B., Miller, R. J., and Nelson, L. A., Temperature-dependent anomalies on the growth of microorganisms, *J. Bacteriol.,* 91, 1827, 1966.
87. Miller, R. J. and Davey, C. B., The apparent effect of water structures on K uptake by plants, *Soil Sci. Soc. Am. Proc.,* 31, 286, 1967.
88. Drost-Hansen, W., Temperature anomalies and biological temperature optima in the process of evolution, *Naturwissenschaften,* 43, 512, 1956.
89. Drost-Hansen, W., The effects on biological systems of higher-order phase transitions in water, *Ann. N.Y. Acad. Sci.,* 125, 471, 1965.
90. Oppenheimer, C. H. and Drost-Hansen, W., A relationship between multiple temperature optima for biological systems and the properties of water, *J. Bacteriol.,* 80, 21, 1960.
91. Falk, M. and Kell, G. S., Thermal properties of water: discontinuities questioned, *Science,* 154, 1013, 1966.
92. Némethy, G., Structure of water and of aqueous solutions, *Cryobiology,* 3, 19, 1966.
93. Farrel, J. and Rose, A., Temperature effects on microorganisms, *Annu. Rev. Microbiol.,* 21, 101, 1967.
94. Myers, G. H., Morrow, M. B., Wyss, O., and Littlepage, J. L., Antarctica: The microbiology of an unfrozen saline pond, *Science,* 138, 1103, 1962.
95. Salisbury, F. B., Martian biology, *Science,* 136, 17, 1962.
96. Hawrylewicz, E., Gowdy, B., and Ehrlich, R., Micro-organisms under a simulated Martian environment, *Nature,* 193, 497, 1962.
97. Young, R. S., Deal, P. H., and Whitfield, O., Response of soil bacteria to high temperature and diurnal freezing and thawing, *Nature,* 216, 355, 1967.
98. Curtis, C. R., Response of fungi to diurnal temperature extremes, *Nature,* 213, 738, 1967.
99. Becquerel, P., La suspension de la vie des spores des Bactéries et des Moisissures desséchées dans le vide, vers le zéro absolu. Ses conséquences pour la dissémination et la conservation de la vie dans l'Univers, *Compt. Rend.,* 231, 1392, 1950.
100. Goos, R. D., Davis, E. E., and Butterfield, W., Effect of warming rates on the viability of frozen fungus spores, *Mycologia,* 59, 58, 1967.
101. Hotchin, J., Lorenz, P., and Hemenway, C., Survival of microorganisms in space, *Nature,* 206, 442, 1965.
102. Horowitz, N. H., Sharp, R. P., and Davies, R. W., Planetary contamination. I. The problem and the agreements, *Science,* 155, 1501, 1967.
103. Sherman, J. M. and Cameron, G. M., Lethal environmental factors within the natural range of growth, *J. Bacteriol.,* 27, 341, 1934.
104. Lamanna, C. and Mallette, M. F., *Basic Bacteriology,* Williams & Wilkins, Baltimore, 1965.
105. O'Donovan, G. A., Kearney, C. L., and Ingraham, J. L., Mutants of *Escherichia coli* with high minimal temperatures of growth, *J. Bacteriol.,* 90, 611, 1965.
106. Hastings, J. W. and Sweeney, B. M., On the mechanism of temperature independence in a biological clock, *Proc. Natl. Acad. Sci. USA,* 43, 804, 1957.
107. Bruce, V. G. and Pittendrigh, C. S., Temperature independence in a unicellular "clock," *Proc. Natl. Acad. Sci. USA,* 42, 676, 1956.
108. Stotzky, G. and Rem, L. T., Influence of clay minerals on microorganisms: I. Montmorillonite and kaolinite on bacteria, *Can. J. Microbiol.,* 12, 547, 1966.
109. Stotzky, G., Influence of clay minerals on microorganisms: II. Effect of various clay species, homoionic clays, and other particles on bacteria, *Can. J. Microbiol.,* 12, 831, 1966.
110. Carlucci, A. F. and Pramer, D., Factors affecting the survival of bacteria in sea water, *Appl. Microbiol.,* 7, 388, 1959.
111. Flannery, W. L., Current status of knowledge of halophilic bacteria, *Bacteriol. Rev.,* 20, 49, 1956.
112. Henis, Y. and Eren, J., Preliminary studies on the microflora of a highly saline soil, *Can. J. Microbiol.,* 9, 902, 1963.
113. Taber, W. A., Evidence for the existence of acid-sensitive actinomycetes in soil, *Can. J. Microbiol.,* 6, 503, 1960.
114. Johnson, T. W. and Sparrow, F. K., *Fungi in Oceans and Estuaries,* Carmer, Weinheim, 1961.
115. McLeod, R. A., The question of the existence of specific marine bacteria, *Bacteriol. Rev.,* 29, 9, 1965.
116. Portner, D. M., Spiner, D. R., Hoffman, R. K., and Phillips, C. R., Effects of ultrahigh vacuum on viability of microorganisms, *Science,* 134, 2047, 1961.
117. Sussman, A. S. and Halvorson, H. A., *Spores, Their Dormancy and Germination,* Harper & Row, New York, 1966.

118. **Siegel, S. M., Renwick, G., Daly, O., Giumarro, C., Davis, G., and Halpern, L.,** The survival capabilities and the performance of earth organisms in simulated extraterrestrial environments, in *Current Aspects of Exobiology,* Mamikunian, G. and Briggs, M. H., Eds., Pergamon, New York, 1965, Chap. 4.

119. **Zobell, C. E. and Morita, R. Y.,** Barophilic bacteria in some deep sea sediments, *J. Bacteriol.,* 73, 563, 1957.

120. **Kriss, A. E.,** *Marine Microbiology, Deep Sea,* Oliver & Boyd, Edinburgh, 1963.

121. **Heden, C.-G.,** Effects of hydrostatic pressure on microbial systems, *Bacteriol. Rev.,* 28, 14, 1964.

122. **Zobell, C. E. and Johnson, F. H.,** The influence of hydrostatic pressure on the growth and viability of terrestrial and marine bacteria, *J. Bacteriol.,* 57, 179, 1949.

123. **Rovira, A. D.,** Interactions between plant roots and soil microorganisms, *Annu. Rev. Microbiol.,* 19, 241, 1965.

124. **Stotzky, G.,** A simple method for the determination of the respiratory quotient of soils, *Can. J. Microbiol.,* 6, 439, 1960.

125. **Greenwood, D. J.,** The effect of oxygen concentration on the decomposition of organic materials in soil, *Plant Soil,* 14, 360, 1961.

126. **Greenwood, D. J.,** Nitrification and nitrate dissimilation in soil. II. Effect of oxygen concentration, *Plant Soil,* 17, 365, 1962.

127. **Stotzky, G. and Goos, R.,** Effect of high CO_2 and low O_2 tensions on the soil microbiota, *Can. J. Microbiol.,* 11, 853, 1965.

128. **Stotzky, G. and Goos, R. D.,** Adaptation of the soil microflora to low O_2 and high CO_2 tensions, *Can. J. Microbiol.,* 12, 849, 1966.

129. **Sherwood, R. T. and Hagedorn, D. J.,** Effect of oxygen tension on growth of *Aphanomyces euteiches, Phytopathology,* 51, 492, 1961.

130. **Gundersen, K.,** Growth of *Fomes annosus* under reduced oxygen pressure and the effect of carbon dioxide, *Nature,* 190, 649, 1961.

131. **Stover, R. H. and Freiberg, S. R.,** Effects of carbon dioxide on multiplication of *Fusarium* in soil, *Nature,* 181, 788, 1958.

132. **Klotz, L. J., Stolzy, L. H., and DeWolfe, T. A.,** Oxygen requirement of three root-rotting fungi in a liquid medium, *Phytopathology,* 53, 303, 1963.

133. **Durbin, R. D.,** Straight-line functions of growth of microorganisms at toxic levels of carbon dioxide, *Science,* 121, 734, 1955.

134. **Durbin, R. D.,** Factors affecting the vertical distribution of *Rhizoctonia solani* with special reference to carbon dioxide concentrations, *Am. J. Bot.,* 46, 22, 1959.

135. **Bartnicki-Garcia, S.,** Mold-yeast dimorphism of *Mucor, Bacteriol. Rev.,* 27, 293, 1963.

136. **Cantino, E. C. and Lovett, J. S.,** Non-filamentous aquatic fungi: Model systems for biochemical studies of morphological differentiation, *Adv. Morphog.,* 3, 33, 1964.

137. **Duguid, J. P. and Wilkinson, J. F.,** Environmentally induced changes in bacterial morphology, in *Microbial Reaction to Environment,* Meynell, G. G. and Gooder, H., Eds., Cambridge University Press, Cambridge, 1961, 69.

138. **Grelet, N.,** Growth limitation and sporulation, *J. Appl. Bacteriol.,* 20, 315, 1957.

139. **Knaipi, G.,** A study of environmental factors which control endospore formation by a strain of *Bacillus mycoides, J. Bacteriol.,* 49, 473, 1945.

140. **Woodruff, H. B.,** Antibiotic production as an expression of environment, in *Microbial Reaction to Environment,* Meynell, G. G. and Gooder, H., Eds., Cambridge University Press, Cambridge, 1961, 317.

141. **Averner, M. and Fulton, C.,** Carbon dioxide: Signal for excystment of *Naegleria gruberi, J. Gen. Microbiol.,* 42, 245, 1966.

142. **Siegel, S. M. and Giumarro, C.,** Survival and growth of terrestrial microorganisms in ammonia-rich atmospheres, *Icarus,* 4, 37, 1965.

143. **Siegel, S. M. and Giumarro, C.,** On the culture of a microorganism similar to the Precambrian microfossil *Kakabekia umbellata* Barghoorn in ammonia-rich atmospheres, *Proc. Natl. Acad. Sci. USA,* 55, 349, 1966.

144. **Siegel, S. M., Roberts, K., Nathan, H., and Daly, O.,** Living relative of the microfossil *Kakabekia, Science,* 156, 1231, 1967.

145. **Moore-Landecker, E. and Stotzky, G.,** Inhibition of fungal growth and sporulation by volatile metabolites from bacteria, *Can. J. Microbiol.,* 18, 957, 1972.

146. **Moore-Landecker, E. and Stotzky, G.,** Morphological abnormalities of fungi induced by volatile microbial metabolites, *Mycologia* (in press).

147. **Vancura, V. and Stotzky, G.,** Excretions of germinating plant seeds, *Folia Microbiol.,* 16, 512, 1971.

148. **Babich, H. and Stotzky, G.,** Ecologic ramifications of air pollution, *Proc. Internatl. Conf. on Transportation and the Environment,* Soc. Automotive Eng., New York, 1972, 198.

149. **Hibben, C. R. and Stotzky, G.,** Effects of ozone on the germination of fungal spores, *Can. J. Microbiol.,* 15, 1187, 1969.

150. **Abeles, F. B., Craker, L. E., Forrence, L. E., and Leather, G. R.,** Fate of air pollutants: Removal of ethylene, sulfur dioxide, and nitrogen dioxide by soil, *Science,* 173, 914, 1971.

151. **Inman, R. E., Ingersoll, R. B., and Levy, E. A.,** Soil: A natural sink for carbon monoxide, *Science,* 172, 1229, 1971.

152. **Inman, R. E. and Ingersoll, R. B.,** Uptake of carbon monoxide by soil fungi, *J. Air Pollut. Control Assoc.,* 21, 646, 1971.
153. **Rasmussen, R. A.,** Isoprene: Identified as a forest-type emission to the atmosphere, *Environ. Sci. Technol.,* 4, 4667, 1970.
154. **Volesky, B. and Zajic, J. E.,** Ethane and natural gas oxidation by fungi, *Dev. Ind. Microbiol.,* 11, 184, 1970.
155. **Taber, W. A.,** Identification of an alkaline-dependent streptomyces as *Streptomyces caeruleus* Baldacci and characterization of the species under controlled conditions, *Can. J. Microbiol.,* 5, 335, 1959.
156. **Taber, W. A.,** Evidence for the existence of acid-sensitive actinomycetes in soil, *Can. J. Microbiol.,* 6, 503, 1960.
157. **Corke, C. T. and Chase, F. E.,** Comparative studies of actinomycete populations in acid podzolic and neutral mull forest soils, *Soil Sci. Soc. Am. Proc.,* 28, 68, 1964.
158. **Holding, A. J. and Jeffrey, D. C.,** Effects of metallic ions on soil bacteria, in *The Ecology of Soil Bacteria,* Gray, T. R. G. and Parkinson, D., Eds., Liverpool University Press, Liverpool, 1968, 516.
159. **Hartley, G. S. and Roe, J. W.,** Ionic concentrations at surfaces, *Trans. Faraday Soc.,* 36, 101, 1940.
160. **Mortland, M. M.,** Urea complexes with montmorillonite: An infrared absorption study, *Clay Minerals,* 6, 143, 1966.
161. **Harter, R. D. and Alrichs, J. L.,** Determination of clay surface acidity by infrared spectroscopy, *Soil Sci. Soc. Am. Proc.,* 31, 30, 1967.
162. **Mortland, M. M.,** Protonation of compounds at clay mineral surfaces, *Trans. 9th Int. Congr. Soil Sci.,* Adelaide, 1, 691, 1968.
163. **Mortland, M. M. and Raman, K. V.,** Surface acidity of smectites in relation to hydration, exchangeable cation and structure, *Clays Clay Min.,* 16, 393, 1968.
164. **Swoboda, A. R. and Kunze, G. W.,** Reactivity of montmorillonite surfaces with weak organic bases, *Soil Sci. Soc. Am. Proc.,* 32, 806, 1968.
165. **McLaren, A. D. and Peterson, G. H.,** Physical chemistry and biological chemistry of clay mineral-organic nitrogen complexes, in *Soil Nitrogen,* Bartholomew, W. V. and Clark, F. E., Eds., Am. Soc. Agron., Madison, 1965, Chap. 10.
166. **McLaren, A. D.,** Enzyme action in structurally restricted systems, *Enzymologia,* 21, 356, 1960.
167. **Tahoun, S. A. and Mortland, M. M.,** Complexes of montmorillonite with primary, secondary, and tertiary amides. I. Protonation of amides on the surface of montmorillonite, *Soil Sci.,* 102, 248, 1966.
168. **McLaren, A. D. and Skujins, J.,** The physical environment of microorganisms in soil, in *The Ecology of Soil Bacteria,* Gray, T. R. G. and Parkinson, D., Eds., Liverpool University Press, Liverpool, 1968, 3.
169. **McLaren, A. D. and Skujins, J.,** Nitrification by *Nitrobacter agilis* on surfaces and in soil with respect to hydrogen ion concentration, *Can. J. Microbiol.,* 9, 729, 1963.
170. **Hattori, T. and Furusaka, C.,** Chemical activities of *E. coli* adsorbed on a resin, *J. Biochem.,* 48, 831, 1960.
171. **Hattori, T. and Furusaka, C.,** Chemical activities of *Azotobacter agile* adsorbed on a resin, *J. Biochem.,* 50, 312, 1961.
172. **McLaren, A. D. and Estermann, E. F.,** Influence of pH on the activity of chymotrypsin at a solid-liquid interface, *Arch. Biochem. Biophys.,* 68, 157, 1957.
173. **Weber, D. F. and Gainey, P. L.,** Relative sensitivity of nitrifying organisms to hydrogen ions in soils and in solutions, *Soil Sci.,* 94, 138, 1962.
174. **Hewitt, L. F.,** *Oxidation-Reduction Potentials in Bacteriology and Biochemistry,* 6th ed., E. & S. Livingston, Ltd., Edinburgh, 1950.
175. **Dolin, M.,** Survey of microbial electron transport mechanisms, in *The Bacteria,* Vol. II, Gunsalus, I. C. and Stanier, R. Y., Eds., Academic Press, New York, 1961, 6.
176. **Takai, Y. and Kamura, T.,** The mechanism of reduction in waterlogged paddy soil, *Folia Microbiol.,* 11, 304, 1966.
177. **Rodrigo, D. M.,** Redox potential — with special reference to rice culture, *Trop. Agric.* (Ceylon), 199, 85, 1963.
178. **Patrick, W. H., Jr. and Mahapatra, I. C.,** Transformation and availability to rice of nitrogen and phosphorous in waterlogged soils, *Adv. Agron.,* 20, 323, 1968.
179. **Clark, W. M.,** *Oxidation-Reduction Potentials of Organic Systems,* Williams & Wilkins, Baltimore, 1960.
180. **Barrow, E. S. G.,** Bacterial oxidations in *Bacterial Physiology,* Werkman, C. H. and Wilson, P. W., Eds., Academic Press, New York, 1951, 325.
181. **Stumm, W.,** Redox potentials as an environmental parameter; conceptual significance and operational limitation, *Trans. 3rd Internatl. Conf. Water Pollut. Res.,* Munich, 1966, Chap. 13.
182. **Zobell, C. E.,** Studies on redox potentials of marine sediments, *Bull. Am. Assoc. Petrol. Geol.,* 30, 477, 1946.
183. **Ponnamperuma, F. N.,** Dynamic aspects of flooded soils and the nutrition of the rice plant, in *The Mineral Nutrition of the Rice Plant,* Johns Hopkins Press, Baltimore, 1965.
184. **Patrick, W. H., Jr.,** Apparatus for controlling oxidation-reduction potentials of waterlogged soils, *Nature,* 212, 1278, 1966.
185. **Aomine, S.,** A review of research on redox potentials of paddy soils in Japan, *Soil Sci.,* 94, 6, 1961.
186. **Ehrlich, H.,** Microbial transformations of minerals, in *Principles and Applications in Aquatic Microbiology,* Heukelekian, H. and Dondero, N. C., Eds., John Wiley & Sons, New York, 1964, 43.
187. **Patrick, W. H., Jr.,** Effects of redox potentials on manganese transformation in waterlogged soils, *Nature,* 220, 5166, 1968.
188. **Connell, W. E. and Patrick, W. H., Jr.,** Sulfate reduction in soil: effects of redox potential and pH, *Science,* 159, 86, 1968.

189. Baas-Becking, L. G. M., Kaplan, I. R., and Moore, D., Limits of the natural environment in terms of pH and oxidation-reduction potentials, *J. Geol.*, 68, 243, 1960.
190. Postgate, J. R., Sulphate reduction by bacteria, *Annu. Rev. Microbiol.*, 13, 505, 1959.
191. Turner, F. T. and Patrick, W. H., Jr., Chemical changes in waterlogged soils as a result of oxygen depletion, *Trans. 9th Internatl. Congr. Soil Sci.*, Adelaide, 4, 53, 1968.
192. Patrick, W. H., Jr. and Wyatt, R., Soil nitrogen loss as a result of alternate submergence and drying, *Soil Sci. Soc. Am. Proc.*, 28, 647, 1964.
193. Patrick, W. H., Jr., Extractable iron and phosphorous in a submerged soil at controlled redox potentials, *Trans 8th Internatl. Congr. Soil Sci.*, Bucharest, 4, 605, 1964.
194. Vallentyne, J. R., Environmental biophysics and microbial ubiquity, *Annu. N.Y. Acad. Sci.*, 108, 342, 1963.
195. Kononova, M. M., *Soil Organic Matter: Its Nature, Its Role in Soil Formation and in Soil Fertility*, Pergamon Press, New York, 1966.
196. van Olphen, H., *An Introduction to Clay Colloid Chemistry*, John Wiley & Sons, New York, 1963.
197. Marshall, C. E., *The Physical Chemistry and Mineralogy of Soils, Soil Materials*, Vol. I, John Wiley & Sons, New York, 1964.
198. Estermann, E. F. and McLaren, A. D., Stimulation of bacterial proteolysis by adsorbents, *J. Soil Sci.*, 10, 64, 1959.
199. Allison, F. E., Soil aggregation – some facts and fallacies as seen by a microbiologist, *Soil Sci.*, 106, 136, 1968.
200. Chesters, G., Attoe, O. J., and Allen, O. N., Soil aggregation in relation to various soil constituents, *Soil Sci. Soc. Am. Proc.*, 21, 272, 1957.
201. Martin, J. P., Martin, W. P., Page, J. B., Raney, W. A., and DeMent, J. D., Soil aggregation, *Adv. Agron.*, 7, 1, 1955.
202. McCalla, T. M., Physico-chemical behavior of soil bacteria in relation to the soil colloid, *J. Bacteriol.*, 40, 33, 1940.
203. Ellis, J. H., Barnhisel, R. I., and Phillips, R. E., The diffusion of copper, manganese, and zinc as affected by concentration, clay mineralogy, and associated anions, *Soil Sci. Soc. Am. Proc.*, 34, 866, 1970.
204. Pramer, D., The persistence and biological effects of antibiotics in soil, *Appl. Microbiol.*, 6, 221, 1958.
205. Soulides, D. A., Antibiotic tolerance of the soil microflora in relation to type of clay minerals, *Soil Sci.*, 107, 105, 1969.
206. Bailey, G. W. and White, J. L., Review of adsorption and desorption of organic pesticides by soil colloids, with implications concerning pesticide bioactivity, *J. Agric. Food Chem.*, 12, 324, 1964.
207. Bailey, G. W. and White, J. L., Herbicides: A comparison of their physical, chemical, and biological properties, *Residue Rev.*, 10, 97, 1965.
208. Marshall, K. C., Survival of root-nodule bacteria in dry soils exposed to high temperatures, *Austral. J. Agr. Res.*, 15, 273, 1964.
209. Müller, H. P. and Schmidt, L., Kontinuierliche Atmungemessungen an *Azotobacter chroococcum* Beij. in Montmorillonit unter chronischer Röntgenbestrahlung, *Arch. Mikrobiol.*, 54, 70, 1966.
210. Miyamoto, Y., Further evidence for the longevity of soil-borne plant viruses adsorbed on soil particles, *Virology*, 9, 920, 1959.
211. Harrison, B. D., The biology of soil-borne plant viruses, *Adv. Virus Res.*, 7, 131, 1960.
212. Hendrix, J. W., Soil transmission of tobacco ringspot virus, *Phytopathology*, 51, 194, 1961.
213. Murphy, W. H., Jr., Eylar, O. R., Schmidt, E. L., and Syverton, J. T., Adsorption and translocation of mammalian viruses by plants. I. Survival of mouse encephalomyelitis and poliomyelitis viruses in soil and plant root environment, *Virology*, 6, 612, 1958.
214. Fildes, P. and Kay, D., The conditions which govern the adsorption of tryptophan-dependent bacteriophage to kaolin and bacteria, *J. Gen. Microbiol.*, 30, 183, 1963.
215. Robinson, J. and Corke, C., Preliminary studies on the distribution of actinophages in soil, *Can. J. Microbiol.*, 5, 479, 1959.
216. Rudolfs, W., Falk, L. L., and Rogotzkie, R. A., Literature review on the occurrence and survival of enteric, pathogenic, and related organisms in soil, water, sewage and sludges and on vegetation, *Sewage Ind. Wastes*, 22, 1261, 1950.
217. Glathe, H., Knoll, K. H., and Makawi, A. A. M., Die Lebensfähigkeit von *Escherichia coli* in verschiedenen Bodenarten, *Z. Pflanz. Düng. Bodenk.*, 100, 142, 1963.
218. Glathe, H., Knoll, K. H., and Makawi, A. A. M., Die Verhalten von *Salmonellen* in verschiedenen Bodenarten, *Z. Pflanz. Düng. Bodenk.*, 100, 224, 1963.
219. Welshimer, H. J., Survival of *Listeria monocytogenes* in soil, *J. Bacteriol.*, 80, 316, 1960.
220. Wentz, M. W., Scott, R. A., and Vennes, J. W., *Clostridium botulinum* type F: Seasonal inhibition by *Bacillus licheniformis*, *Science*, 155, 89, 1967.
221. van Dijk, H., Colloid chemical properties of humic matter, in *Soil Biochemistry*, Vol. 2, McLaren, A. D. and Skujins, J., Eds., Marcel Dekker, Inc., New York, 1971, Chap. 2.
222. Bail, O., Ergebnisse experimenteller Populations forschung, *Z. Immun. Forsch.*, 60, 1, 1929.
223. Tyrell, E., MacDonald, R., and Gerhardt, P., Biphasic system for growing bacteria in concentrated culture, *J. Bacteriol.*, 75, 1, 1958.
224. Contois, D. E., Kinetics of bacterial growth: Relationship between population density and specific growth rate of continuous cultures, *J. Gen. Microbiol.*, 21, 40, 1959.

225. **Sinclair, N. A. and Stokes, J. Z.**, Factors which control maximal growth of bacteria, *J. Bacteriol.*, 83, 1147, 1962.
226. **Freter, R. and Ozawa, A.**, Explanation for limitation of populations of *Escherichia coli* in broth culture, *J. Bacteriol.*, 86, 904, 1963.
227. **Gallup, D. M. and Gerhardt, P.**, Dialysis fermentor systems for concentrated culture of microorganisms, *Appl. Microbiol.*, 11, 506, 1963.
228. **Ecker, R. E. and Schaechter, M.**, Bacterial growth under conditions of limited nutrition, *Annu. N.Y. Acad. Sci.*, 102, 549, 1963.
229. **Schultz, J. S. and Gerhardt, P.**, Dialysis culture of microorganisms: Design, theory and results, *Bacteriol. Rev.*, 33, 1, 1969.
230. **Martin, J. P., Ervin, J. O., and Sheperd, R. A.**, Decomposition and aggregating effect of fungus cell material in soil, *Soil Sci. Soc. Am. Proc.*, 23, 217, 1959.
231. **Martin, J. P. and Richards, S. J.**, Decomposition and binding action of a polysaccharide from *Chromobacterium violaceum* in soil, *J. Bacteriol.*, 85, 7288, 1963.
232. **Martin, J. P., Ervin, J. O., and Sheperd, R. A.**, Decomposition of iron, aluminum, zinc, and copper salts or complexes of some microbial and plant polysaccharides in soil, *Soil Sci. Soc. Am. Proc.*, 30, 196, 1966.
233. **McLaren, A. D. and Skujins, J.**, The physical environment of microoorganisms in soil, in *The Ecology of Soil Bacteria*, Gray, T. R. G. and Parkinson, D., Eds., Liverpool University Press, Liverpool, 1968, 3.
234. **Griffin, D. M.**, Observations on fungi growing in a translucent particulate matrix, *Trans. Br. Mycol. Soc.*, 51, 319, 1968.
235. **Park, D.**, Antagonism – The background to soil fungi, in *The Ecology of Soil Fungi*, Parkinson, D. and Waid, J. S., Eds., Liverpool University Press, Liverpool, 1960, 148.
236. **Rovira, A. D. and Greacen, E. L.**, The effect of aggregate disruption on the activity of soil microorganisms in the soil, *Austral. J. Agric. Res.*, 8, 659, 1957.
237. **Greenwood, D. J.**, Measurement of microbial metabolism in soil, in *The Ecology of Soil Bacteria*, Gray, T. R. G. and Parkinson, D., Eds., Liverpool University Press, Liverpool, 1968, 138.
238. **McLaren, A. D. and Skujins, J.**, Trends in the biochemistry of terrestrial soils, in *Soil Biochemistry*, Vol. 2, McLaren, A. D. and Skujins, J., Eds., Marcel Dekker, New York, 1971, Chap. 1.
239. **McLaren, A. D. and Ardakani, M. S.**, A vector biochemical approach to consecutive reactions in soil, in *Modern Methods in the Study of Microbial Ecology*, Uppsala (in press).
240. **Brock, T. D.**, Microbial growth rates in nature, *Bacteriol. Rev.*, 35, 39, 1971.
241. **Gray, T. R. G. and Williams, S. T.**, Microbial productivity of soil, in *Microbes and Biological Activity*, Hughes, D. E. and Rose, A. H., Eds., Cambridge University Press, Cambridge, 1971, 225.
242. **Morrill, L. G. and Dawson, J. E.**, Growth rates of nitrifying chemoautotrophs in soil, *J. Bacteriol.*, 83, 205, 1962.
243. **Babiuk, L. A. and Paul, E. A.**, The use of fluorescein isothiocyanate in the determination of the bacterial biomass of grassland soil, *Can. J. Microbiol.*, 16, 57, 1970.
244. **Murrell, W. G.**, Spore formation and germination as a microbial reaction to the environment, in *Microbial Reaction to Environment*, Meynell, G. G. and Gooder, H., Eds., Cambridge University Press, Cambridge, 1961, 100.
245. **Stotzky, G., Goos, R. D., and Timonin, M. I.**, Microbial changes occurring in soil as a result of storage, *Plant Soil*, 16, 1, 1962.
246. **Sussman, A. S.**, Dormancy of soil microorganisms in relation to survival, in *Ecology of Soil-Borne Plant Pathogens*, Baker, K. I. and Snyder, W. C., Eds., University Calif. Press, Berkeley, 1965, 99.
247. **Odum, E. P.**, *Fundamentals of Ecology*, 2nd ed., Saunders, Philadelphia, 1959.
248. **Brady, R. J., Can, E. C. S., and Pelczar, M. J., Jr.**, Sporulation of *Bacillus sphaericus* grown in association with *Erwinia atroseptica*, *J. Bacteriol.*, 81, 725, 1961.
249. **Sonneborn, T. M.**, The metagon: RNA and cytoplasmic inheritance, *Am. Naturalist*, 99, 279, 1965.
250. **Marshall, K. C. and Alexander, M.**, Competition between soil bacteria and *Fusarium*, *Plant Soil*, 12, 143, 1960.
251. **Finstein, S. and Alexander, M.**, Competition for carbon and nitrogen between *Fusarium* and bacteria, *Soil Sci.*, 94, 334, 1962.
252. **Stotzky, G.**, Replica plating technique for studying microbial interactions in soil, *Can. J. Microbiol.*, 11, 629, 1965.
253. **Stotzky, G. and Post, A. H.**, Growth rates of microorganisms in soil, *Agron. Abstr.*, 89, 1965.
254. **Wright, J. M.**, The production of antibiotics in soil. III. Production of gliotoxin in wheat straw buried in soil, *Ann. Appl. Biol.*, 44, 461, 1956.
255. **Stallings, J. H.**, Soil produced antibiotics – plant disease and insect control, *Bacteriol. Rev.*, 18, 131, 1954.
256. **Stevenson, I. L.**, Antibiotic activity of actinomycetes in soil as demonstrated by direct observation techniques, *J. Gen. Microbiol.*, 15, 372, 1956.
257. **Pramer, D. and Starkey, R. L.**, Decomposition of streptomycin, *Science*, 113, 127, 1951.
258. **Pramer, D. and Starkey, R. L.**, Determination of streptomycin in soil and the effect of soil colloidal material in its activity, *Soil Sci.*, 94, 48, 1962.
259. **Lockwood, J. L.**, Soil fungistasis, *Annu. Rev. Phytopath.*, 2, 341, 1964.
260. **Park, D.**, The importance of antibiotics and inhibiting substances, in *Soil Biology*, Burges, A. and Raw, F., Eds., Academic Press, New York, 1967, 435, Chap. 14.
261. **Fredericq, P.**, On the nature of colicinogenic factors: A review, *J. Theor. Biol.*, 4, 159, 1963.

262. **Reeves, P.,** The bacteriocins, *Bacteriol. Rev.,* 29, 24, 1965.
263. **Nomura, M.,** Colicins and related bacteriocins, *Annu. Rev. Microbiol.,* 21, 257, 1967.
263a. **Novick, R. P.,** Extrachromosomal inheritance in bacteria, *Bacteriol. Rev.,* 33, 210, 1969.
264. **Lingappa, B. T. and Lingappa, Y.,** The nature of self-inhibition of germination of conidia of *Glomerella cingulata, J. Gen. Microbiol.,* 43, 91, 1966.
265. **Lingappa, B. T. and Lingappa, Y.,** Role of auto-inhibitors on mycelial growth and dimorphism of *Glomerella cingulata, J. Gen. Microbiol.,* 56, 35, 1969.
266. **Woods, D. R. and Bevon, E. A.,** Studies on the nature of the killer factor produced by *Saccharomyces cerevisiae, J. Gen. Microbiol.,* 51, 115, 1968.
267. **Beale, G. H. and Jurand, A.,** The classes of endosymbiont of *Paramecium aurelia, J. Cell Sci.,* 5, 65, 1969.
268. **Sonneborn, T. M.,** Kappa and related particles in paramecium, *Adv. Virus Res.,* 6, 229, 1959.
269. **Mitchell, R. and Alexander, M.,** Lysis of soil fungi by bacteria, *Can. J. Microbiol.,* 9, 169, 1963.
270. **Stolp, H. and Starr, M. P.,** Bacteriolysis, *Annu. Rev. Microbiol.,* 19, 79, 1965.
271. **Strominger, J. L. and Ghuysen, J. M.,** Mechanisms of enzymatic bacteriolysis, *Science,* 156, 213, 1967.
272. **Bloomfield, B. J. and Alexander, M.,** Melanins and resistance of fungi to lysis, *J. Bacteriol.,* 93, 1276, 1967.
273. **Birch, L. C. and Clarck, D. P.,** Forest soil as an ecological community, with special reference to the fauna, *Q. Rev. Biol.,* 28, 13, 1953.
274. **Duddington, C. L.,** The predacious fungi and their place in microbial ecology, in *Microbial Ecology,* Williams, R. E O. and Spicer, C. C., Eds., Cambridge University Press, Cambridge, 1957, 218.
275. **Jenkins, W. R. and Taylor, D. P.,** *Plant Nematology,* Rheinhold, New York, 1967.
276. **Pramer, D.,** Fungal parasites of insects and nematodes, *Bacteriol. Rev.,* 29, 382, 1965.
277. **Pramer, D.,** Nematode-trapping fungi, *Science,* 144, 382, 1964.
278. **Barnett, H. L.,** Mycoparasitism, *Mycologia,* 56, 1, 1964.
279. **Anderson, E. S.,** The relations of bacteriophages to bacterial ecology, in *Microbial Ecology,* Williams, R. E. O. and Spicer, C. C., Eds., Cambridge University Press, Cambridge, 1957, 189.
280. **Kleczkowska, J.,** A study of phage-resistant mutants of *Rhizobium trifolii, J. Gen. Microbiol.,* 4, 298, 1950.
281. **Hollings, M. and Stone, O. M.,** Viruses that infect fungi, *Annu. Rev. Phytopath.,* 9, 93, 1971.
282. **Safferman, R. S. and Morris, M. E.,** Growth characteristics of the blue-green algal virus LPP-1, *J. Bacteriol.,* 88, 771, 1964.
283. **Safferman, R. S.,** Virus diseases in blue-green algae, in *Algae, Man, and the Environment,* Jackson, D. F., Ed., Syracuse University Press, Syracuse, 1968, 429.
284. **Safferman, R. S., Diener, T. O., Desjardins, P. R., and Morris, M. E.,** Isolation and characterization of AS-1, a phycovirus infecting the blue-green algae, *Anacystis nidulans* and *Synechococcus cedrorum, Virology,* 47, 105, 1972.
285. **Schuster, F. L.,** Intranuclear virus-like bodies in the amoeboflagellate *Naegleria gruberi, J. Protozool.,* 16, 724, 1969.
286. **Starr, M. P. and Seidler, R. J.,** The bdellovibrios, *Annu. Rev. Microbiol.,* 25, 649, 1971.
287. **Klein, D. A. and Casida, L. E., Jr.,** Occurrence and enumeration of *Bdellovibrio bacteriovorus* in soil capable of parasitizing *Escherichia coli* and indigenous soil bacteria, *Can. J. Microbiol.,* 13, 1235, 1967.
288. **Sullivan, C. W. and Casida, L. E., Jr.,** Parasitism of *Azotobacter* and *Rhizobium* species by *Bdellovibrio bacteriovorus, Antonie van Leeuwenhoek,* 34, 188, 1968.
289. **Parker, C. A. and Grove, P. L.,** *Bdellovibrio bacteriovorus* parasitizing *Rhizobium* in Western Australia, *J. Appl. Bacteriol.,* 33, 253, 1970.
290. **Mishustin, E. N. and Nikitina, A.,** Infection and lysis of Gram-negative bacteria by parasitic bacteria, *Bdellovibrio bacteriovorus, Izv. Akad. Nauk. SSSR (Biol.),* 3, 423, 1970.
291. **Varon, M. and Shilo, M.,** Interaction of *Bdellovibrio bacteriovorus* and host bacteria. I. Kinetic studies of attachment and invasion of *Escherichia coli* B by *Bdellovibrio bacteriovorus, J. Bacteriol.,* 95, 744, 1968.
292. **Stotzky, G.,** Relevance of soil microbiology to search for life on other planets, *Adv. Appl. Microbiol.,* 10, 17, 1968.
293. **Schmidt, E. L. and Bankole R. O.,** Detection of *Aspergillus flavus* in soil by immunofluorescent staining, *Science,* 136, 776, 1962.
294. **Eren, J. and Pramer, D.,** Application of immunofluorescent staining to studies of the ecology of soil microorganisms, *Soil Sci.,* 101, 39, 1966.
295. **Hill, I. R. and Gray, T. R. G.,** Application of the fluorescent-antibody technique to an ecological study of bacteria in soil, *J. Bacteriol.,* 93, 1888, 1967.
296. **Schmidt, E. L., Bankole, R., and Bohlool, B.,** Fluorescent antibody approach to study of rhizobia in soil, *J. Bacteriol.,* 95, 1987, 1968.
297. **Schmidt, E. L.,** Fluorescent antibody techniques in the study of microbial ecology, in *Modern Methods in the Study of Microbial Ecology,* Uppsala (in press).
298. **Gray, T. R. G., Baxby, P., Hill, I. R., and Goodfellow, M.,** Direct observation of bacteria in soil, in *The Ecology of Soil Bacteria,* Gray, T. R. G. and Parkinson, D., Eds., Liverpool University Press, Liverpool, 1968, 171.
299. **Strugger, S.,** Fluorescence microscope examination of bacteria in soil, *Can. J. Res.,* C, 26, 188, 1948.
300. **Trolldenier, G.,** The use of fluorescence microscopy for counting microorganisms, in *Modern Methods in the Study of Microbial Ecology,* Uppsala (in press).

301. Casida, L. E., Jr., Observations of microorganism in soil and other natural habitats, *Appl. Microbiol.*, 18, 1065, 1968.
302. Casida, L. E., Jr., On the isolation and growth of individual microbial cells from soil, *Can. J. Microbiol.*, 8, 115, 1962.
303. Casida, L. E., Jr., Infrared color photography: selective demonstration of bacteria, *Science*, 159, 199, 1968.
304. Marshall, K. C., Sorptive interactions between soil particles and microorganisms, in *Soil Biochemistry*, Vol. 2, McLaren, A. D. and Skujins, J., Eds., Marcel Dekker, New York, 1971, Chap. 14.
305. Hagen C. A., Hawrylewicz, E. J., Anderson, B. T., Tolkacz, V. K., and Cephus, M. J., Use of the scanning microscope for viewing bacteria in soil, *Appl. Microbiol.*, 16, 932, 1968.
306. Jenny, H. and Grossenbacher, K., Root-soil boundary zones as seen by the electron microscope, *Soil Sci. Soc. Am. Proc.*, 27, 273, 1963.
307. Gray, T. R. G., Stereoscan electron microscopy of soil microorganisms, *Science*, 155, 1668, 1967.
308. Stotzky, G. and Bystricky, V., Electron microscopic observations of surface interactions between clay minerals and microorganisms, *Bacteriol. Proc.*, A93, 1969.
309. Harter, R. D. and Stotzky, G., X-ray diffraction, electron microscopy, electrophoretic mobility, and pH of some stable smectite-protein complexes, *Soil Sci. Soc. Am. Proc.*, 1972 (in press).
310. Kubiena, W. L., *Micropedology*, Collegiate Press, Ames, 1938.
311. Alexander, F. E. S. and Jackson, R. M., Examination of soil microorganisms in their natural environment, *Nature*, 174, 750, 1954.
312. Hepple, S. and Burges, A., Sectioning of soil, *Nature*, 177, 1186, 1956.
313. Burges, A. and Nicholas, D. P., Use of soil section in studying amounts of fungal hyphae in soil, *Soil Sci.*, 92, 25, 1961.
314. Jones, D. and Griffiths, E., The use of soil sections for the study of soil microorganisms, *Plant Soil*, 20, 232, 1964.
315. Dobbs, C. C. and Hinson, W. H., Some observations on fungal spores in soil, in *The Ecology of Soil Fungi*, Parkinson, D. and Waid, J. S., Eds., Liverpool University Press, Liverpool, 1960, 33.
316. Cholodny, N., Über eine neue Methode zur Untersuchung der Boden mikroflora, *Arch. Mikrobiol.*, 1, 620, 1930.
317. Chesters, C. G. A., A method of isolating soil fungi, *Trans. Br. Mycol. Soc.*, 24, 352, 1941.
318. Thornton, R. H., The screened-immersion plate: a method of isolating soil microorganisms, *Research*, 5, 190, 1952.
319. Tribe, H. T., Ecology of microorganisms in soils as observed during their development upon buried cellulose films, in *Microbial Ecology*, Williams, R. E. O. and Spicer, C. C., Eds., Cambridge University Press, Cambridge, 1957, 287.
320. Waid, J. S. and Woodman, M. J., A method of estimating hyphal activity in soil, *Pédologie*, 7, 155, 1957.
321. Warcup, J. H., Methods for isolation and estimation of activity of fungi in soil, in *The Ecology of Soil Fungi*, Parkinson, D. and Waid, J. S., Eds., Liverpool University Press, Liverpool, 1960, 3.
322. Sewell, G. W. F., A slide-trap method for the isolation of soil fungi, *Nature*, 177, 708, 1956.
323. Mueller, K. E. and Durrell, L. W., Sampling tubes for soil fungi, *Phytopathology*, 47, 243, 1957.
324. Perfil'ev, B. V. and Gabe, D. R., *Capillary Methods of Studying Microorganisms*, Akad. Nauk SSSR, Moscow, 1969. English translation, Oliver & Boyd, Edinburgh.
325. Perfil'ev, B. V. and Gabe, D. R., The capillary microbial-landscape method in geomicrobiology, and the use of the microbial-landscape method to investigate bacteria which concentrate manganese and iron in bottom deposits, in *Applied Capillary Microscopy: The Role of Microorganisms in the Formation of Iron-Manganese Deposits*, Gurevich, M.S., Ed., Sinclair, F. L., Transl., Consultants Bureau, New York, 1965, 1 and 9
326. Aristovskaya, T. V., The use of capillary techniques in ecological studies of microorganisms, in *Modern Methods in the Study of Microbial Ecology*, Uppsala (in press).
327. Aristovskaya, T. V. and Zavarzin, G. A., Biochemistry of iron in soil, in *Soil Biochemistry*, Vol. 2, McLaren, A. D. and Skujins, J., Eds., Marcel Dekker, New York, 1971, Chap. 13.
328. Nikitin, D. I. and Makarieva, E. D., Use of electron microscope for the calculation of microorganisms in soil suspensions, *Pochvovedenie*, 10, 51, 1970.
329. Williams, S. T., Parkinson, D., and Burges, A., An examination of the soil washing technique by its application to several soils, *Plant Soil*, 22, 167, 1965.
330. Chou, C. K. and Stephen, R. C., Soil fungi, their occurrence, distribution, and associations with different microhabitats, together with a comparative study of isolation techniques, *Nova Hedwigia*, 15, 393, 1968.
331. Parkinson, D., Techniques for the study of soil fungi, in *Modern Methods in the Study of Microbial Ecology*, Uppsala (in press).
332. Stotzky, G., Microbial respiration, in *Methods of Soil Analysis. II. Chemical and Microbiological Properties*, Black, C. A. et al., Eds., American Society of Agronomy, Madison, 1965, 1550.
333. Stolp, H. and Petzold, H., Untersuchungen über einen obligat parasitischen Mikroorganismus mit lytischer Aktivität für *Pseudomonas*-Bakterien, *Phytopath. Z.*, 45, 364, 1962.
334. Casida, L. E., Jr., An abundant microorganism in soil, *Appl. Microbiol.*, 13, 327, 1965.
335. Furcolow, M. L. and Chick, E. W., Eds., *Histoplasmosis*, Charles C Thomas, Springfield, 1971.
336. Nikitin, D. I., Lokhmacheva, R. A., and Vasil'eva, L. V., Microorganisms growing on soil fulvic acids, *Proc. 9th Internatl. Cong. Microbiol.*, Moscow, 1966, c 2/15, 269.
337. Casida, L. E., Jr. and Wood, R. T., Isolation and characterization of a highly pleomorphic soil bacillus, *Bacteriol. Proc.*, A12, 1967.

338. Barghoorn, E. S. and Tyler, S. A., Microorganisms from the Gunflint chart, *Science*, 147, 563, 1965.
339. Schopf, J. W., and Barghoorn, E. S., Alga-like fossils from the early Precambrian of South Africa, *Science*, 156, 508, 1967.
340. Barghoorn, E. S. and Schopf, J. W., Microorganisms from the late Precambrian of Central Australia, *Science*, 150, 337, 1965.
341. Barghoorn, E. S. and Schopf, J. W., Microorganisms three billion years old from the Precambrian of South Africa, *Science*, 152, 758, 1966.
342. Cloud, P. E., Jr., Significance of the Gunflint (Precambrian) microflora, *Science*, 148, 27, 1965.
343. Jackson, T. A., Fossil actinomycetes in middle Precambrian glacial varves, *Science*, 155, 1003, 1967.
344. Schopf, J. W., Barghoorn, E. S., Maser, M. D., and Gordon, R. D., Electron microscopy of fossil bacteria two billion years old, *Science*, 149, 1365, 1965.
345. Schopf, J. W., Ehlers, E. G., Stiles, D. V., and Birle, J. D., Fossil iron bacteria preserved in pyrite, *Proc. Am. Phil. Soc.*, 109, 228, 1965.
346. Urey, H. C., Biological material in meteorites: A review, *Science*, 151, 157, 1966.
347. Bunt, J. S. and Rovira, A. D., The effect of temperature and heat treatment on soil metabolism, *J. Soil Sci.*, 6, 129, 1955.
348. Chase, F. E. and Gray, P. H. H., Application of the Warburg respirometer in studying respiratory activity in soil, *Can. J. Microbiol.*, 3, 335, 1957.
349. Runkles, J. R., Scott, A. D., and Nakayama, F. S., Oxygen sorption by moist soils and vermiculite, *Soil Sci. Soc. Am. Proc.*, 22, 15, 1958.
350. Peterson, G. H., Respiration of soil sterilized by ionizing radiations, *Soil Sci.*, 94, 71, 1962.
351. Domsch, K. H., Bodenatmung, Sammelbericht über Methoden und Ergebnisse, *Zentralbl. Bakteriol.*, Abt. II, 116, 33, 1962.
352. Black, C. A., Evans, D. D., White, J. L., Ensminger, L. E., and Clark, F. E., Eds., *Methods of Soil Analysis*, Vols. I and II, American Society of Agronomy, Inc., Madison, 1965.
353. Audus, L. H., A new soil perfusion apparatus, *Nature*, 158, 419, 1946.
354. Audus, L. H., The biological detoxication of hormone herbicides in soil, *Plant Soil*, 3, 170, 1957.
355. Greenwood, D. J. and Lees, H., An electrolytic rocking percolator, *Plant Soil*, 11, 87, 1959.
356. Lees, H., The soil percolation technique, *Plant Soil*, 1, 221, 1949.
357. Quastel, J. H. and Scholefield, P. G., Biochemistry of nitrification in soil, *Bacteriol. Rev.*, 15, 1, 1951.
358. Macura, J. and Malek, J., Continuous-flow method for the study of microbiological processes in soil samples, *Nature*, 182, 1796, 1958.
359. Briggs, M. H. and Spedding, D. J., Soil Enzymes, *Sci. Progr.*, 51, 217, 1963.
360. Durand, G., Les enzymes dans le sol, *Rev. Ecol. Biol. Sol.*, 2, 141, 1965.
361. Hofmann, E. and Hoffmann, G., Die bestimmung der biologischen tätigkeit in Böden mit enzymmethoden, *Adv. Enzymol.*, 28, 365, 1966.
362. Porter, L. K., Enzymes, in *Methods of Soil Analysis, II. Chemical and Microbiological Properties*, Black, C. A. et al., Eds., American Society of Agronomy, Madison, 1965, Chap. 112.
363. Skujins, J., Enzymes in soil, in *Soil Biochemistry*, Vol. 1, McLaren, A. D. and Peterson, G. H., Eds., Marcel Dekker, New York, 1967, 371.
364. Strickland, J. D. H., Microbial activity in aquatic environments, in *Microbes and Biological Productivity*, Hughes, D. E. and Rose, A. H., Eds., Cambridge University Press, Cambridge, 1971, 231.
365. Stotzky, G., Dawson, J. E., Martin, R. T., and ter Kuile, G. H. H., Soil mineralogy as a factor in the spread of *Fusarium* wilt of banana, *Science*, 133, 1483, 1961.
366. Stotzky, G. and Martin, R. T., Soil mineralogy in relation to the spread of *Fusarium* wilt of banana in Central America, *Plant Soil*, 18, 317, 1963.
367. Stotzky, G., Clay minerals and microbial ecology, *Trans. N.Y. Acad. Sci.*, II, 30, 11, 1967.
368. Stotzky, G., Techniques to study interactions between microorganisms and clay minerals in vivo and in vitro, in *Modern Methods in the Study of Microbial Ecology*, Uppsala (in press).
369. Jackson, M. L., Soil clay mineralogical analysis, in *Soil Clay Mineralogy*, Rich, C. I. and Kunze, G. W., Eds., University North Carolina Press, Chapel Hill, N.C., 1964, Chap. 8.
370. Ajello, L., Comparative ecology of respiratory mycotic disease agents, *Bacteriol. Rev.*, 31, 6, 1967.
371. Sweany, H. C., Ed., *Histoplasmosis*, Charles C Thomas, Springfield, 1960.
372. Stotzky, G. and Post, A. H., Soil mineralogy as possible factor in geographic distribution of *Histoplasma capsulatum*, Can. J. Microbiol., 13, 1, 1967.
373. Goos, R. D., Growth and survival of *Histoplasma capsulatum* in soil, *Can. J. Microbiol.*, 11, 979, 1965.
374. McKinnon, J. E., Histoplasmosis in Latin America, in *Histoplasmosis*, Furcolow, M. L. and Chick, E. W., Eds., Charles C Thomas, Springfield, 1971, Chap. 17.
375. Stotzky, G., Influence of clay minerals on microorganisms: III. Effect of particle size, cation exchange capacity, and surface area on bacteria, *Can. J. Microbiol.*, 12, 1235, 1966.
376. Stotzky, G. and Rem, L. T., Influence of clay minerals on microorganisms. IV. Montmorillonite and kaolinite on fungi, *Can. J. Microbiol.*, 13, 1535, 1967.

377. **Santoro, T., Stotzky, G., and Rem, L. T.,** The electrical sensing zone particle analyzer for measuring germination of fungal spores in the presence of other particles, *Appl. Microbiol.,* 15, 935, 1967.

378. **Smith, C. D.,** Nutritional factors that are required for the growth and sporulation of *Histoplasma capsulatum,* in *Histoplasmosis,* Furcolow, M. L. and Chick, E. W., Eds., Charles C Thomas, Springfield, 1971, Chap. 10.

379. **Garrett, S. D.,** *Biology of Root-Infecting Fungi,* Cambridge University Press, Cambridge, 1956.

380. **Stotzky, G. and De Mumbrum, L. E.,** Effect of autoclaving on X-ray characteristics of clay minerals, *Soil Sci. Soc. Am. Proc.,* 29, 225, 1965.

381. **Kunc, F. and Stotzky, G.,** Influence of clay minerals on heterotrophic microbial activity in soil, *Agron. Abs.,* 1968.

382. **Kunc, F. and Stotzky, G.,** Breakdown of some aldehydes in soils with different amounts of montmorillonite and kaolinite, *Folia Microbiol.,* 15, 216, 1970.

383. **Kunc, F.,** Decomposition of vanillin by soil microorganisms, *Folia Microbiol.,* 16, 41, 1971.

384. **Santoro, T. and Stotzky, G.,** Effect of electrolyte composition and pH on the particle size distribution of microorganisms and clay minerals as determined by the electrical sensing zone method, *Arch. Biochem. Biophys.,* 122, 664, 1967.

385. **Santoro, T. and Stotzky, G.,** Sorption between microorganisms and clay minerals as determined by the electrical sensing zone particle analyzer, *Can. J. Microbiol.,* 14, 299, 1968.

386. **Santoro, T. and Stotzky, G.,** Influence of cations on flocculation of clay minerals by microbial metabolites as determined by the electrical sensing zone particle analyzer, *Soil Sci. Soc. Am. Proc.,* 31, 761, 1967.

387. **Santoro, T. and Stotzky, G.,** Effect of cations and pH on the electrophoretic mobility of microbial cells and clay minerals, *Bacteriol. Proc.,* A15, 1967.

388. **Santoro, T. and Stotzky, G.,** Further observations on electrophoretic mobility of microorganisms and clay minerals, *Bacteriol. Proc.,* A24, 1968.

389. **Harter, R. D. and Stotzky, G.,** Formation of clay-protein complexes, *Soil Sci. Soc. Am. Proc.,* 35, 383, 1971.

390. **Edwards, P. Q.,** Histoplasmin sensitivity patterns around the world, in *Histoplasmosis,* Furcolow, M. L. and Chick, E. W., Eds., Charles C Thomas, Springfield, 1971, Chap. 14.

391. **Stotzky, G.,** Ecological eradication of fungi — dream or reality, in *Histoplasmosis,* Furcolow, M. L. and Chick, E. W., Eds., Charles C Thomas, Springfield, 1971, Chap. 62.

GROWTH OF BACTERIA IN MIXED CULTURES

Author: **J. L. Meers**
Imperial Chemical Industries Ltd.
Billingham, Teesside
England

INTRODUCTION

Most natural environments (which may be defined as any environment other than a laboratory culture vessel) are constantly subject to invasion by a wide variety of microorganisms. If physicochemical conditions such as temperature, pH, hydration, etc. are within specified limits and the environment contains an available energy source plus the other nutrients required for microbial growth, a microbial community will develop. However, not all species which by chance are transferred to a given environment will become stable members of the community, despite the fact that when grown in axenic (i.e., pure) laboratory culture many more species appear to be capable of growth in a given environment than are in fact found in that environment. Conversely, some species that are found in nature are very difficult to grow as pure laboratory cultures. These apparent contradictions may be reconciled when the importance of the interactions between the component species of mixed microbial populations is appreciated. It is the twofold purpose of this review to discuss the ways in which the growth of one microbial species can affect the growth of another and then to examine specific examples of

natural ecosystems with a view to assessing the extent to which recent work has led to an understanding of the complex interrelationships that occur in natural environments.

With the above terms of reference, the bibliography could become voluminous, therefore only selected references are given, from which various topics may be followed up in greater detail. Although it is evident from the title that the central theme of this review concerns the interactions between bacterial species, some discussion of how the growth of other types of microorganisms affect bacterial growth is included.

TYPES OF MICROBIAL INTERACTION

Although many microbial interrelationships do not fall neatly into categories, it is convenient to classify the different types of microbial interaction. Several different classification schemes have been proposed, but none are universally accepted and a variety of alternative terms have been used by different authors. Microorganisms can be considered to interact in six distinct ways. The following terms will be used here.

Competition occurs when the populations of two or more species are mutually limiting because of their joint dependence on a common factor or factors external to them. The term *commensalism* describes a situation in which the growth of one species is promoted by the presence of a second species in a community, the growth of this second species being unaffected by the presence of the first. *Mutualism* is similar to commensalism but in this case the growth of each species is promoted by the presence of the other. In a *symbiotic* relationship the two species have an obligate reliance on each other in an environment where they are invariably found growing together. If the mutualistic relationship is of a looser nature, the term *protocooperation* is sometimes used. *Synergism* is a third type of mutualism in which the formation of specific products is greater in mixed than in pure populations. In an *amensal* relationship the growth of one species is repressed because of the presence of a toxic substance or substances produced by another.

The distinction between *predation* and *parasitism* is particularly vague when discussing microbial interrelationships. *Predation* occurs when one organism totally engulfs and digests another organism which thereby loses the ability to reproduce itself. *Parasitism* occurs when one organism feeds or reproduces at the expense of the tissues or body fluids of another, usually larger organism that is necessarily damaged by the relationship.

A distinction is often made between open and closed environments. A microbial community may grow in an environment for an indefinite period because nutrients flow continually into the environment and end products and cells flow continually out. Such an environment, designated an open environment, leads to population stability or homeostasis, although oscillations in the density of the component species may be observed. These changes in population may be due either to periodic fluctuations in the environment or to interrelationships between microbial species which lead to oscillations of population density. Open systems are typified in the laboratory by various types of chemostat or continuous culture apparatus and in natural environments by the rumen or an activated sludge plant. In a closed environment, growth of a microbial community continues only for a finite time because the organisms bring about changes in the environment which exclude the possibility of further growth. Microbial growth often exhausts the supply of a limited quantity of

an essential nutrient, but such changes as accumulation of toxic metabolites or lowering of pH levels may also eventually result in the death of the whole population. In natural closed environments (such as compost heaps or decaying carcasses), a succession of dominant species occurs before the environment becomes unsuitable for further growth. In the laboratory the conventional batch culture is an example of a closed system.

The distinction between open and closed systems is important because the factors influencing the survival of given species are different in the two types of environment.

Neutralism

In order to show that two species have no effect on one another, it must be demonstrated that the growth of the populations of the two species is the same in either axenic or mixed cultures. Data that demonstrate the existence of neutralism are sparse; the paper of Lewis[115] represents one example. In her paper, Lewis describes how strains of *Lactobacillus* and *Streptococcus* were established as both pure and mixed cultures in chemostats. The population densities were almost the same in the pure and mixed cultures, which indicated a state of neutralism (Figure 1). Lewis noted the existence of considerable oscillations of the populations which she ascribed to fluctuations in the pH levels of the system (which lacked pH control). However, such oscillations do not seem to be particularly marked in the published results.

In a closed system during the initial stages of colonization when all nutrients are in excess, it seems likely that examples of neutralism can be demonstrated. However, as time proceeds and nutrient levels are depleted, only those species whose growth is limited by the lack of different essential nutrients can be expected to have no effect on one another. The same would be true in an open system. Therefore, it follows that only species with dissimilar nutrient requirements can be expected to exhibit neutralism. However, since a number of nutrients are considered to be required by all bacteria (e.g., magnesium, sulfur), neutralism is not an immutable feature of the relationship between any two species, but only between particular species in particular environments; if the environments were changed such that a nutrient that was essential to both species

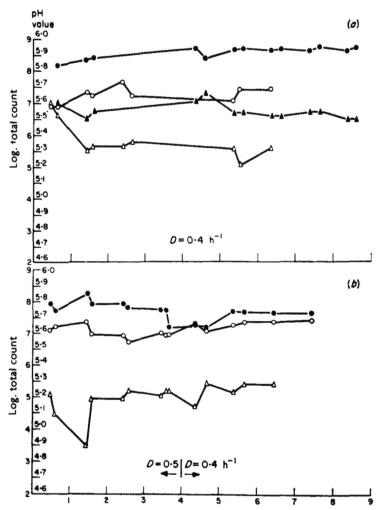

FIGURE 1. Total counts and pH values of chemostat cultures of *Lactobacillus* sp. and *Streptococcus* sp. Open circles, counts of *Lactobacilli*; closed circles, counts of *Streptococci*. (a) Pure cultures of each species. Open triangles, pH values of *Lactobacillus* culture; closed triangles, pH values of *Streptococcus* culture. (b) Two membered culture. Triangles, pH values.[115] (From Lewis, P. M., *J. Appl. Bacteriol.*, 30, 408, 1967. With permission.)

became deficient, then a competitive situation would occur. A neutral relationship could be envisaged between populations of two species, both of which were growth limited by a deficiency of different nutrients which was produced by a third species, the former two species having a commensal relationship with the latter. A similar neutral relationship can occur in which the growth of the two populations was not limited by nutrient depletion, but by other adverse changes in the environment (e.g., toxin production or pH value change) brought about by the growth of a third species in the ecosystem. Other hypothetical neutral relationships can be proposed, but

this area remains almost unexplored. While neutralism seems likely to be of importance in natural environments, whether or not this is so remains unknown. Pipes[180] could be correct in suggesting that the microbial community in an activated sludge plant may be considered to be in a state of synergism of the whole population. This suggestion would exclude the possibility of the existence of neutralism in this environment. If neutralism in natural environments is rare, then the value of many laboratory experiments on pure cultures is markedly decreased. Clearly it would be of value to find out if neutralism is common in natural environments.

Competition

The *phenomenon of competition between* different species is frequently studied by ecologists. This interest probably stems from the importance, stressed by Darwin,[49] of competition in the natural selection of species. However, Darwin failed to define the term "competition" unambiguously and it can be argued that he used the term synonymously with the struggle for existence (see Darwin,[49] Chapter 3). Since Darwin's time, the biological meaning of this term has become confused and is used to describe struggles for food, and strife against enemies, parasites, severe climatic conditions, hunger, and disease. It is recommended that the term "competition" be restricted to the situation in which the populations of two species are mutually limiting because of their joint dependence on a common factor or factors external to them. Haldane[76] stated that "the fitness of plants in the Darwinian sense must be tested with plants grown in competition." This statement implies the existence in plants (and presumably other life forms) of heritable differences in competitive ability. It is legitimate then to ask whether or not this competitive ability is associated with any particular physiological property. A number of such properties have been cited as being of importance in conferring competitive advantage, but rapid growth rate is most commonly assumed to be of prime importance.

In developing his thesis on microbial growth rates, Smith[130] assumed that natural selection will always favor species or mutants with rapid maximum growth rates. Hence, it was argued that mutants with decreased generation times will invariably outgrow the parent population and, over an extended period, organisms will be selected which will grow as rapidly as the maximum possible rate of duplication of specific macromolecules. Smith[120] concluded that growth rate in bacteria is limited by the rate of protein synthesis. From this argument it is predicted that all bacterial species would tend to have similar short doubling times. Although I will argue in subsequent paragraphs that a rapid growth rate is not invariably advantageous, Smith's sole theoretical line of argument is supported, for example, by the data of Oberhofer and Frazier.[164] These authors showed that in axenic batch cultures, *Escherichia coli* and *Staphylococcus aureus* grew to similar cell densities, but that the former had the shorter generation time (Figure 2). When both species were grown together in the same medium, *E. coli* grew rapidly to the cell density achieved in ascenic culture and (presumably) exhausted the supply of an essential nutrient before the *S. aureus* population had grown extensively (Figure 2). Since *E. coli* did not produce a toxin active against its competitor, it appears that the advantage of a rapid growth rate has been demonstrated when two species compete for a limited supply of nutrient in a closed system. In this system one species had such a distinct advantage over the other that its growth was virtually unaffected by the presence of the competitor. If, however, the growth rates of the two competing species were similar, it would be predicted that the growth of both species would be restricted in a mixed culture. Such a situation has been observed by Annear.[7] If a proportion of the mixed culture from an experiment, such as that illustrated in Figure 2, was transferred to fresh medium, allowed to grow, and then serially subcultured in the same way, one would predict that the faster growing organisms would form an increasingly large proportion of the mixed community. After a sufficiently long period of time, the slower growing population would

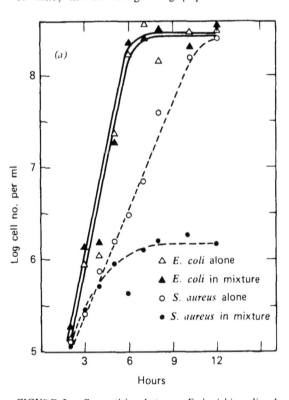

FIGURE 2. Competition between *Escherichia coli* and *Staphylococcus aureus* when the species were grown either alone or as mixtures in batch cultures.[164]

disappear from the mixture. This kind of intermittent growth can occur in natural environments and it follows that slow growing organisms require some physiological advantages in order to survive in the longer term against faster growing competitors. Several such alternative advantageous traits can be proposed. For example, the ability to remain dormant but viable between periods of active growth would be a considerable alternative advantage to a rapid growth rate. Slow growing organisms with effective means of dispersal could also compete effectively against faster growing organisms which were less effective in reaching a new growth situation.

Braun[17] reported results that showed the existence of small but significant differences in the generation times of "wild-type" and nutritionally deficient mutants in favor of the latter. It is, therefore, possible to appreciate from Smith's[130] argument why auxotrophic mutants become established in natural environments where a plentiful supply of growth factors exists. Alternatively, in nutritionally poorer environments, prototrophs would clearly be at a competitive advantage. Thus, Smith's proposal that competitive pressures tend to favor the selection of populations with short generation times has been shown to be justified in some simple closed systems.

Powell[187] and Moser[152] have provided mathematical analyses of the fate of contaminants or of mutants in continuous flow systems. An important assumption made in these analyses is that there is no interaction between the two varieties of organisms other than their competition for the available nutrients. For a rigorous mathematical treatment the paper by Powell[187] is recommended, but the essential points regarding nutrient competition may be summarized as follows. The nomenclature and symbols used are primarily those recommended by an international commission at the Second International Symposium on the Continuous Culture of Microorganisms, Prague, 1962 (see pages 379-382 in proceedings).

Consider the growth of a population of organisms in a steady state in a chemostat. Its change in concentration with time is given by:[83]

$$\frac{dx}{dt} = X \left[\mu_m \frac{S_1}{K_s + S_1} - D \right] \tag{1a}$$

where t = time

Z or X = concentration of organisms by mass

S = concentration of limiting substrate

λ_m or μ_m = maximum mass growth rate constant for the organism and medium concerned.

L_s or K_s = saturation constant for the given organism and substrate

D = dilution rate

But, in the steady state $\frac{dx}{dt} = 0$ and therefore

$$D = \mu_m \left(\frac{S_1}{K_s + S_1} \right) = \mu \tag{2a}$$

Analogous equations can be derived for a second organism growing under identical growth conditions in a second chemostat

$$\frac{dz}{dt} = Z \left[\lambda_m \left(\frac{S_2}{L_s + S_2} \right) - D \right] \tag{1b}$$

When $\frac{dz}{dt} = 0$

then

$$D = \lambda_m \left(\frac{S_2}{L_s + S_2} \right) = \lambda \tag{2b}$$

Suppose now that a small volume of the culture of organism 2 is introduced into the culture of organism 1. The initial change in concentration with time of organism 2 is given by

$$\frac{dz}{dt} = Z \left[\lambda_m \left(\frac{S_1}{L_s + S_1} \right) - D \right] \tag{1c}$$

The population of organism 2 is unlikely to be in a steady state and hence $\frac{dz}{dt}$ is not likely to be zero. For $\frac{dz}{dt}$ to be positive the following relationship must exist

$$\lambda_m \frac{S_1}{L_s + S_1} > \mu_m \left(\frac{S_1}{K_s + S_1} \right) \tag{3a}$$

i.e., the growth rate of organism 2 at a substrate concentration of S_1 must be greater than D. Under these circumstances organism 2 will displace organism 1 at a rate depending on the extent to which the constants in Equation 3a differ from one another. If, however,

$$\lambda_m \left(\frac{S_1}{L_s + S_1} \right) < \mu_m \left(\frac{S_1}{K_s + S_1} \right) \qquad (3b)$$

organism 2 will not replace organism 1, but will itself be washed out of the culture.

It is conceivable that

$$\lambda_m \left(\frac{S_1}{L_s + S_1} \right) = \mu_m \left(\frac{S_1}{K_s + S_1} \right) \qquad (3c)$$

and under these conditions populations of organisms 1 and 2 could theoretically coexist indefinitely. However, since the equilibrium described by Equation 3c is not stable and small changes in parameters such as temperature are likely to have different effects on the growth of the two species, this type of equilibrium is unlikely to be maintained over prolonged periods except in the most rigorously controlled laboratory environments. The most likely case in which Equation 3c might describe a stable equilibrium is that in which a mutant population could coexist with the parent organisms. The prerequisite for such a situation would be that the nature of the mutation distinguishing the mutants from the parent organisms would be such that the growth parameters K_s and μ_m were not altered.

In Figure 3 hypothetical curves are plotted showing how the growth rates of organisms A and B might vary with the limiting substrate concentration. During the course of an extensive series of experiments in which *Aerobacter aerogenes*, *Pseudomonas fluorescens*, *Bacillus subtilis*, *B. megaterium*, *Staphylococcus epidermidis*, and *Torula utilis* were grown in mixed magnesium-, potassium-, or carbon-limited mixed cultures, no combination was observed in which two species coexisted in chemostat cultures.[37] This implies that the curves illustrated in Figure 3 are rarely superimposable. In the experiments mentioned above, the course of events which most commonly followed the mixing of populations of two species may be outlined as follows. Organisms A and B in Figure 3 may be used as model examples.

If a small number of cells of organism A were transferred to a chemostat containing a steady-state population of organism B, then (assuming no lag period) the population of organism A would begin growing at a rate μ_A (since the growth-limiting substrate concentration would be S_B: Figure 3). Because μ_A is greater than D, the concentration of the transferred species would

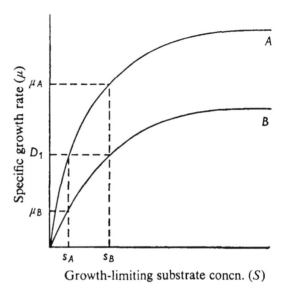

Growth-limiting substrate concn. (S)

FIGURE 3. Hypothetical saturation curves for two microbial species (A and B) growing in substrate-limited chemostat cultures. At a fixed dilution rate (D_1) the growth-limiting substrate concentration in a culture of A organisms would be s_A and of B organisms s_B. It is assumed that the maximum growth rate (μ_m) for A organisms is greater than for B organisms, and that the saturation constants (K_s values) bear a similar relationship.[133] (From Meers, J. L. and Tempest, D. W., *J. Gen. Microbiol.*, 52, 309, 1968. With permission.)

increase. Its growth would cause the growth-limiting substrate concentration to decrease toward the value S_A; at this latter concentration organism B would grow only at a rate μ_B which, being less than D_1, would result in B's being washed completely from the culture. Thus, from this theory, one would predict that organism A would totally displace organism B from the chemostat culture and this should occur irrespective of either the initial concentration of each organism in the mixed culture, or, in the case illustrated in Figure 3. It should be noted, however, that if only very few type A organisms were transferred, there is a finite chance that they could be washed from the culture before establishing themselves.[187]

Tempest et al.[219] described an experiment in which *Aerobacter aerogenes* outgrew several gram-positive organisms in a manner predicted by the above theory. It was subsequently shown that the physiological basis for the competitive supremacy of gram-negative species over gram-positive species in magnesium-limited environments was due to differences in cell wall structure.[135] Thus, the gram-negative species contained polymers on the cell surfaces which had a strong affinity for cations, whereas the teichoic acids, which were largely responsible for magnesium adsorption in *Bacillus subtilis*, had a lower affinity for cations, despite having a greater capacity for magnesium adsorption. The competition between *A. aerogenes* and various *Bacillus* species for magnesium is probably one of the best understood competitive situations studied to date, because the basis of the competitive advantage of *A. aerogenes* can be explained clearly in terms of physiological differences between this and the other species.[135]

Figure 4 illustrates the results of previously unpublished experiments in which *A. aerogenes* and *Torula utilis* competed for growth-limiting quantities of potassium at two different dilution rates. Two useful points are illustrated by these figures. First, at high dilution rates, the replacement of the yeast by the bacterium was more rapid than at low dilution rates. This result does not necessarily follow from the analysis of Powell,[187] but is consistent with theory and was observed in all combinations of organisms which I have studied. Second, since the population of *T. utilis* decreased at a rate almost exactly the same as the theoretical wash out rate for nongrowing organisms (dashed lines in Figure 4), the difference between the growth parameters μ_m and K_s for these two species must be great (see Equation 3a). In many experiments it was observed that following cross-inoculation the population of the species that was ultimately displaced from the culture initially increased in number (Figure 4). Changes in cell size commonly accompany changes in growth rate[84] and changes in growth rate necessarily followed the mixing of populations. Hence, the anomalous population increases were probably due to changes in cell size and did not indicate changes in the dry weight of the culture. When *A. aerogenes* and *B. subtilis* were competing for a limited supply of potassium, the rate of displacement of *B. subtilis* was not as rapid as the rate of displacement of *T. utilis* by *A. aerogenes* at the same dilution rate. From these results, one could predict that *B. subtilis* would replace *T. utilis* in a potassium-limited environment, which was found to be the case. Thus, the rate of replacement of one species by another gave an indication of the relative shapes of the saturation curves (Figure 3) for the two species.

Using the same type of argument, based on experimental data, the results of a series of carbon-limited mixed-culture experiments could be summarized by the following listing. A species could replace any of those below it in the list, but was replaced by any of the species above it.

1. *A. aerogenes*
2. *B. subtilis*
3. *B. megaterium*
4. *T. utilis*

The point of special interest in these experiments (which were all conducted at a dilution rate of 0.3 hr^{-1}) was the slow rate of replacement of one species by another. This indicated that the extent to which all of the species decreased the extracellular glucose concentration was similar at a dilution rate of 0.3 hr^{-1}. The results are representative of a large number of mixed culture experiments that have been completed by the author in various nutrient-limited environments. In the large majority of experiments it was found

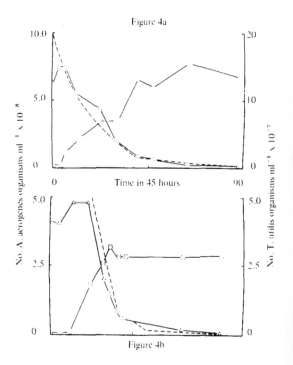

FIGURE 4. The growth of *Aerobacter aerogenes* (□) and *Torula utilis* (∧) in a potassium-limited simple salts medium in a chemostat. The dilution rate was set at 0.05 hr^{-1} (a) and 0.3 hr^{-1} (b). The temperature was 33° and the pH value was 6.4. The dotted lines show the theoretical washout rates for nongrowing organisms (previously unpublished data).

that one species replaced another in a way similar to that illustrated by Figure 4. This was the type of result predicted by Powell[187] and Moser.[152] Exceptions were observed when some species were found to secrete substances which stimulated growth.

Brunner et al.[22] described experiments in which the results were quite different from those described previously, although the experimental procedure was similar. These authors found that when *E. coli* and *Serratia marcescens* were grown as glucose-limited mixed cultures both species persisted in constant proportions at a variety of dilution rates. Clearly such a result is inconsistent with the theories proposed, unless both species had identical saturation curves (Figure 3). Such a possibility was, however, excluded by Brunner et al.,[22] who plotted the saturation curves in order to calculate the dilution rate at which the two species would be expected to coexist (Equation 3c). The paradox presented by their results was appreciated by Brunner et al.[22] who proposed two ideas to overcome the problem. The first explanation may be paraphrased as follows. As the dilution rate was increased, the density of *E. coli* decreased, leading to a concomitant increase in the concentration of growth-limiting substrate (glucose) as described by Herbert et al.[83] The faster growing *S. marcescens* could then have utilized this "extra" glucose that was not used by the *E. coli*. Thus, it was concluded that the steady-state values of substrate concentration in the heterogenous population were determined by the organism with the higher maximum growth rate. Such an explanation, however, neglects to take into account the fact that the growth of *S. marcescens* will decrease the glucose concentration to a level lower than that at which *E.coli* would be predicted to grow and maintain their population in the chemostat. The other explanation for their results proposed by Brunner et al.[22] was based on the phenomenon of variability of generation times in a population of microorganisms. Although it has been shown that not all of the organisms in a population have the same generation time, it is difficult to see how such a phenomenon can adequately explain the results under discussion. One is finally forced to conclude that the results reported by Brunner et al.[22] must be explained by some kind of interrelationship other than competition.

Veldkamp[231] has reviewed work on enrich-ment in continuous culture. Of special note in this connection is the work of Jannasch,[100] who showed that if samples of sea water were used as an inoculum for chemostat cultures, then the species which became dominant depended on the dilution rate used during the experiment. These results implied that the curves for specific growth rate vs. limiting nutrient concentration (c.f. Figure 3) crossed. Such a situation was anticipated by Powell[187] and was demonstrated more recently by Meers.[139] The result of an experiment in which *Bacillus subtilis* var. *niger* and *Torula utilis* were growing together under magnesium-limited conditions in a chemostat is shown in Figure 5. The dilution rate was alternated between 0.05 and 0.08 hr^{-1}, both dilution rates being well below those at which the organisms would otherwise be washed from the culture (i.e., 0.7 and 0.5 hr^{-1}, respectively). *B. subtilis* replaced the yeast at the higher dilution rate, but the reverse was true at the lower dilution rate. Thus, complementary fluctuations, or oscillations, in the density of the bacterial and yeast populations occurred in response to changes in the flow rate. In subsequent experiments the dilution rate was not alternated, but was kept constant at a value of 0.05 or 0.08 hr^{-1}. It was then observed that either *B. subtilis* or *T. utilis* continued to be washed from the fermenter at an exponential rate until, after approximately six days, cultures were obtained that on microscopic examination appeared to be pure. Even after this length of time, if the dilution rate was changed from 0.08 to 0.05 hr^{-1}, the undetectably small population of *T. utilis* gradually increased in number. But, because of the predictable exponential of this increase, the yeast population only became large enough to influence the growth of the bacterial population after a period of several days. Hence, one could predict the type of asymptotic curves that are illustrated by Figure 5. The significance of this observation will be apparent later when the reasons for oscillations in microbial populations are discussed. But it is worth noting at this stage that *T. utilis* would replace *B. subtilis* in a system that was operating with a mean residence time of 20 hr, despite the fact that the bacterium had the higher maximum growth rate. Thus, the contention of Smith[130] that natural selection will favor organisms with rapid maximum growth rates cannot be assumed to be universally true.

The results discussed above concern

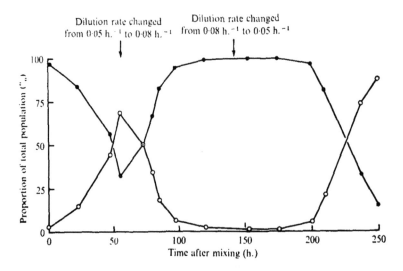

FIGURE 5. Growth of *Bacillus subtilis* (●) and *Torula utilis* (○) in a magnesium-limited simple salts medium in a chemostat. The initial concentration of *B. subtilis* was 6 x 10⁸ organisms/ml and of *T. utilis* 1 x 10⁷ organisms/ml. The pH was 6.4 ± 0.1 throughout the experiment and the temperature was 33°. At the start of the experiment the dilution rate was set at 0.5 hr⁻¹. This rate was increased to 0.08 after 55 hr and returned to 0.05 hr⁻¹ after 143 hr.[139] (From Meers, J. L., *J. Gen. Microbiol.*, 67, 359, 1967. With permission.)

competition between different species of organisms. Analogous competitive situations would be expected to be observed when mutant strains arise in continuous flow systems. Novick[162] has studied the changeover of strains during the long-term cultivation of *E. coli* in a chemostat. He was able to demonstrate that 5 or 6 such changeovers took place during 450 generations and that each succeeding strain was able to reduce further the extracellular concentration of growth-limiting substrate to a lower level, at a constant dilution rate, than the preceding dominant strain. Schlegel and Jannasch[198] have reviewed work on the selection of variants in continuous culture.

A natural development of the work on competition for a single growth-limiting substrate is to study competition for two substrates in a chemostat. Mateles et al.,[127] Chian and Mateles,[38] and Mateles and Chian[128] have made such investigations. In the series of experiments conducted by Mateles and his co-workers, two substrates were supplied in growth-limiting quantities as the carbon sources for pure and mixed cultures of *E. coli* and *Pseudomonas fluorescens.* As a general rule, above certain dilution rates only one substrate was used in the pure cultures. This

effect is analogous to the diauxic behavior observed by Monod[149] and is probably caused by catabolite repression.[121] In mixed cultures, however, the populations of both species were maintained indefinitely in proportions which varied with the dilution rate and the kinetics of substrate uptake was different from that observed in the pure cultures. Clearly, further work would be required to explain unequivocally the precise nature of the interrelationships between the species studied by Mateles and Chian.[128] However, the results do seem to indicate that the two species might have been coexisting in the cultures because they were each utilizing the alternative substrates. For such a situation to be observed, one has to conclude that each species had a competitive advantage over the other for one of the growth-limiting substrates. But the possibility that commensalism or mutualism was occurring in the cultures was not excluded by Mateles and Chian.[128] One interesting point mentioned by the authors was that adaptation occurred in the cultures and that mutants were selected which were better able to utilize both substrates simultaneously. Mueller[154] studied a somewhat similar situation in which the preferential utilization of the methane fraction of natural gas was demonstrated even though a mixed culture of bacteria was used, and in this case no selective enrichment of ethane or propane utilizers was observed. Dworkin and Foster,[55] however, found that their cultures did become enriched with ethane oxidizers when growing on natural gas.

Cooney and Mateles[41] have discussed the concept of dual limitation in pure cultures and have provided evidence that *Aerobacter aerogenes* can grow in a fermenter where there was one primary limiting nutrient while a second nutrient was available to the cell in an amount less than the cell was capable of utilizing. Such a fermenter is similar to that observed when *Torula utilis* and *Pseudomonas fluorescens* were competing for limited supplies of both magnesium and oxygen.[137] Under fully aerobic magnesium-limited growth conditions *P. fluorescens* competitively displaced the yeast population at a dilution rate of 0.3 hr^{-1}. If, however, the air supply to the mixed culture was carefully adjusted so that only partially aerobic growth was possible, then a stable, self-regulating mixed population could be established in which essentially all of the available magnesium and oxygen was metabolized by the microorganisms. The most likely explanation of this result is that the diminished oxygen supply limited the growth of the *P. fluorescens* population without allowing the supply of magnesium to be completely exhausted by this species. At the same time, the *T. utilis* population was growing anaerobically and metabolizing the remaining magnesium, which under fully aerobic growth conditions would have been utilized by the bacterial population. Under these circumstances, one would predict that if extra magnesium was added to the medium, the yeast population would increase while the bacterial population remained steady. If, on the other hand, extra oxygen was supplied, the bacterial population would increase at the expense of the yeast population. This sort of complex situation could easily exist in, for example, sewage sludge tanks, where variations in oxygen supply alter the microbial flora,[103] and many analogous situations in which two substrates limit the growth of mixed populations can be envisaged.

Apart from competition for essential nutrients, bacterial competition for light and for space is possible. It is reasonable to assume that competition for light plays a role in selecting photosynthetic bacterial strains in aquatic communities and this point will be returned to in a later section.

Bail[9] proposed the concept of a bacterial space requirement to account for inhibition of growth at high population densities. But this belief is not now generally accepted as being of importance in natural environments, because the population density of most species can be raised to levels far higher than those observed in nature by the addition of appropriate nutrients. In laboratory liquid cultures there seems to be no prima facie reason why physical overcrowding should prevent the growth of bacteria to the population densities that are normally observed in colonies growing on agar plates, where the possibility of competition for space seems more likely.

Dias et al.[53] have shown that there were considerable fluctuations in the distribution of the species growing as a slime layer in continuous flow devices. Thus, low concentrations of ammonium chloride or oxygen or high flow rates resulted in attached populations that were dominated by the species *Sphaerotilus natans*, but the reasons for this observation were unclear. Munson and Bridges[155] have reported an experiment in which a culture of a tryptophan-requiring mutant of *E. coli* was displaced by a prototrophic mutant of the parent population that became attached to the vessel surfaces. Bungay and Bungay[27] also stated that adherence to vessel walls was a key factor influencing the outcome of competition experiments. This result was caused by slower growing species persisting in appreciable numbers in competition with rapidly growing organisms if the former species were continually reinoculated from the wall growth into the main culture. The potential importance of wall growth can be well appreciated from the report by Larsen and Dimmick[112] that up to 90% of the cells in culture fluid can be the progeny of adhering organisms when the surface area in continuous cultures of *Serratia marcescens* was increased by the addition of glass wool. The competitive advantage of species that are able to become attached to surfaces would be particularly marked in open environments with a rapid dilution rate. Even if the dilution rate exceeded the maximum growth rate of the organisms, the attached population could maintain itself. This is a further example of a situation in which a species with a low maximum growth rate could compete effectively with a species capable of rapid multiplication.

In summarizing work on competition between bacterial populations there seems to be good evidence to support the claim that if two species compete for a limited supply of a nutrient in an identical ecological niche, one species would be expected to replace the other. This is in effect a restatement of a well-known, but disputed, dictum in ecology.[66,146,158] More complex competitive interrelationships have also been observed and in recent years progress has been made toward obtaining an understanding of the interrelationships between competitive species which do not occupy identical ecological niches.

Amensalism

When bacterial species grow, the environment may be altered in such a way that the growth of other species is inhibited. This kind of harmful interrelationship may be caused either by the removal of essential nutrients or by the formation of toxic products, and it is not always clear from published results on the subject which type of antagonism has occurred. The toxic products likely to cause amensalism may be divided into two groups: viz., inorganic substances which usually are of low potency and organic compounds which are often toxic at low concentrations.

During growth, bacteria commonly excrete organic acids. This leads to a lowering of pH which can either inhibit or promote the growth of other species. Thus, Contois and Yango[40] described an interrelationship in which *Lactobacillus* strains were growing in a chemostat that did not have pH control. One species grew well and caused a lowering of the pH to levels that were inhibitory to itself, but not to another strain. The second strain then started to grow, but did not produce acid and therefore the pH in the culture rose. The first strain again became dominant and the cycle repeated itself. The converse of this situation was illustrated by experiments in which *Bacillus subtilis* and *Aerobacter aerogenes* were competing for limited supplies of potassium.[137] The interesting point about these experiments was that *B. subtilis* became nonviable when *A. aerogenes* was present at high concentration in the cultures. This result was caused by the production of acids by *A. aerogenes* which lowered the pH value in the culture vessels to 4.5. Subsidiary experiments showed that *A. aerogense* grew well between the pH limits 4.3 to 8.5, but *B. subtilis* could only grow well between the pH limits 6.5 to 8.5. Therefore, it was concluded that the negative growth rate of *B. subtilis* was caused by the pH change.

Control of growth by acid production may not simply be due to pH value changes. Some organic acids have specific growth inhibitory effects which are independent of pH. For example, Hentges[82] showed that the growth of *Shigella flexneri* was inhibited by the formic and acetic acids produced by an *Aerobacter* species. Tramer[224] and Ichikawa and Kitamoto[96] have also shown that organic acids have growth inhibitory effects which are not due to pH changes. Neronova et al.[156] have published details of an interesting population effect with *Propionibacterium shermanii*. In this pure culture system, excreted propionate and acetate were shown to inhibit the growth of the bacteria. Since the quantity of propionate excreted was proportional to population density, a population effect was observed. Therefore, these workers proposed the following equation to describe the relationship between growth rate μ and population density:

$$\mu = \frac{\mu_o K_p}{K_p + P} \qquad (4)$$

in which μ_o is the specific growth rate in a defined nutrient medium in the total absence of growth inhibitors (propionate in the case considered); P is the concentration of growth inhibitor; K_p is a constant, numerically equal to the propionate concentration at which $\mu = \frac{\mu_o}{2}$ Although Neronova et al.[156] produced evidence to support the validity of Equation 4 only in pure cultures, it seems reasonable to suggest that such an equation could also describe the effect of the population density of an inhibitor-producing strain on a second strain which was susceptible to the inhibitor.

The same might be said with regard to the equation of Contois[39] who adapted Monod's equation[149] (Equation 2) so as to include a population factor (P) as follows:

$$\mu = \frac{\mu_m S}{K_s P + S} \qquad (5)$$

However, the conclusion that the growth of *Aerobacter aerogenes* is inhibited by increased population density was not substantiated by Meers and Tempest,[133] and Powell[188] was able to show (using a turbidostat) that population density influenced μ_m — a result contrary to the findings of

Contois.[39] The turbidostat gave unequivocal results in these experiments and it seems doubtful that the equations of Contois[39] are of general applicability as claimed.

Ammonia can be released in large quantities by organisms using amino acids as an energy source.[140] Chemoautotrophs such as *Nitrobacter* are known to be inhibited by small quantities of ammonia and therefore the production of this compound probably causes amensalism in some environments.[213]

Microbial growth may bring about a reduction in redox potential which can result in commensalism or in competition for oxygen, but since anaerobiosis often also causes the production of toxic compounds such as hydrogen sulfide and organic acids, this situation can be quite complex.[143] Conversely, the production of oxygen by photosynthetic organisms may inhibit the growth of anaerobic species. Thus, Jannasch[99] has shown that the anaerobic growth of *Pseudomonas stutzeri* was inhibited by the presence of *Chlorella*, which produced oxygen.

Hydrogen peroxide production has been reported to be important in inhibiting bacterial growth,[109] but since most bacteria produce the enzymes catalase or peroxidase, the ecological importance of the production of this compound remains to be established.[4]

A particularly neat way of demonstrating amensalism was used by Parker and Snyder.[170] Pure cultures of *Streptococcus salivarium* and *Veillonella alcalescens* were grown in separate chemostats and continuously fed into a mixed culture vessel together with fresh medium. The dilution rate of this third vessel was set at the value greater than the maximum growth rate of either of the species. Since the growth of each species was slower in the mixed than in the pure culture vessels, it appeared that amensalism had been demonstrated.

An antibiotic is by definition a substance produced by one species of microorganism which at low concentrations kills or inhibits the growth of another microorganism. Bacteria both produce and are susceptible to antibiotics. In the laboratory the classical test for antibiotic production is performed by allowing the antibiotic produced by one strain to diffuse through agar and inhibit the growth of an antibiotic-sensitive species. A clear inhibition zone surrounds the colony of antibiotic-producing strains, and the phenomenon of amensalism has been demonstrated. From the massive accumulation of data concerning antibiotic activity, it is clear that antibiotic production confers a selective advantage on certain strains of bacteria, fungi, actinomycetes, and algae when grown in laboratory cultures. However, the ecological significance of antibiotic production in natural environments is less clear and it can be argued that antibiotics are simply one of a number of waste products excreted by microorganisms, and that their antagonistic activity is coincidental. At the other extreme, Waksman[235] has proposed that antibiotic production is a purposeful behavior on the part of the antibiotic-producing organism. Evidence in favor of the ecological importance of antibiotics has been discussed at length by Brian[18] and it is a widely held view that such toxins generally account for the successful establishment of many species in natural ecosystems. A number of examples that provide evidence for the importance of antibiotics in some specific environments will be cited in later sections, but before wider generalizations are made, several points should be borne in mind.

A successful invasion of an established ecosystem by a small population of an antibiotic-producing species is unlikely to be brought about by amensalism. The reason for this conclusion is that the initially small population of the invading species would be unlikely to produce sufficient quantities of antibiotic to affect the outcome of the invasion. An argument that can be made against such a thesis is that in many natural environments the struggles for survival occur in microhabitats containing only a few organisms. If so, amensalism could be an important factor in ensuring successful colonization of new habitats. Since the soil environment is probably composed of a series of microhabitats, it is reasonable to expect to isolate antibiotic-producing species from soil samples. But again, if this is so, one is forced to wonder why antibiotic-resistant populations are not more abundant in well-studied natural environments.[4] If antibiotic production is ecologically important, antibiotic-resistant strains would be expected to be selected in the way observed in hospitals where antibiotic therapy has obvious ecological significance. Hill[242] has argued that this is precisely what happens when penicillin β-lactamase-producing organisms grow in the presence of *Penicillium chrysogenum* and organisms unable to produce the penicillin-degrading enzyme. It is also worth noting that antibiotics are produced in the laboratory by the growth of organisms in specially designed media. In natural environments, conditions may well not be conducive to antibiotic production and indeed in unsupplemented soil most microorganisms fail to produce antibiotics.[18] However, this does not exclude the possibility that antibiotic production occurs in those microhabitats which contain high nutrient concentrations. Wright[237] showed that toxins were produced in higher concentrations in areas of soil adjacent to decaying matter than in unsupplemented soil and Hill[242] has recently shown that antibiotic production by *P. chryso-genum* was of ecological significance in soil supplemented with maize seeds. Some antibiotics have been ascribed roles distinct from those associated with amensalism. Thus, the peptide antibiotics of *Bacillus* species have been shown to be excreted as a preliminary to sporulation; their production is associated with changes in cell surface structure.[201] Thus, these substances at least may have functions other than as weapons to be used in an antagonistic struggle with other species. Any ecological significance of these peptides may be incidental. Populations of antibiotic-excreting cells do not necessarily become dominant in many environments despite their presumed advantage.[4] This again does not necessarily provide an argument against the ecological significance of antibiotic production, but it does indicate that at least this phenomenon is not of overriding importance in antagonistic relationships in natural ecosystems.

Thus, although the effect of antibiotic production on bacterial interrelationships in laboratory experiments and in therapeutic applications has been abundantly demonstrated, the functional role of these substances in nature is not altogether clear. It is often proposed that antibiotics are produced in order to confer a selective advantage on the producers, which they probably often do. However, Waksman's proposals[235] that this is a purposeful behavior are still lacking unequivocal support.

Bacteriocines are bacterial antibiotics that are distinctive because they are active only against strains closely related to the producing organism. Since competition is likely to be most intense between closely related species, bacteriocine production could be significant in such competition. Pohunck[182] suggested that streptococci produce bacteriocines that are active against lactobacilli and it is this property that enables the former species to colonize the vagina. However, Hamon and Péron[78] rejected the idea of bacteriocines being produced in order to influence the outcome of competitive situations and concluded that they probably serve as fertility factors aiding cell-to-cell contact. This does not exclude the possibility that bacteriocines also cause amensalism and thereby influence the outcome of competition between closely related bacterial strains.

A number of bacteria produce enzymes that decompose cell wall polymers and cause lysis. Cells may either cause the depolymerization of their own cell walls (autolysis) or the cell walls of organisms of another species (heterolysis). Enzymes that attack the ubiquitous bacterial mucopeptides are important lytic agents, because the mucopeptides are thought to give the bacterial cell wall its strength and rigidity. Mucopeptides contain a "backbone" of repeating units of glucosamine and muramic acid to which various peptide moieties may be attached. Various enzymes attack different points of the polymer molecule.[194] Following heterolysis, the enzyme-producing organisms may derive nutrients from the remains of the lysed organisms and amensalism then merges into parasitism or predation.

The ability to lyse cohabitants confers two advantages on a population of microorganisms. The population of potential competitors is reduced and in addition extra nutrients are released from the lysed cells. This phenomenon has been studied most thoroughly in soil environments where organisms with the ability to produce extracellular depolymerizing enzymes are widespread.[3]

Predation

Bacteria are preyed upon by certain species of *Myxobacteria* and *Myxomycetes* and the majority of protozoan species belonging to the classes *Mastigophora*, *Ciliata*, and *Sarcodina* consume bacteria as their usual typical source of nutrients. Thus, bacteria are the first members of important food chains in aquatic and terrestrial environments. Bacterial predators are also known, and Chet et al.[37] have described a motile predatious bacterium that consumed *Pythium debaryanum*. It could well be argued that the bacterial species *Bdellovibrio bacteriovorus* is a predator and some authors have described it as such.[85] However, this species is most commonly considered to be a parasite. The relationship between phagocytes and invading bacteria in metazoan vascular systems is analogous to a predator-prey system and some of the results of experiments describing bacterial/protozoan interactions in chemostats could be of relevance to medical research.

Gause[65] was the first to study systematically the competition for food between different species of protozoa and the interrelationships between ciliates and their prey. He showed that the increase in the number of individuals of either of two ciliate species was greater in axenic than in mixed cultures. Thus, Gause obtained data for predatory ciliates that were very similar to those illustrated by Figure 2. He also showed that damped oscillations in the populations of predator and prey occurred in batch cultures. No significant advances in obtaining an understanding of the kinetics of ciliate/bacteria interactions were achieved until recent years when several workers began studying the growth of ciliates in continuous flow systems. The work of Gause remains the best documented description of predator-prey systems in closed laboratory environments.

Contois and Yango[40] reported the results of an experiment in which a mixed population of the myxamoeba *Dictyostelium discoideum* and the bacterium *Aerobacter aerogenes* persisted in continuous cultivation. Insufficient enumerations were made to detect any oscillations in the populations. Bungay[26] also grew a species of myxamoeba with bacteria and observed oscillations in the density of the populations. Curds[46] found that populations of ciliates and bacteria oscillated for prolonged periods when the two types of organisms were grown together as a continuous culture. These experiments seemed to indicate that the predators grew and consumed bacteria until they cut off their own food supply. The ciliate population then decreased in numbers. Reduced predation allowed the bacterial population to increase again and the cycle was repeated.

In the early mathematical models of Lotka[119] and Volterra[234] the assumption was made that the growth rate of a predator species was a function of the concentration of the prey organisms

$$\frac{dH}{dt} = \mu_h H - k_1 HP \tag{6}$$

$$\frac{dP}{dt} = k_2 HP - \delta P \tag{7}$$

where H = concentration of prey; t = time; μ_h = specific growth rate of prey; k_1 and k_2 = constants; P = concentration of predators; δP = specific death rate of predators.

In open systems such as a chemostat, Equations 6 and 7 become:

$$\frac{dH}{dt} = \mu_h H - DH - k_1 HP \tag{8}$$

$$\frac{dP}{dt} = k_2 HP - \delta P - DP \tag{9}$$

where D = dilution rate.

Although Equations 5 to 9 predict oscillatory population levels, as are observed in real systems, they do not indicate the existence of maximum growth rates of either predator or prey, and no account is taken of the fact that the specific growth rate of the prey is a function of the concentration of nutrients. If it is assumed that the growth rates of both predator and prey are functions of substrate concentration in the way suggested by Monod,[149] then one obtains the more refined models described by Bungay and Bungay[27] which have been developed by Canale.[31]

$$\frac{dH}{dt} = \mu_m \frac{SH}{K_s + S} - DH - \frac{\eta_m}{w} \frac{HP}{L_s + H} \tag{10}$$

$$\frac{dP}{dt} = \eta_m \frac{HP}{L_s + H} - DP \tag{11}$$

where μ_m and η_m = maximum growth rates for prey and predator, respectively; K_s and L_s = saturation constants for prey and predator, respectively; s = concentration of growth limiting nutrient of prey; and w = predator yield constant, i.e., predators produced/hosts consumed.

However, despite their weaknesses, the Volterra/Lotka models fit observed data as well as the models that incorporate the Monod relationship.[27] Investigations designed to define more precisely the relationship between the growth rate of *Tetrahymena pyriformis* as a function of prey (*Aerobacter aerogenes*) organisms have been reported by Curds and Cockburn.[45,47] In the earlier paper it was found that the individual feeding rate of a ciliate was governed by the concentration of ciliates as well as the concentration of bacteria present. From these observations a model for ciliate feeding was derived in which the equation of Contois[39] was adapted to describe the population effect found by Curds and Cockburn.[45]

Thus

$$f = \frac{f_m \bar{b}}{C\bar{n} + \bar{b}} \qquad (12)$$

where f = ciliate feeding rate coefficient; f_m = maximum ciliate feeding rate; $\bar{b}$ = mean bacterial concentration; C = a growth parameter (constant); and $\bar{n}$ = mean number of ciliates.

Equation 12 may be converted into

$$\frac{\bar{b}}{\bar{n}\,f} = \frac{\bar{b}}{\bar{n}}\frac{1}{f_m} + \frac{C}{f_m} \qquad (13)$$

When $\frac{\bar{b}}{\bar{n}\,f}$ was plotted against $\frac{\bar{b}}{\bar{n}}$ a straight line was obtained. Therefore, it was concluded that Equation 13 was correct and that the slope of the line was equal to $\frac{1}{f_m}$ and the intercept on the absissa was equal to the constant C.

In more recent work, Curds and Cockburn[47] designed a two-stage continuous culture apparatus in which they studied the growth and feeding kinetics of T. pyriformis. Steady states were obtained in the second stage fermenter by pumping a pure culture of A. aerogenes from the first stage into the second stage which also contained the ciliate population. Contrary to the earlier findings of these authors, it was found that no population effect existed, but that the ciliate feeding rate was dependent upon ciliate cell size and growth rate. Cell size in turn depended on growth rate and hence a more complicated analysis of predator/prey growth kinetics was possible.

Canale[31] applied methods of nonlinear analysis to characterize predator/prey models. Three types of situation were predicted which were dependent on the kinetic constants of the two organisms. These were (1) limit cycle oscillations in a model in which predator growth rate was a function of the concentration of prey, (2) a system where damped oscillations occurred, and (3) a non-oscillatory response in which the two populations asymptotically approached constant or steady-state values. Curds[48] has used a digital computer simulation technique to analyze the stability of

bacterial and ciliate populations growing together in a single-stage continuous culture apparatus. Curds[48] used the kinetic constants from Curds and Cockburn[47] and assumed that the growth of T. pyriformis obeyed simple Michaelis-Menten kinetics. The results of a computer model are compared in Figure 6 with observed results. The computer model agrees with observed results qualitatively, but not quantitatively. When the equations of Curds and Cockburn[47] were used for the computer program, rather than the Michaelis-Menten equations, the agreement was no better, despite the fact that these equations took account of observed variations in ciliate cell size. Curds[48] used the computer system to examine the effects of varying various parameters on the equilibrium between bacteria and ciliates. It was predicted that if the dilution rate of the continuous fermenter was increased to a value approaching the maximum growth rate of the ciliate, then damped oscillations would occur and eventually steady population densities could be expected. At slightly higher dilution rates, steady-state populations of bacteria and protozoa were obtained without initial oscillations in population density. With changes in a variety of other parameters, the computer model predicted the amplitude and frequency of the oscillations.

At the moment there are several areas of uncertainty regarding the kinetics of growth of protozoa. For example, how quickly do protozoa respond to increases in bacterial population density? It may well be wrong to assume that this response is instantaneous. Gaudy et al.[64] have shown that the growth rate of bacterial species does not respond rapidly to changes in substrate concentration and it seems reasonable to expect ciliates to behave similarly. Also, it seems unlikely (despite the marked progress made recently by Curds and his co-workers) that a true description of ciliate mass growth rate as a function of bacterial population density has yet been obtained.

Various factors are important in determining whether or not particular bacterial species will be selected as prey. Alexander[4] has provided a list showing the food choices of various microbial predators. Chemical composition, particularly of cell surfaces, may be important in prey selection. Extracellular slime or capsular polysaccharides are generally considered to contribute to making certain bacterial strains resistant to phagocytosis

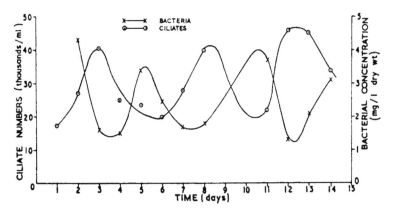

FIGURE 6A. Fluctuations in the populations of *Aerobacter aerogenes* and *Tetrahymena pyriformis* organisms during cultivation in a single-stage continuous culture apparatus.[236]

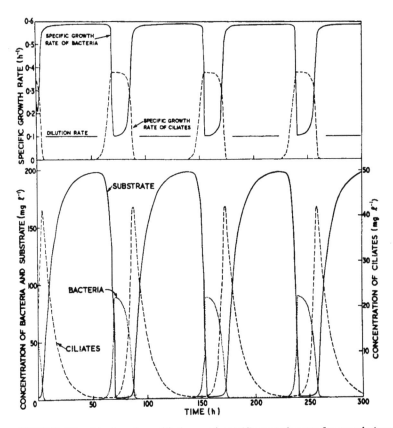

FIGURE 6B. Limit cycle oscillations and specific growth rates for populations of bacteria and ciliates as predicted by computer analysis. The kinetic constants assumed were as follows: concentration of growth-limiting substrate (sucrose) in the in-flowing medium 200 mg l^{-1}; dilution rate, 0.1 hr^{-1}; maximum specific growth rates 0.6 and 0.43 hr^{-1}; saturation constants 4.0 and 12.0 mg l^{-1}, and yield coefficients 0.45 and 0.54 for bacteria and ciliates, respectively.[47] (From Curds, C. R., *Water Res.*, 5, 793, 1971. With permission.)

and the same phenomenon may be important in bacteria/protozoa interactions. Toxins are produced by a number of bacterial species. Thus, the pigments of *Serratia marcescens* and *Chromobacterium violaceum* have been reported to protect these species from protozoal ingestion,[241] and *Pseudomonas fluorescens* produces extracellular substances which are toxic to protozoa.[243] On the other hand, the predatory bacterium described by Chet et al.[37] was attracted to its prey (*Pythium debaryanum*) by substances released into the medium by the fungus. A nonpredatory bacterium was not attracted by these substances, which appeared to be cellulose, and its oligopolymers, which in turn are known to be components of the cell wall of *Pythium*. Large size could be important in protecting some bacteria from predation. Bacteria may also be able to escape capture by being motile. Gause[65] described an experiment in which motile *Paramecium caudatum* cells survived predation far better in an environment containing a refuge than in a homogenous environment. Such a situation may also be important for the survival of bacteria. Some bacteria, particularly as spores, may be able to resist digestion and thus escape predation. But, as Alexander[4] points out, it is not usually possible to explain why particular bacterial species are not consumed by specific protozoan species, although they appear to be nontoxic.

Parasitism

Bacteria are parasitized by two types of organisms — the bacteriophages and the bacterial species *Bdellovibrio bacteriovorus*. Conversely, there are bacterial species that grow and multiply within protozoal cells. These have physiological characteristics that enable them to avoid being digested by the enzymes that decompose those bacterial species that are the usual prey of protozoa (see review by Ball[10]). Fungi and algae are also subject to attack by bacterial endoparasites.

B. bacteriovorus is a unique species. The cells penetrate the cell wall of the host bacterium and then grow in the periplasm until they surround the host cell, which is ultimately lysed. At one time it was believed that *Bdellovibrio* strains were obligate parasites, but Seidler and Starr[199] have recently shown that host-independent *Bdellovibrio* species can be isolated. Stolp and Starr[214] studied the growth of *B. bacteriovorus* on *Erwinia amylovora*

(Figure 7) and showed how the decline in the *Erwinia* population was followed by a decrease in the parasite population. Varon and Shilo[229] have studied the kinetics of *Bdellovibrio* invasions of *Escherichia coli* B in some detail. These workers demonstrated that a series of saturation curves (c.f. Figure 3) described the attachment of *B. bacteriovorus* to *E. coli*, the slope and saturation level of the curves varying with experimental conditions. One important factor determining both the attachment rate and the saturation level was the physiological state of the *Bdellovibrio* cells. The majority of the actively motile cells were attached after 20 min of incubation with the host. Inhibitors of *Bdellovibrio* motility (chelating agents, NaN_3, and low pH) inhibited attachment, as did anaerobiosis. Inhibitors of protein synthesis (streptomycin, puromycin, and chloramphenicol) had no effect on attachment, but inhibited invasion of the host cells. Penicillin, however, affected neither attachment nor invasion. It appears likely from these results that the synthesis of cell wall-degrading enzymes is required before invasion can be completed successfully. The results of Varon and Shilo[229] support earlier suggestions by Stolp and Starr,[214] who stressed the importance of the motility of *B. bacteriovorus* in its attack on the host. They proposed that the physical impact between motile parasite and its prey caused damage to the host cell wall which was a prerequisite for attachment. The main advantage of motility may, however, simply be that

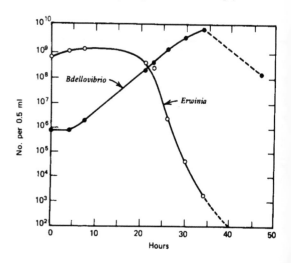

FIGURE 7. The growth of *Bdellovibrio bacteriovorus* on *Erwinia amylovora* in batch culture.[214] (From Stolp, H. and Starr, M. P., *Antonie van Leeuwenhoek*, 29, 217, 1963. With permission.)

the *Bdellovibrio* cells can seek out their hosts more effectively if they are in a healthy motile state. Varon and Shilo[229] found that the average number of cells attached to *E. coli.* varied with the parasite/host ratio, and also that the host population was heterogenous in its binding capacity. Thus, those organisms that were attacked within seconds by tens of *Bdellovibrio* cells lysed before a new generation of parasites could be produced, whereas other host cells bound only a few parasites, which were able to reproduce before the host lysed.

The action of virulent phage on bacteria shows several similarities to the action of *Bdellovibrio* species. Successful invasion is preceded in both cases by attachment to the bacterial cell surface. In the absence of phage receptor sites, phage infection cannot take place and it is probable that the development of phage resistance is commonly associated with a selection of mutants with altered cell surface characteristics. Figure 8 illustrates the results of an experiment showing how a phage attack is often followed by secondary growth of phage resistant cells. In addition to phage resistance brought about by the mutational loss of phage receptor sites, cells may produce a capsule which acts as a physical barrier between the virus and the receptor sites.[174] Paynter and Bungay[173] have studied the kinetics of the interaction between virulent bacteriophages and their hosts in continuous flow systems. In such a system, if only sensitive bacteria and virulent phage persisted, one would predict oscillations of host and parasite populations similar to the predator/prey oscillations described in the previous section. Oscillations were indeed observed by Paynter and Bungay,[173] but the situation was complicated by the presence of phage resistant cells. High substrate concentrations should be reached after the populations of sensitive cells first declined and the selective pressure for the growth of resistant organisms would be high. However, the population of sensitive cells remained in the growth vessel and showed periodic bursts of growth. The most obvious explanation for this observation is that the sensitive cells had a competitive advantage over the resistant mutants in competing for the undefined growth-limiting nutrient. When the density of the phage population fell (due to lack of receptive hosts and washout of the nongrowing phage particles from the fermenter), the transitory period of growth of the sensitive bacterial popula-

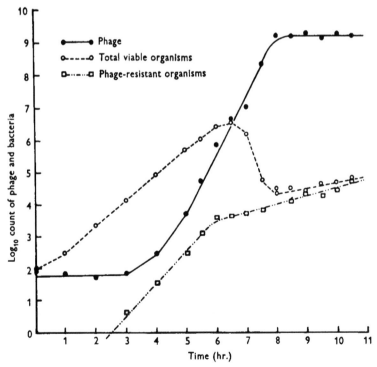

FIGURE 8. The growth of Vi-phage A and Vi-type A of *Salmonella typhi* from small initial inocula in batch culture.[6] (From Anderson, E. S., in *Microbial Ecology,* 7th Symp. Soc. Gen. Microbiol., Williams, R. E. O. and Spicer, C. C., Eds., Cambridge University Press, London, 1957, 189. With permission.)

tions was possible. The density of the phage particles would then again increase and the cycle would repeat itself. The population of sensitive cells would never be completely removed from the system because phage adsorption follows the kinetics of a first order reaction[212] and is determined by the concentration of both phage particles and cells.[6] Therefore, at low concentrations of sensitive bacteria, phage infection would not occur and the population of phage-sensitive organisms could maintain itself in the culture between the periods of active growth.

The other reason cited by Paynter and Bungay[173] for the persistence of the sensitive cells in the presence of virulent phage was "peaceful co-existence." Such a phenomenon was first conceived by Hunter[94] and, although this appears to be a somewhat vague concept, Horne[90] has more recently described results which are consistent with it. In this work, coevolution of both phage and host organisms occurred during a 2-day transition period, after which a stable relationship was established and maintained in continuous culture vessels for 52 weeks. Only small fluctuations in the bacterial and phage populations were observed and the bacterial population reestablished itself at a similar level to that which occurred prior to phage infection. Plaques formed by phages taken from the vessels after stabilization and plated onto lawns of control *E. coli* were reduced in size, and the latent period was extended when compared with the phage type used to inoculate the fermenter. The bacteria were resistant to both the original and the evolved phage after stabilization of the population. Such an evolution parallels the appearance of attenuated viruses in natural environments and one has then to discover why the less virulent forms are at a competitive advantage. In a heterogenous system, highly virulent parasites kill all of their hosts and preclude the possibility of their own further growth. Attenuated strains may persist indefinitely with their host population in another habitat, from where they could recolonize areas previously occupied by the highly virulent strains.

Contois and Yango[40] also studied the interaction between *E. coli* and a coliphage in continuous culture. They found that the phage titer rose rapidly to a maximum and then declined to a level that was constant with time. The bacterial population dropped to a low level at which it persisted indefinitely.

Three groups of workers[70,90,173] studied the interaction between *E. coli* and coliphages in continuous cultures and obtained contrasting results. More recently, Paynter and Bungay[176] challenged some of Horne's[90] conclusions, and Horne[91] replied to the criticisms. This exchange seems to add little of substance to the previous work of these authors, in all of which the data seem to be too incomplete to resolve the conflicts of opinion on such topics as the significance of differences in phage types, plaque morphology, or dilution rates. The intriguing probability that mutations of both the phage and bacterial populations take place in continuous flow systems leads to the conclusion that such a dynamic situation could be very complex. More detailed experimentation is required to elucidate this type of interaction which may be occurring between constantly changing populations that may never be expected to reach a stable steady state.

Several groups have studied the continuous growth of lysogenic cultures and have observed the spontaneous or induced appearance of virulent bacteriophages. Noack[160] has provided a theoretical analysis which predicts that for a given set of growth conditions, two stable steady concentrations of free phages and lysogenic cells can exist at a given dilution rate in a continuous flow system. The crux of Noack's argument is that the interaction between the injected phage DNA molecules and the immunity repression system of the host cell leads to either an increased or a decreased induction rate for vegetative development of both the prophage and the superinfected phage. If the superinfection rate is high enough, then all of the immunity repressor molecules will be neutralized and an autocatalytic production of phage will occur. On the other hand, if the superinfection rate is low, then a decreased induction rate of phage replication will occur and a different ratio of the number of free phage particles to lysogenic cells will be established. Noack's analysis predicts that transitions between the two steady states should be caused either by permanent variations in dilution rate or by a temporary disturbance of the regulated equilibrium growth conditions. Such a disturbance could be brought about by irradiation or by heating a temperature sensitive lysogenic complex. Noack[160a] has presented some preliminary data in which the steady-state concentration of an unnamed bacterial species was decreased following an induction heating at $43°C$ for a period of 30 min. This result was consistent with the theory described above. Paynter and Bungay[175] completed a similar experiment with *E. coli* and found that for a given steady-state bacterial population there could be more than one steady-state level of free phage. The results published to support this conclusion were at best only suggestive. Noack[160a] found that following heat induction there was a steady fall in bacterial population density to a lower steady-state value, whereas the bacterial populations observed by Paynter and Bungay[175] returned to their pre-induction levels.

Northrop[161] studied the appearance of virulent bacterial phages in lysogenic *E. coli* cultures after continuous growth in the presence of the mutagen triethylene melamine. These experiments were conceptually similar to those of Noack[160a] and Paynter and Bungay,[175] but Northrop showed that following the induction of the formation of virulent virus particles, bacterial colonies that were resistant to this virus appeared. The virus titer remained high despite the fact that apparently all of the cells became phage resistant. Northrop concluded that virus production continued because of the constant formation of virus-sensitive cells by back mutation from the resistant population.

The groups of workers who have studied the kinetics of the interaction between phage and bacteria have obtained different results, even when the experimental procedures have been similar. Either alternative strains and species behave in a distinct way or small differences in the growth conditions have large effects on the interrelationships. Therefore, possibly no general result can be predicted to follow phage attack of continuous cultures of bacteria. Alternatively, some of the published work in this area may not reveal the full nature of the interactions that were occurring. This question is of more than academic interest in situations where large investments are being made in commercial continuous culture systems for the production of bacterial protein for animal feed.

Commensalism

Many bacteria produce substances that promote the growth of other species. Such a phenomenon is probably the most widespread form of commensalism. However, there are other ways by which the growth of one population may stimulate the growth of a cohabitant species. For example, one section of the population may remove a substance which is toxic to another or one species may convert an otherwise unavailable substrate into a form which is useful to its commensalistic partner.

The formation of extracellular products is a common if not invariable feature of microbial growth. These compounds are potential substrates for the growth of other species, although they may be little more than the waste products of the organisms excreting them. There are a very large number of substances produced by microorganisms,[145] and the combinations of possible commensal relationships are legion. Some of these compounds may be produced in large quantities and could serve as the carbon and energy sources for other species. Thus, ethanol produced by yeast cultures growing on glucose as a carbon source may be used by *Acetobacter* strains to produce acetic acid. Other organisms in turn could use this acetic acid as an energy source. Thus, a succession of species may convert a given carbon compound into water and carbon dioxide in a series of distinct stages. In closed environments the succession will occur over a period of time during which different populations will become dominant in turn. Most of the work on succession has been done with organisms growing in natural environments and, although a good deal of descriptive data exists, the factors determining prominence in the succession are poorly defined.[4] Nevertheless, it is reasonable to assume that successive conversion of available energy sources is important.

Vitamins and growth factors are often excreted by bacteria in small quantities and have been shown to be used by commensal species. A common way of indicating colonies that produce copious quantities of a given compound is to observe the growth of an auxotrophic species which has a requirement for the desired compound. If the auxotroph grows around any particular colony, then that colony may usually be expected to be a mutant producing an excess of the compound under consideration. Owen[167] has found that mixed cultures of methane- and methanol-utilizing organisms are often difficult to separate and grow as axenic cultures because one species has a commensal relationship with the other. Thus, some *Hyphomicrobium* species only grew well on methane as a carbon source in the presence of *Pseudomonas* species. The same was true when the cultures were grown in continuous culture.[166] Further work has led to the suggestion[167] that the *Hyphomicrobium* species were growing on methanol produced by the pseudomonads. Table 1 indicates the variety of types of compounds that have been shown to serve as the basis of commensal relationships. The results of Demain et al.[51] are of special interest since they demonstrate how the cross-feeding of purineless mutants by wild-type *Bacillus subtilis* can occur in a heterogenous population of a single species. Halmann et al.[77] have also shown how the growth of a single species can be promoted by the activities of a proportion of the population. The latter authors conclude that in cultures of *Pasturella tularensis* a small number of bacteria in the population excrete a growth-promoting substance that is essential for the growth of the major proportion of the total population. This phenomenon was considered to be a normal characteristic of the growth of *P. tularensis,* whereas the results of Demain et al.[51] illustrated a situation contrived in the laboratory. Halmann et al.[77] suggested that the population dynamics of the different strains of *P. tularensis* were determined on the one hand by a high mutation rate toward the organisms producing the growth promoter and on the other hand by the lack of selective pressure in favor of growth-promoter producing organisms. Although the concept proposed by Halmann et al.[77] is intriguing, only preliminary data were presented to support it, and corroborative data are needed.

Several investigators have studied commensalism in the laboratory using chemostats to provide an open system. Established theory would predict that such cultures should coexist indefinitely without oscillations in the population densities and that the density of the species producing growth-promoting substances should be unaffected by the presence of the commensal species. If the concentration of the growth-limiting substrate was altered, then the population densities of the two species should alter proportionately. The results of Shindala et al.[204] illustrate a commensalism interaction. *Saccharomyces cerevisiae* and *Proteus vulgaris* were grown as continuous mixed cultures in which the yeast produced niacin or a related compound that was required by the bacterium. However, detailed study of the results reveals anomalies that seem to contradict some of the main conclusions of the paper. For example, a figure is provided that shows absolute agreement between the population densities of *S. cerevisiae* in pure and mixed cultures, but other data seem to suggest that the steady-state population density of the yeast was sometimes more than halved by the presence of its commensal (i.e., in a pure culture growing at a dilution rate of $0.2 \ hr^{-1}$ the pure culture population density of yeast organisms was 10.3×10^6 organisms ml^{-1}, but in mixed culture the yeast population was 4.8×10^6 organisms ml^{-1} in the

TABLE 1

Examples of Compounds Serving as the Basis of Commensal Relationships

Compound	Species producing compound	Species requiring compound	Reference
Purine	*Bacillus subtilis*	*B. subtilis* auxotrophs	51
Organic acid	*Aerobacter cloacae*	Unnamed bacterium	14
Isobutyrate	*Diphtheroid*	*Treponema microdentium*	80
Nicotinic acid	*Saccharomyces cerevisiae*	*Proteus vulgaris*	204
Vitamin K	*Staphylococcus aureus*	*Bacteroides melaninogenicus*	67
Nitrite ions	*Nitrosomonas*	*Nitrobacter*	132
Hydrogen sulfide	*Desulphovibrio*	Sulfur bacteria	19
Water	*Bacillus mesentericus*	*Clostridium botulinum*	104
Polysaccharides	algae	bacteria	114
Hydrogen	Rumen bacteria	*Methanobacterium ruminatium*	85
Methane	Anaerobic methane bacteria	methane-oxidizing bacteria	19
Ammonium ions	Many heterotrophs	*Nitrosomonas*	—
Nitrite	*Nitrosomonas*	*Nitrobacter*	—
Nitrate	*Nitrobacter*	Denitrifying bacteria	—
Acetyl phosphate	*Diphtheroid*	*Borrelia vincenti*	157
Fructose	*Acetobacter suboxydans*	*Saccharomyces carlsbergensis*	36

same medium; compare Figures 2 and 4 of Shindala et al.[204]). Unfortunately, the substance limiting the growth of the yeast population was not identified. But since it was shown not to be a vitamin-like factor and ample glucose was supplied for the population densities obtained, it seems likely that it was a trace element that was required by both species. When niacinamide was added to a mixed culture, the bacterial population density increased and at precisely the same time the density of the yeast population decreased. This result would seem to imply that a competitive situation existed even before the niacin was added. If this were so, then these results would be explained if the increase in bacterial population brought about a lowering of the concentration of the substrate limiting the growth of the yeast organisms — a substrate that the bacteria also required and for which they had the greater affinity. In order to demonstrate rigorously the existence of commensalism in an open system, one must be able to show that the dependent species does not utilize the substrate limiting the growth of its commensalistic partner. Tsuchiya and Fredrickson[247] have reported the results of continuous culture experiments in which what was described as competition plus commensalism was observed in stable mixed cultures of *S. cerevisiae* and *Lactobacillus casei*. Competition arose because the two organisms competed for glucose. Bergter and Noack[14] grew *Aerobacter aerogenes* and another gram-negative organism (unnamed) in a chemostat under nitrogen-limited conditions with glucose as the carbon and energy source. The unnamed species competed for the limited supply of nitrogen, but was dependent on *A. aerogenes* for a supply of organic acids which were essential for its growth. Oscillations in population density were observed in these experiments and a mathematical model was presented to account for the results.

Commensalism involving organic growth factors in open systems may often be associated with competition. To set up a system in which true commensalism occurs in a defined medium may require careful contrivance. This may indicate that true commensalism, as strictly defined, may be rare and that the dependent species in commensal relationships may rarely have absolutely no effect on the associated species.[204] Commensalism seems more likely to occur in suitable open environments containing mixed populations of species with widely dissimilar nutrient requirements. Such relationships are probably often loose and transient in nature and may not always involve set pairs of species.

The environment may be altered by one species in such a way that a second species can grow because of changes in pH value or redox potential. For example, the growth of lactic acid bacteria may reduce the pH value of media so that yeast species can grow and it is often observed that the growth of fastidious anaerobes is facilitated by the presence of aerobes which can remove remaining traces of oxygen from the environment. Conversely, photosynthetic organisms may produce oxygen that is required for the growth of aerobes.

The phenomenon of one species removing substances that would otherwise prevent the growth of a second species is widespread. Examples of this type of commensalism are shown in Table 2. Sometimes the situation can become quite complex. For example, certain *Pseudomonads* produce the compound *N*-heptyl-4-hydroxyquinoline-*N*-oxide, which is an inhibitor of oxidative metabolism.[42,116] Streptomycin is more active against aerobically growing cells, and hence in the presence of *N*-heptyl-4-hydroxyquinoline-*N*-oxide-secreting *Pseudomonads* other species become streptomycin resistant because they grow anaerobically despite the presence of oxygen.

Jannasch and co-workers[117,227,232] have studied the growth of a species of *Thiocystis* (purple) and a species of *Chromatium* (brown) in batch and chemostat mixed cultures. Both species required sulfide, but their growth was inhibited by high sulfide concentrations. In a batch culture in which the initial sulfide concentration was 2mM, *Thiocystis* outgrew *Chromatium*, whereas the reverse was true in a sulfide-limited mixed culture experiment that was performed in a chemostat. This apparent anomaly is readily explained by the relationships between specific growth rate and sulfide concentration (Figure 9). At an initial sulfide concentration of 2mM *Chromatium* could grow relatively quickly only after the *Thiocystis* had metabolized much of the sulfide in the batch culture medium. However, in the continuous culture experiments, the steady-state sulfide concentration was low and under these conditions the *Chromatium* had the growth advantage for the reasons described previously. One intriguing point that arises from Figure 9 is that a range of dilution

TABLE 2

Examples of Compounds Which are Toxic to one Species
but Which are Removed from the Environment by a Second Species

Compound	Details of interrelationship	Reference
Concentrated sugar solutions	Osmophilic yeasts metabolize the sugar and thereby reduce the osmolarity so allowing the growth of species which are sensitive to high osmotic pressures.	153
Oxygen	Aerobic organisms may reduce the oxygen tension, thus allowing anaerobes to grow.	153
Hydrogen sulfide	Toxic H_2S is oxidized by photosynthetic sulfur bacteria, and the growth of other species is then possible.	227
Food preservatives	The growth inhibitors benzoate and sulfur dioxide are destroyed biologically.	4
Lactic acid	The fungus *Geotrichum candidum* metabolizes the lactic acid produced by *Streptococcus lactis*; the acid would otherwise accumulate and inhibit the growth of the bacteria.	58
Mercury-containing germicides	*Desulphovibrio* species form H_2S from sulfate, and the sulfide combines with mercury-containing germicides and permits bacterial growth.	216
Antibiotics	Enzymes are produced by some species of bacteria which break down antibiotics; thus the growth of antibiotic sensitive species is allowed.	197
Phenols	Some bacteria can oxidize phenols and thereby allow other species to grow.	153
Trichlorophenol	A number of gram-negative bacteria can absorb trichlorophenol in their cell wall lipids and thereby protect *Staphylococcus aureus* from its action.	226

rates exists between which the outcome of a competition experiment would be determined by the initial extracellular concentration of sulfide in the growth vessel. For example, at a dilution rate of 0.12 hr^{-1}, a small population of *Thiocystis* introduced into the *Chromatium* culture would be displaced because the sulfide concentration would be too low for the former species to maintain itself in the growth vessel. On the other hand, a small population of *Chromatium* could not maintain itself when introduced into a steady-state culture of *Thiocystis* because the high sulfide concentration would be inhibitory to growth. This work shows how a commensal type of relationship observed in batch cultures can become a quite different type of interaction in an open system. Sulfide is not the only substrate that might be expected to cause the type of phenomenon described by Van Gemerden and Jannasch[227] because a number of compounds give the type of inhibition curves described by these authors. Meers and Tempest[133] also described experiments in which the outcome of a competition experiment depended on the initial concentrations of the organisms in a mixed population. However, the cause of this result was quite different from that described by Van Gemerden and Jannasch[227] which will be described.

A form of commensalism, an explanation for which is elusive, has been observed when a population of one species extends the temperature tolerance range of a second species. For example, the growth of a mesophilic *Bacillus* at 65°C was only possible in the presence of a thermophilic

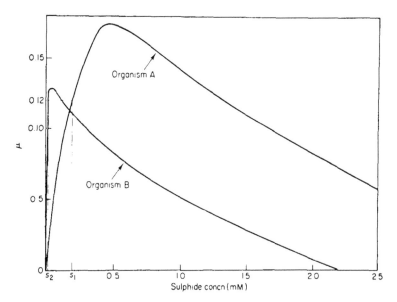

FIGURE 9. Relationship between specific growth rate and sulfide concentration of two purple sulfur bacteria. Curve B (*Chromatium* sp.) was determined experimentally; curve A (*Thiocystis* sp.) is hypothetical.[227] (From Veldkamp, R. H. and Jannasch, H. W., *J. Appl. Chem. Biotechnol.*, 22, 105, 1972. With permission.)

species.[245] Similar types of results have been obtained with certain methanol oxidizing bacteria which grow as mixed cultures at 50°C, but as pure cultures will not grow at temperatures above 40°C.[167]

Alexander[4] has discussed a phenomenon in which one species stimulates a neighbor to enter a resting stage when nutrients are depleted or when inhibitors accumulate. An example of this process is endospore formation in *Bacillus* species. Endospore formation may be induced by metabolic products of a partner in the interspecific association.[4] Data illustrating this phenomenon are meager and it could be that the onset of sporulation is hastened in mixed cultures simply because of an increased rate of depletion of nutrients when several species grow together.

A large number of bacterial species produce extracellular enzymes that depolymerize potential substrates into metabolizable units. The organisms producing the extracellular enzymes have no special claim to the metabolizable substrates so formed and it is probably quite common in natural environments for many species to depend on the enzymes produced by benevolent neighbors. An example of such a phenomenon has been observed in the laboratory where nonamylolytic species grew well on starch-containing media in the presence of amylase-producing *Bacillus subtilis*,

but only poorly on the same medium as pure cultures.[136] Such a relationship may well have been competitive as well as commensalistic, but was not studied in detail.

Mutualism

The types of relationships from which both species benefit are particularly varied. This type of interaction can range from a loose cooperation, which is not essential for the survival of the interacting species (sometimes termed protocooperation), to an obligate association on which both species depend for their continued survival (symbiosis). Alternatively, two species may produce more of a given product when grown together than when grown alone. Such a relationship is termed synergism. Observed interactions often do not fall neatly into these categories and usually show the characteristics of more than one type of interrelationship.

One form of protocooperation is that involving the exchange of growth factors between two species. Such a phenomenon, which is usually termed syntrophism, has been observed on several occasions and may well be common in natural environments. Nurmikko[163] demonstrated that *Lactobacillus plantarum* produced folic acid which was required by *Streptococcus faecalis*. These species in turn produced phenylalanine which was required by *L. plantarum*. Figure 10 illustrates one of Nurmikko's batch culture experiments. Yeoh et al.[238] showed that *Proteus vulgaris* and *Bacillus polymyxa* grew as continuous mixed cultures, but that neither species grew independently in the medium of Nurmikko.[163] *P. vulgaris* required biotin for growth, whereas *B. polymyxa* required niacin, and it was concluded that each species provided the vitamin required by the other. However, the total population of organisms oscillated in concentration — a phenomenon which would not be predicted if syntrophism was the only interrelationship between the species. Further experiments indicated that *P. vulgaris* also produced a proteinaceous compound that inhibited the growth of *B. polymyxa*.

Burkholder[28] has shown that two marine bacterial species can grow together in a thiamine-free medium despite the fact that both species required thiamine when grown as axenic cultures. The explanation for this result was that each species excreted intermediates (riboflavin and pantothenate) needed by the other species for the biosynthesis of this vitamin.

Sometimes the substances exchanged in proto-cooperative relationships are energy sources rather than growth factors. For example, Okuda and Kobayashi[165] have suggested that the growth and nitrogen fixation of *Azotobacter vinlandii* and *Rhodopseudomonas capsulatus* were stimulated in mixed cultures of the two species due to the symbiotic relationship illustrated by Figure 11. In other cases only one species provides an energy source, whereas the other produces growth factors. This kind of relationship may exist between bacteria and auxotrophic marine algae;[19] the bacteria provide vitamins that are essential for the algae which in turn release organic molecules required by the heterotroph. Mutualism is generally considered as a relationship between organisms growing in close proximity to one another. However, the exchange of oxygen and carbon dioxide between photosynthetic and heterotrophic organisms is a form of mutualism and it is not difficult to design a multitude of closed loop cycles showing how almost all species are involved in mutualistic relationships in natural environments.

Another form of mutualism is that in which one species destroys a toxin that would otherwise suppress the growth of its associate species. This second species in turn produces substrates required for the growth of the former species. The example par excellence of such a symbiotic relationship is provided by the "species" *Methanobacillus*

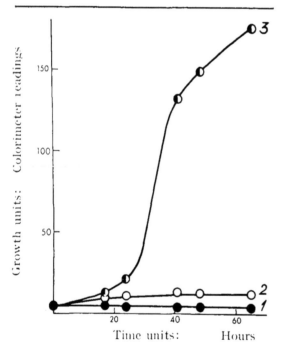

FIGURE 10. Growth of *Lactobacillus arabinosus* 17-5 (phenylalanine-requiring strain) and *Streptococcus faecalis* R (folic acid-requiring strain) in a synthetic medium lacking phenylalanine and folic acid. Curve 1, *S. faecalis*; curve 2, *L. arabinosus*; curve 3, both strains together growing in symbiosis.[163] (From Nurmikko, V., *Experientia*, 12, 245, 1956. With permission.)

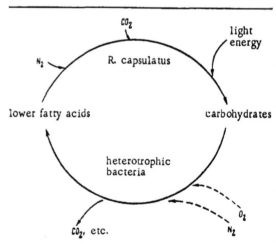

FIGURE 11. Proposed exchange of substrates in mixed cultures of *Rhodopseudomonas capsulatus* and *Azotobacter vinlandii*. The symbiotic relationship was demonstrated in laboratory experiments and was suggested to occur widely among the microorganisms isolated from rice fields.[165] (From Okuda, A. and Kobayashi, M., *Mikrobiologiya*, 32, 797, 1963. With permission.)

omelianskii. Recently, Bryant et al.[24] have shown that this presumed species, which had been the subject of many investigations, was in fact a mixture of two species of rod shaped organisms growing in a symbiotic relationship. One of these species (the S organism) was a gram-negative motile anaerobic rod which fermented ethanol with the formation of hydrogen and acetate

$$CH_3 CH_2 OH + H_2 O \rightarrow CH_3 COO^- + H^+ + 2 H_2$$

However, this species was inhibited by hydrogen which explained why growth of the pure culture was poor on an ethanol medium. The second species involved in the symbiotic association was a gram-variable, nonmotile, anaerobic rod which could utilize hydrogen but not ethanol for growth and methane formation

$$4H_2 + CO_2 \rightarrow CH_4 + 2H_2 O$$

Thus, this second species oxidized hydrogen, produced by the S organism, which would otherwise have accumulated to inhibitory levels. It is surprising that *M. omelianskii*, which is one of the most abundant bacteria in anaerobic sludge, should have been studied for so many years before it was shown to be a mixed culture. This indicates how close is the interdependence between the species. It is possible that other supposed pure cultures are in fact symbiotic mixtures. Pollock[184] has described another mutualistic relationship in which the autoinhibitory end products of the metabolism of one species (long-chain fatty acids) were used as carbon sources by a second species.

Synergism is a term which is often applied to describe situations that may be caused by commensalism or symbiosis. Often the basis for the increased productivity is unclear. Table 3 provides examples of synergistic relationships.

Meers and Tempest demonstrated that organisms of an axenic culture can produce autostimulatory substances.[133] This observation is difficult to categorize, but can be considered as either mutualism between members of a given species or as one aspect of competition. The phenomenon was discovered during experiments in which *Bacillus* species and *Torula utilis* organisms were competing for limited quantities of magnesium in chemostat cultures (Figure 12). The outcome of these experiments depended on the initial con-

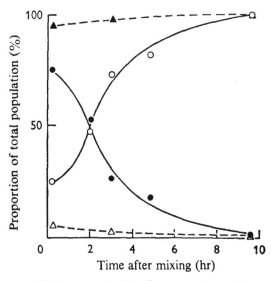

FIGURE 12. Growth of *Bacillus megaterium* and *Torula utilis* in magnesium-limited simple salts medium in a chemostat. In the experiment represented by the solid lines the initial concentration of *B. megaterium* (•) was 9 x 10^7 organisms/ml and *T. utilis* (○) 3 x 10^7 organisms/ml, i.e., 75% and 25% of the initial mixed population, respectively. Also recorded (broken lines) are the results of a similar experiment in which the initial concentration of *B. megaterium* (▲) was increased to 1.8 x 10^8 organisms/ml and that of *T. utilis* (△) lowered to 1 x 10^7 organisms/ml, i.e., 95% and 5% of the total population, respectively. In both experiments the dilution rate was 0.3 hr^{-1} and the temperature was 33°. The pH value was not controlled, but did not vary beyond the range 6.1 to 6.5 in each experiment.[133] (From Meers, J. L. and Tempest, D. W., *J. Gen. Microbiol.*, 52, 309, 1968. With permission.)

centration of organisms in the mixed inoculum. A study of the effect of cell-free supernatant fluids from *B. subtilis*, *B. megaterium*, and *T. utilis* cultures on the growth of these organisms showed that the growth of the *Bacillus* species was promoted to the greatest extent by their own culture liquids. These results indicated that the relationship between specific growth rate and magnesium concentration was dependent on the presence of extracellular products that promoted magnesium uptake. Further experiments confirmed this conclusion, which is of special interest because it indicates that the classical equations describing microbial growth are unsatisfactory when substances are produced which affect the growth rate of other organisms in the community. Meers and Tempest[133] suggested an alternative equation to that described by Monod[149] in order to describe the relationship between growth rate

TABLE 3

Synergistic Relationships Involving Bacteria

Substance produced	Organisms involved	Remarks	Reference
Cellulose	*Corynebacterium sp.* and *Pseudomonas caryophylli*	The mixed population rapidly assimilated inhibitors of enzyme synthesis.	16
Bacitracin	*Bacillus subtilis* and *Pseudomonas sp.*	May be due to prolonged sporogenesis in mixed cultures.	217
Biomass from methane	Not defined	Mixed cultures have several advantages for biomass production.	202
Putrescine	*Escherichia coli* and *Streptococcus faecalis*	*S. faecalis* produced ornithine which *E. coli* converted into putrescine.	62
Lecithinase	*Pseudomonas* strains	Each strain produces a portion of the active enzyme.	12
Prodigiosin	Strains of *Serratia marcescens*	—	205
Pathogenesis	*Staphylococcus aureus* and *Proteus vulgaris*	The two weakly pathogenic species did more harm to mice as mixtures than as pure cultures.	8
Gas production	Two types of Bacteria	This phenomenon can lead to false positives in water analysis.	89
Acid production during yogurt manufacture	*Lactobacillus bulgaricus* and *Streptococcus thermophilus*	Good example of symbiosis.	50
Cheese flavor	Lactic acid bacteria	—	—
Ginger beer	*Saccharomyces pyriformis* and *Lactobacillus vermiformis*	The yeast grows in the gelatinous sheath of the bacteria.	25

and substrate concentration when excreted metabolites promoted growth

$$\mu = \mu_m \frac{(1 + \beta P)}{(1 + P)} \left(\frac{s}{K_s + s}\right) \quad (14)$$

where P represents the concentration of extracellular growth-promoting metabolite, β is a constant, and μ_m is defined as the maximum growth rate when P = O. The other symbols are defined elsewhere. Equation 14 might also be used to describe the growth of an organism that is stimulated by the presence of a cohabitant. The interrelationship could be either commensalism or mutualism. Under constant growth conditions P is probably directly proportional to the concentration of the organisms that produce the growth-stimulating substance. But P might also vary with such parameters as growth rate, pH, or temperature, and it is therefore preferable not to make P equal to population density. The constant β is different for each species and reflects the extent to which the maximum specific growth rate is altered by the presence of the growth-promoting products. Equation 14 can be converted into Equation 15 where $\beta_p = \beta - 1 = $ a constant.[137]

$$\mu = \mu_m \left(\frac{S}{K_s + S}\right) + \mu_m \left[\left(\frac{S}{K_s + S}\right)\left(\frac{\beta_p P}{1 + p}\right)\right] \quad (15)$$

Equation 15 is in fact more wieldy than might at first sight appear. When the growth-promoting substance is absent (i.e., P = O) the equation reduces to the Monod type. This is to say that the expression $\mu_m \left(\frac{S}{K_s + S}\right)$ describes the rate of growth in the absence of promoting substances, whereas $\mu_m \left[\left(\frac{S}{K_s + S}\right)\left(\frac{\beta_p\ P}{I + P}\right)\right]$ describes the effect of growth-promoting substance on growth rate. The subject of what Powell has termed "hypertrophic growth" (i.e., stimulation of the growth of a species by metabolites excreted by that species) has been discussed at length elsewhere.[137,188] This phenomenon probably explains why Mian et al.[142] found that when the second stage of a two-stage continuous culture apparatus was fed both with the effluent from the first stage and with additional fresh medium, the growth rate of organisms in this second stage could exceed the maximum growth rate which had been measured previously.

Parker[171] fed steady-state populations of continuously grown *Streptococcus salivarious, Veillonella alcalescens*, and *Staphylococcus aureus* into a common mixed culture vessel and measured the minimum doubling times of the organisms in the different vessels. It was found that the generation times of the organisms in the mixed culture vessel were often comparatively short. This observation appears to be similar to hypertrophic growth except that in this case the growth-promoting substances are exchanged between members of populations of different species, whereas in the cases observed by Meers and Tempest[133] and Mian et al.[142] axenic cultures were used.

MICROBIAL INTERACTIONS IN SOME SPECIFIC NATURAL ENVIRONMENTS

In the preceding sections different types of microbial interrelationships have been discussed in the light of published data obtained from laboratory experiments. Most of the laboratory experiments were conducted with defined systems containing only two different species. However, most natural ecosystems are far more complex and contain many different species. Nonetheless, it is reasonable to suggest that the same kinds of interaction occur in nature as have been described in the laboratory experiments and it is the object of the second part of this review to discuss how these same interactions determine the growth of microorganisms in several specific natural environments.

Anaerobic Sludge Digestion

The microbiology of anaerobic sludge digestion has been reviewed recently by Toerien and Hattingh[223] and by Kirsch and Sykes.[106] More reviewers conclude that although progress has been made in waste treatment technology, a complete ecological analysis of anaerobic digestion has not been obtained. The precise relevance of the many different individual species that have been isolated from sludge may be poorly understood, but it is clear that different groups of organisms combine to digest effectively a wide variety of compounds. Two major groups of organisms can be distinguished, the methanogenic and the nonmethanogenic bacteria.

Raw sewage contains a wide variety of compounds which the nonmethanogenic bacteria convert into volatile fatty acids, hydrogen, and carbon dioxide. The predominant bacterial species bringing about this primary transformation are gram-negative non-spore-forming obligate anaerobes. However, facultative anaerobes are present in low numbers and those organisms have been shown to persist even when sterile medium is used to feed a digester.[222] This indicates that facultatively anaerobic species are not present only because they are constantly seeded from the raw sewage. In an experiment in which a sudden load of a readily fermentable carbon source was added to anaerobic sludge, a transient surge of the growth of facultative anaerobes occurred.[122] The most obvious explanation of this result is that an increase in the population of these organisms in the sludge was prevented under steady-state conditions because the obligate anaerobes competed more effectively for the limited supply of available carbon. Although the obligate anaerobes appear to have this competitive advantage over facultative species in sludge digesters, the latter group fill a small ecological niche which is not as yet defined. Other types of organisms that seem to play a part in anaerobic sludge digestion are the proteolytic, cellulolytic, lipolytic, sulfate-reducing, and denitrifying bacteria.[223]

Although a very wide variety of nonmethanogenic bacteria have been isolated from anaerobic digesters, the interrelationship between these species remains obscure. Since many of these obligate anaerobes have complex nutritional requirements, it is reasonable to suggest that commensalism and mutualism play an important part in regulating their populations.

Species of *Methanobacterium, Methanobacillus, Methanococcus,* and *Methanosarcina* convert hydrogen and various organic acids and alcohols into methane. These species depend on the activities of the nonmethanogenic bacteria in anaerobic sludge for a supply of these compounds, and thus a commensal or perhaps mutualistic relationship exists between the two groups of species. This relationship is delicately balanced because the slow-growing methanogenic organisms are sensitive to changes in such parameters as pH, temperature, or the organic load on the digester. Once the methanogenic bacteria stop growing at a rate sufficient to assimilate the constantly formed organic acids, the pH of the digester falls and thus the chances of recovery of the methane producers are diminished. Such a situation may be caused by various factors and the general remedy for curing a so-called stuck or sour digester is to increase the pH by the addition of ammonia or lime. Complete recovery may take many weeks and stuck digesters often have to be almost completely emptied and refilled with fresh sludge. It is probable that present techniques would be greatly improved with a better understanding of the interrelationships between the microbial species in the sludge. It seems likely that if the diagnosis of the cause of stuck fermenters could be made in terms of the state of the microbial population, an appropriate remedy could be suggested that would speed the recovery of normal operations. It may turn out that the nonmethanogenic bacteria have different growth optima from the methanogenic species, and that the whole operation is best carried out in a two-stage system.

Methanogenic organisms isolated from anaerobic sludge digesters provide excellent examples of mutualism. Smith[210] found an unexpectedly high proportion of hydrogen-oxidizing species among the methanogenic bacteria isolated from anaerobic sludge, despite the belief that most of the methane produced in a fermenter can be traced to the fermentation of organic acids. This point, as well as the biochemistry of methane formation, has been discussed in detail recently by Kirsch and Sykes,[106] who suggest the following reactions for the oxidation of valeric acid by *Methanobacterium suboxydans* and *M. propionicum:*

M. suboxydans

$$2CH_3CH_2CH_2CH_2COOH + 4H_2O = 2CH_3CH_2COOH + 2CH_3COOH + 8H$$
$$CO_2 + 8H = CH_4 + 2H_2O$$

M. propionicum

$$4CH_3CH_2COOH + 8H_2O = 4CH_3COOH + 4CO_2 + 24H$$

$$4CO_2 + 24H = 3CH_4 + CO_2 + 6H_2O$$

The acetate produced by *M. propionicum* could then be decomposed by another species. The important question[106] is whether species such as *M. suboxydans* or *M. propionicum* are in fact true species or are they mixtures of symbionts such as *M. omelianskii* proved to be. If they are mixtures, then carbon dioxide reduction by hydrogen may be one of the main reactions brought about by the truly methanogenic species and may also be the reason why high proportions of hydrogen-utilizing species can be isolated from anaerobic sludge.

Aerobic Sludge Digestion

The aerobic digestion plants that are in general use are of three basic types. These are percolating filters, activated sludge, and activated algae.

Percolating filters have been in use for a number of years and although a good deal of empirical knowledge has accumulated with regard to their design and operation, the understanding of the interactions between the fauna and flora that grow on the filter beds is only rudimentary. As the effluent passes through the bed, the concentration of specific nutrients changes and some substrates become converted into compounds that commensal species may be able to metabolize. These circumstances produce a succession community in which the species are arranged in space rather than in time. Thus, in the top layers of the bed heterotrophs utilize organic matter, the concentration of which rapidly diminishes as the fluid trickles to the lower strata. Here autotrophic species such as *Nitrobacter* and *Nitrosomonas* oxidize ammonia in layers where the dissolved oxygen tension is suitably high. Protozoa, fungi, and algae are to be found as well as small animals that utilize the solid matter on the bed. Clearly this is a challenging but neglected area of microbiology where no doubt most if not all of the previously mentioned types of interrelationships take place.

An activated sludge plant is in essence a continuous culture apparatus with the important feature of feedback of much of the cell yield.

Although the culture vessel is necessarily non-aseptic and is continuously reinoculated with microorganisms from the incoming sewage, only those species which can successfully compete for the growth-limiting nutrients will form part of the dominant flora. Thus, *E. coli* is normally found in raw sewage, but it is not dominant in activated sludge.[54] The nature of the growth-limiting substrate affects the outcome of competition in activated sludge. When the concentration of nitrogen or phosphorus limits growth, the fungi *Geotrichum* and *Sphaerotilus* have a competitive advantage over flocculating bacterial species.[53,179] The presence of these fungal species causes bulking of activated sludge; growth conditions can be selected that place them at a competitive disadvantage over species that are more desirable from the operator's point of view. However, many of the important factors involved in determining the outcome of competitive situations in activated sludge remain unclear. Pipes[180] has suggested the importance in activated sludge of two factors that are commonly assumed to influence survival in competitive situations. Organisms able to form storage products or with low requirements for the growth-limiting substrates would be considered to have a better chance of survival than organisms lacking these properties. Although these suggestions are reasonable, neither is a truism (as is sometimes assumed) and both lack experimental support. Much of the empirical data relating changes in nutrient concentration to properties such as reduction in biological oxygen demand (BOD) could be at least partly explained in terms of population shifts caused by competition.

Species that flocculate readily make for easy operation of activated sludge plants and the presence of predatory protozoa has long been known to be an important factor influencing floc formation. The reason for this has been recently explained by Curds and Cockburn,[45,47] who have suggested that protozoa ingest dispersed bacteria in effluents and thereby exert a selective pressure in favor of flocculating species which can be considered to have a refuge (c.f. Gause's experiments[65] mentioned elsewhere). This idea supersedes earlier vague concepts of protozoan-induced flocculation. Passage through each trophic level dissipates energy and carbon dioxide and one might suggest that the more predation the better. However, as has been discussed in a previous section, excessive predation could lower the concentration of bacteria to a lower level than would be required for digestion of the primary substrate. Therefore, the phenomenon of floc formation may be a useful buffer regulating the predator/prey relationship between the bacteria and protozoa in activated sludge. Pike and Curds[181] have discussed the importance of protozoa in activated sludge at length. They have also discussed the succession of microorganisms that is observed during the establishment of activated sludge.

Phage and *Bdellovibrio* may be isolated from activated sludge, but it is reported that neither is important in removing bacteria from sludge.[54]

Commensalism and mutualism have been reported to occur in effluent treatment plants. A well-documented industrial application of the use of a type of commensal relationship has been described by Barta.[11] In this process a two-stage continuous culture system is used to decontaminate industrial effluents. This type of interrelationship, in which the successive growth of microbial species brings about a breakdown of complex molecules, is probably common in effluent treatment plants.[29,103]

The microbial ecology of the activated sludge process has been reviewed by Pipes[180] and Pike and Curds.[181]

Activated algae is a term that has been used to describe a process in which a mixed algal and bacterial culture may be used to treat waste water. The interrelationship between the two types of organism is mutualistic with *Chlorella* using carbon dioxide produced by the bacteria and the bacteria in turn using oxygen produced by the photosynthetic algae. Humenik and Hanna[92] have studied the respiratory relationships in such a culture and have shown that stable populations can be established from which excess biomass may be withdrawn daily. Such a system has several advantages from an effluent treatment point of view and may be developed in tropical countries into a process for the production of biomass for animal feed. Activated algae is effectively an updated version of the system in which settled sewage is run into oxidation ponds and held there for about a month in order to decrease the amount of organic residue.

Wine and Beer Production

During the production of ethanol by strains of *Saccharomyces* species several types of microbial interrelationship can be distinguished. A number of species may compete for the available nutrients in the wort or must, whereas other species depend on the products of yeast metabolism. Amensalism has been well established and neutralism seems likely.

A number of studies have been made of the flora found on the skin of grapes and the commonest species isolated were of the genera *Saccharomyces, Hansenula, Hanseniaspora, Pichia, Torulopsis, Candida,* and *Kloeckera.*[5] These species remain during the grape crushing process and therefore are also to be found in the must. In traditional processes (which are still used by most European wine producers) these species form the fermentation inoculum. The natural yeast flora varies from year to year[178] and from one area to another.[35] Hence, the mixture of yeasts used for a traditional type of fermentation varies widely and this is said to influence the taste of the wine produced. Evidence on the latter point is provided by expert wine tasters, while a detailed understanding of how changes in the yeast population of musts affect taste appears to be lacking. What is clear, however, is that well-documented examples of amensalism occur during the fermentation. Goswell[72] reports that during a typical wine fermentation the following succession of yeast species is likely to be observed. In the initial stages, when little ethanol was present, the yeasts most likely to be found included *Kloeckera apiculata, Hanseniaspora guillermondii,* and *Candida pulcherrima.* As the fermentation continued *Saccharomyces cervisiae* var. *ellipsoideus, S. rosei,* and *S. veronae* were observed in increasing numbers as the former three species declined. During the final stages when the alcohol concentration passed about 10% by volume, *S. cerivisiae* var. *ellipsoideus, S. oviformis, S. chevaleiri,* and *S. italicus* predominated. This succession observed by Goswell[72] may be compared with the observations on the alcohol concentrations produced by axenic cultures of various yeasts grown in the presence of an excess of carbon source (Table 4). It is assumed that further growth of the various yeast species was prevented when the concentration of alcohol became inhibitory. This correlation between ethanol tolerance levels and the succession of yeast species observed during fermentation of musts has been widely reported. Therefore, it seems clear that amensalism is the main cause of this succession.

TABLE 4

**Percent (w/v) of Ethanol Produced by
Various Yeast Species[5]**

	Percent ethanol produced	
Yeast	Minimum	Maximum
Candida pulcherrima	0.1	0.7
Kloeckera apiculata	1.1	7.5
Saccharomyces rosei	4.0	8.5
S. chevalieri	10.0	13.7
S. oviformis	10.9	12.9
S. italicus	10.0	12.5
S. cerevisiae var. *ellipsoideus*	8.5	15.0

Bacteria are commonly found growing during beer or wine fermentations. This growth is sometimes desirable but more often causes spoilage. Grapes contain varying quantities of tartaric and malic acids and during the fermentation small quantities of succinic acid are produced. The yeasts normally used for wine production do not metabolize these acids, although Gandini and Tarditi[63] have reported that *S. pombe* does so and have recommended the use of this species to remove excess malic acid. Benda and Schmitt[13] have, however, reported that the presence of *S. pombe* impairs the quality of wines. It is more normal to make use of the presence of various species of lactic acid bacteria (which form part of the flora found on grape skins) to reduce the titratable acidity due to malic acid. Such a fermentation is considered desirable for the production of Burgundy wines which would otherwise contain excess acidity. To make certain that the malo-lactic fermentation takes place before bottling, the casks of red wine are kept warm during the first winter of maturation, although this fermentation is usually finished during the course of the yeast alcoholic fermentation. A malo-lactic fermentation is most desirable in highly acid wines, which are inherently inhibitory to the growth of the required bacteria. Therefore, chalk is sometimes added to wines to adjust the pH to a level suitable for bacterial growth.[105] Kunkee[108] has suggested other methods that may be used to control the malo-lactic fermentation, all of which have been developed empirically. Although it seems likely that the lactic acid bacteria responsible for the production of lactic acid from malic acid have a neutral relationship with the alcohol-forming yeasts, a clear understanding of the growth of these species is lacking. If such an understanding could be obtained, it is likely that this important reaction could be better controlled and much of the wine, which is at present spoiled because of the difficulty in encouraging or discouraging the malo-lactic fermentation, could be saved.

In addition to promoting the desired malolactic fermentation, species of lactic acid bacteria also produce lactic acid from carbohydrates during the production of beer and wine. There is a great deal of confusion with regard to the taxonomy of the lactic acid bacteria isolated from wine, but species of *Pediococcus, Leuconostoc,* and *Lactobacillus* are considered to be the most important. To define which reactions are brought about by particular species of bacteria is difficult and a great deal of research is required before the precise interrelationships which exist between various species can be defined. However, two types of glucose catabolism have been described by Rainbow[189] and these may be presented as follows:

glucose → lactic acid + ethanol + CO_2

3 glucose → 4 glycerol + 2 acetic acid + 2 CO_2

During the production of "silky" beer the former reaction predominates. Lactic acid bacteria can also catabolize fructose (Rainbow, 1971) and pentose sugars[151] as follows:

3 fructose → 2 mannitol + lactic acid + acetic acid + CO_2

pentose → acetic acid + lactic acid

The bacteria which bring about the above four reactions may be considered to be competing with the various yeast species for available carbohydrates, but it seems likely that these spoilage organisms, which have fastidious nutritional requirements, also depend on nutrients excreted by the yeast population.[15] Strains of *Pediococcus cerevisiae* probably grow in a similar competitive/commensal relationship with yeasts during beer production, and form diacetyl (which spoils the taste of beer) or polysaccharides (which cause "ropiness"[180]). The lactic acid bacteria described above are necessarily acid and ethanol tolerant facultative anaerobes, but in addition aerobic acetic acid bacteria can be isolated from beers and wines. These species belong to the genera *Acetobacter* and *Acetomonas*, and if excess air accumulates during or after fermentation these species convert ethanol or acetaldehyde into acetic acid. Unlike the lactic acid bacteria, the acetic acid bacteria do not have exacting nutritional requirements and if excess air is available they will convert ethanol into acetic acid as rapidly as it is formed. This may be considered a commensal relationship, which is highly desirable during the production of vinegar.

In the old-fashioned Orleans process for the production of high quality vinegar, acetic acid bacteria are allowed to grow slowly on the surface of wine kept in partly filled flasks. During more widely practiced vinegar-making processes, the commensal relationship between alcohol-producing yeasts and acetic acid bacteria is removed in both time and place. Most commonly the alcoholic liquor is allowed to trickle over a packing which provides a support for the acetic acid bacteria. In some modern plants, vinegar is produced by cultivating acetic acid bacteria in submerged culture by batch or continuous processes.[34]

Another useful commensal relationship which can occur when wine is allowed to become partially aerobic leads to the formation of Fino sherry. The yeasts which form the so-called flor belong mostly to the genus *Saccharomyces*, although many genera and species have been isolated from the film which forms during sherry production. It is of interest to note that the sherry characteristics are not much affected by the strain or species.[5] Accumulation of acetaldehyde is the most important reaction promoted by this system when used for sherry production. The use of film yeasts to "age" red wines more rapidly has been reported from Spain and Italy.[32] Cantarelli[32] described a method in which the concentrations of volatile acidity, tannin, and color were decreased, while esters and acetaldehyde were produced by three weeks' growth of a surface culture of *S. oviformis.* Carr et al.[33] have described an apparatus in which sherry may be produced continuously; they note that vinegar could be produced in the same way. Presumably wine could be artificially aged by a suitable continuous process.

In addition to the competitive, amensal, and commensal relationships described above, an example of what appears to be neutralism can be cited. Strains of *S. diasticus* possess a glucoamylase which enables them to ferment polysaccharides which cannot be metabolized by other yeasts found during beer and wine fermentations.[68] Normally these polysaccharides would remain in finished beer and impart body which is lost when beer becomes infected with *S. diasticus.*

Lüthi[120] has suggested that synergism between organisms in wine is important in producing the desired characteristics.

Most of the methods used to produce beers and wines with desirable qualities were developed empirically, made toward understanding the complex microbial relationships which occur during the fermentations. Nonetheless, it remains difficult to promote the often desirable malo-lactic fermentation while preventing other secondary fermentations at the same time. At present it is possible by keeping equipment scrupulously clean to inhibit the growth of undesired organisms with the judicious use of sulfur dioxide. Rainbow[189] has suggested that it may be possible to control the growth of fastidious lactobacilli in beer through a knowledge of their exact nutritional requirements. A recent trend in fermentation is to produce wines in stainless steel plants and use a heavy inoculum of a pure yeast culture.[148] Such an inoculum would probably tend to swamp the growth of "wild yeasts" during a batch process. However, such an approach would not be adequate if continuous flow techniques are to be used in the future. From the previous discussion it would be anticipated that homeostatic mechanisms would be expected to occur and some of the micro-organisms observed in batch processes would be expected to grow in a continuous fermentation in a condition of dynamic equilibrium. The nature of the factors influencing continued survival of species during the two types of fermentation would not be the same and the same species might not predominate in continuous as in batch fermentations. Pure yeast cultures could be used in aseptic continuous cultivation processes for the production of beer, wine, or vinegar. However, under such growth conditions selective pressures would exist which would favor the survival of, for example, mutants which could metabolize the polysaccharides in wort which are considered essential for beer quality. It is possible to predict that after a period of continuous growth of yeast on complex media such as wort or must, a heterogeneous community composed of mutants of the original strain would become established. These mutants could develop so as to occupy similar niches to those discussed above with reference to processes containing different species of organisms. Such population differences would be expected to produce beer or wine of different quality and any change in the quality of these products is undesirable. However, an improved understanding of the interrelationships between the different microbial species that are found in these fermentations will lead to a less empirical approach to problem solving in this industry and it may then become possible to alter process conditions in such a way that product quality is not impaired.

Skin

Marples[125] has described the conditions for microbial life on the human skin. The surface is constantly bathed in an oily emulsion containing a complex mixture of nutrients suitable for bacterial growth. A fraction of the mixed microbial culture that develops is continuously removed by rubbing and cleaning and therefore the surface of the skin is a good example of an open ecosystem.

Strains of *Staphyloccus epidermidis* are the dominant aerobic members of the cutaneous bacterial flora and are found in the greatest numbers near the mouths of the follicular canals. Röckl and Muller[193] did not isolate any bacteria in the ducts of unobstructed sweat or sebaceous glands. Using the figures for rate of sweat production and size of sweat glands quoted by Randall and McClure,[190] the specific growth rate required in order for a free living population to maintain itself in a sweat duct would be over 1.0 hr^{-1}. Since attainment of this growth rate seems unlikely, the potential colonists may simply be washed out from these canals at a greater rate than they can reproduce themselves. This conclusion is consistent with the observation that canals are frequently colonized by *Staphylococci* if the flow of liquid is obstructed, as in the development of acne.[215] The reasons for the success of *S. epidermidis* on the skin and in particular in the area of the follicular canal opening are not completely clear. The best documented explanation implicates amensalism due to the formation of antibiotics. Thus, Shehadeh and Kligmann[203] have shown that the resident gram-positive species of the axilla exerted an inhibitory influence on the growth of gram-negative bacteria which are not normally found in significant numbers in this area. However, when the growth of gram-positive species was inhibited by appropriate antibiotic treatment, the population of gram-negative species in the axilla increased markedly. Such a result does not exclude the possibility that the growth of the gram-negative species was inhibited because *S. epidermidis* competes more effectively for nutrients that are normally in limited supply. When extra nutrients become available (for example, following sweating or skin damage) the concentration of the bacterial flora increases.[125] Hence, it can be argued that competition for nutrients is one factor influencing the dominance of bacterial species on the skin. Marples[125] suggests that, as only small amounts of glucose are present on the normal skin, this compound is an unimportant energy source for the skin population. However, it can be proposed that the observed low concentrations of such a readily metabolizable nutrient on the surface of the skin[195] are a sign that those microorganisms that are able to compete effectively for this substrate have grown to the maximum steady-state populations density possible. It should be noted that sweat contains far greater concentrations of glucose[118] than can be isolated from the skin surface. The survival of *S. aureus* on the skin is of more than academic interest and the reason why this species does not regularly become established on skin remains unclear. It has been established that coagulase negative species prevent the establishment of the potentially pathogenic coagulase positive species. *S. aureus* readily colonizes the sterile surfaces of new-born infants, but is subsequently replaced by coagulase negative species. Despite the fact that Evans et al.[59] have demonstrated the production of antibiotics by *S. epidermidis* species. Rountree and Barbour[196] were unable to show that antibiotic production was responsible for the exclusion of *S. aureus* from the nasal cavity.

Although coagulase negative *Staphylococci* inhibit the growth of *S. aureus*, probably either by competition or amensalism, the presence of several other species promotes its growth. Thus. in wounds and in impetigo lesions. *S. aureus* often grows in the presence of beta hemolytic *Streptococci* with which it has a mutualistic relationship.[125] Marples suggests that since the growth of *S. epidermidis* is inhibited by serum,[57] *S. aureus* can grow successfully in wounds and lesions, which are bathed in this substance,

without having to compete with the former species.

Girard[69] has shown that the presence of *Streptococcal* strains in the buboes of plague results in the gradual disappearance of *Pasturella pestis* organisms. It is an intriguing thought that perhaps a highly lethal disease could be cured by intentional infection by a less virulent pathogen.

The dominant anaerobic species found on skin is *Corynebacterium acnes*, which is generally found in the pilosebaceous canals where a very dense population can develop. The relationship between *C. acnes* and *S. epidermidis*, both of which species grow in the follicular canal area of the skin, is unclear. It may be postulated that *S. epidermidis* reduces the oxygen tension, facilitating the growth of the anaerobic species. If both species compete for nutrients flowing from the sebaceous glands and *C. acnes* has the competitive edge. then a self-regulating mixed population can be envisaged. If the environment becomes aerobic due to starvation of *S. epidermidis*, the growth of *C. acnes* would be inhibited, thus allowing an increase in the *Staphylococcus* population and the concomitant decrease in dissolved oxygen tension. The cycle could then be repeated. Oscillations of the concentrations of the two species would then be observed. This could explain why Pachtman et al.[169] found such wide variations in the population of *S. epidermidis* even in samples taken at different times from the same area of skin from the same individual. Other anaerobic species such as the *Clostridia* probably also depend on other species to decrease the oxygen tension and thereby provide a suitable environment for their growth.

Many skin lesions caused by the presence of bacterial species are subject to secondary infection and a succession of species is often observed. For example. the ulcers caused by the presence of *Spirochaetes* provide a suitable environment for colonization by a variety of organisms which are generally considered to live on the necrotic material caused by the primary infection. *Corynebacterium diphthereae* is a frequent colonizer of the ulcers that are characteristic of yaws. Marples and Bacon[124] have suggested that this species and *Treponema pertenue* (the etiological agent for yaws) have a synergistic relationship with each other. Large numbers of a variety of other bacterial species (particularly staphylococci, streptococci, and diphtheroids) can also be isolated from such ulcers, but the interrelationships between the species are only vaguely defined. The presence of *Mycobacterium leprae* causes such profound changes in the skin that it is not surprising that the skin population of sufferers from leprosy is distinctive.[207]

Marples and Kligman[126] have shown that orally administered antibiotics can substantially alter the microbial flora of the skin. Some species produce antibiotic-degrading enzymes in response to antibiotic treatment and sometimes this phenomenon has unfortunate results. For example, Sanders et al.[197] have demonstrated that penicillinase-producing strains of *S. epidermidis* permit the survival of satellite colonies of *Neissenia gonorrhoea* which appear to be penicillin resistant only because of the activities of the staphylococci. This intriguing type of interrelationship could be more widespread and has obvious important implications. An analogous situation exists when *Corynebacterium pyogenes* produces catalase, which probably helps in the establishment of anaerobic organisms that infect the feet.[192]

Thompson and Shibuya[221] have demonstrated a complex interrelationship in which streptococci inhibit the growth of *C. diphthereae* but the inhibition is neutralized by the presence of staphylococci. The reasons for this observation remain to be elucidated. Roberts[191] has described in some detail the interrelationships between the four species which together cause ovine foot rot (Table 5). At least three of the four microbial species are required to induce foot rot regularly and synergism is said to be the main interrelationship linking the species.

The interrelationships between bacteria and bacteriophage are probably of considerable importance. For example, Maxted[129] has shown that the surviving organisms of a streptococcal population attacked by phage were bacteriophage resistant, possessed increased amounts of M protein, and were more virulent. Hence, phage attack may contribute to the selection of virulent strains of organisms. A similar type of explanation may be proposed to explain why various phage types of *S. aureus* become dominant at regular intervals in vivo.

There are many accounts in the medical literature of the mixed microbial communities which are to be found in normal and diseased skin. However, a great deal remains to be learned about the factors regulating the microbial community of the skin. The study of this environment presents a special challenge because it probably contains several different microhabitats in which the factors governing survival are quite distinct. The study of these individual microenvironments will no doubt continue to test the ingenuity of microbiologists.

Rumen

The control of the growth of rumen microorganisms has been reviewed recently by Hobson.[85,86] Much of the work on this topic has been directed toward explaining the relationship between diet, the microbial flora and fauna of the

TABLE 5

The Interrelationships Between the Four Species Causing Ovine Foot Rot[191]

Organism	Ability to force entry into host tissues	Axenic growth in host tissue	Production of growth factors for other species	Resistance to host defences	Ability to cause tissue damage
Corynebacterium pyogenes	-	+	+	+	-
Fusiformis necrophorus	+	-	-	+	+
Fusiformis nodosus	-	-	+	+	+
Motile fusiform	-	+	+	+	-

rumen, and the production of compounds of food value to the ruminant.

It has often been observed that changes in the ration fed to ruminants causes changes in the proportions of volatile fatty acids (VFAs) and methane in the rumen. Increased concentrations of readily available carbohydrates and decreased roughage lead to a decrease in the proportion of acetic acid and an increase in the proportion of propionic and butyric acids. These results could be caused by a change in the metabolic activity of given species of microorganisms or by changes in the composition of the microbial flora associated with the different rations. Latham et al.[113] have shown that marked differences exist in the microbial flora of cows fed either hay or cereal rations, but they could not establish statistically significant relationships between the concentration of the major VFAs and lactic acid and the viable count of selected microbial species. Quantitative as well as qualitative changes in the diet are important. Thus, Mann[123] found that when the barley ration fed to a heifer was abruptly increased, gross changes in the bacterial flora were observed and the pH value of the rumen fluid fell from 5.7 to 4.5. In general, high starch diets cause decreases in pH values which promote the growth of *Bacteriodes* and various other species which are presumably able to compete effectively at lowered pH levels for the available nutrients. If the pH value falls further. then streptococci, lactobacilli, and yeasts predominate. Thus, it seems to be established that changes in diet bring about alterations in the rumen flora and fauna, but that after a period the rumen flora become readapted to a changed diet and a new set of microbial interrelationships becomes established.

Hobson[85] has argued that competition for limiting nutrients is not important in determining which species will predominate in the rumen. The basis of Hobson's argument is that in natural open systems factors other than nutrient availability will decide which organisms will initially colonize a new environment. Once this growth is established, these organisms will remain dominant and consume a larger proportion of available nutrients than species present in lower concentrations. Hobson goes on to state that competition can not be a decisive factor in determining the population of stable cultures because it seems unlikely that the dilution rate imposed in a natural continuous culture would be one at which the populations of

organisms could, in theory, all coexist in competition for nutrients. Eadie[56] observed the replacement of one protozoan species by another in the rumen and argued. on similar lines to Hobson,[85] that since a large number of organisms were replaced by relatively few of another type, direct competition for food seemed unlikely. This line of argument cannot be reconciled with the principles of competition outlined in the section of this review devoted to this topic and indeed appears to be demonstrably untenable. The observation that many species coexist in, for example, the rumen without displacing one another does not show that competition is unimportant in determining the composition of the rumen flora. This is because it is reasonable to expect that there are many ecological niches in the rumen, each one of which has its dominant species.

Hungate[93] has suggested that the comparatively low concentration of cellulolytic organisms observed in the rumens of animals fed cellulose-containing diets could be due to their inability to compete with other organisms for the sugars made available by cellulolysis. Cellulolytic bacteria require branched-chain acids for growth[209] and depend on noncellulolytic microorganisms to synthesize these compounds from carbohydrates, which may well have been made available by the action of cellulase. This possibility provides excellent scope for a kinetic study of an intriguing type of mutualistic interrelationship.

Commensal relationships have been demonstrated in the rumen. For example. the lactate-fermenting and methanogenic bacteria depend on the activities of other species to produce their nutrients. On occasions when the pH level of the rumen falls, these delicately balanced commensalistic relationships become disturbed and, for example. lactic acid is produced at a rate greater than it can be metabolized by the lactate-fermenting species. Other intermediates in food chains (sulfide, nitrite or ammonia) can accumulate and cause illness or even the death of the host ruminant if commensal relationships in the rumen are upset. Vitamins also form the basis of commensal interrelationships in the rumen.[85]

Amensalism may occur in the rumen. Hollowell and Wolin[88] showed that the growth of *E. coli* was suppressed by an unidentified substance in rumen fluids and concluded that the inhibitor was produced by a resident rumen species. They showed that bacteriophage was not responsible for their results. However, Adams et al.[1] did isolate phage from the rumen. These phages infected *Serratia sp.* and *Streptococcus bovis*, both of which species are found in low numbers in the rumen. No phage was isolated that could infect the dominant rumen species *Bacteroides ruminicola.*

Protozoan species normally grow in the rumen. Kurihara et al.[110] introduced protozoa into unfaunated rumens and observed a decrease in bacterial numbers, which was attributed to predation. Slyter and Wolin[208] showed that when protozoa were removed from the rumen the production of propionic acid increased, whereas methane production decreased. Defaunation had no major effect on the growth and maintenance of the ruminants, although it is clear that protozoa have an effect on the kinetics of bacterial growth in the rumen.

The rumen is commonly thought of as a continuous flow system. However, there are important differences between growth in the rumen and in conventional laboratory continuous culture apparatus.[86] Thus, in the rumen there is an approximately constant flow of basal medium (saliva) through the growth vessel, but an intermittent flow of nutrients. Since the latter variable depends on the not always predictable behavior of the host ruminant, it is not surprising that the rumen population is subject to constant variations. Thus, following a meal taken by the ruminant, the conditions in the rumen may be similar to a batch fermentation during the logarithmic growth phase.[23] When the readily metabolizable nutrients have been assimilated, less readily available nutrients are utilized, and the growth rate of most microbial species would be expected to decrease until the time when the ruminant takes another meal. Under such growth conditions, species with high maximum growth rates would tend to be favored for the reasons proposed by Maynard Smith.[130] Since these growth conditions, which will favor the microbial sprinters, alternate with conditions where slow growth is inevitable and since it has been shown that different species can be favored by these two growth conditions,[139] it follows that complicated oscillations of the rumen populations would be expected and are indeed observed.

Alimentary Canal

Dental plaque is a common feature of two of the most widespread diseases of man — caries and periodontitis. Since plaque is composed of several species of organisms which are thought to be interdependent, a good deal of investigation of the interrelationships between oral microorganisms has taken place. This topic has been reviewed recently.[87,131]

The mouth contains three well-defined sites for bacterial colonization These are the surface of teeth, the gingival crevices, and the tongue surface. The mouths of infants are sterile at birth but a succession of species colonizes the different sites as the infant matures.[43] Marked changes occur following tooth eruption because a number of species require a surface to colonize before becoming established in the mouth. The ability to become attached to a surface in the mouth gives some species a marked competitive advantage in this habitat.[228] Because of this, the kinetics of growth is not that of the perfectly mixed reactor used for most theoretical studies of bacterial growth. The organisms in plaque are packed very closely together and selective pressures have probably led to a situation in which many species of microorganisms exist in commensal or symbiotic relationship with each other.[186] Numerous other examples of commensal and mutualistic relationships have been observed to occur between the species found growing in the mouth.[87,172] These relationships include the previously mentioned phenomenon of aerobic species decreasing the dissolved oxygen tension, allowing anaerobic species to grow.

Amensalism probably occurs in the mouth because of, for example, the production of hydrogen peroxide[81] and volatile fatty acids.[52]

Young et al.[239] observed that *Candida albicans* and *Lactobacillus* species regulated each other's growth because the yeast organisms produced growth factors required by the bacteria. The bacteria in turn produced lactic acid which inhibited the growth of *C. albicans.* These two species are normally found in association in the oral cavity.[117]

It thus appears that commensalism, mutualism, amensalism, and perhaps nutrient competition and neutralism are the most important interrelationships that occur among the oral microorganisms. The ability to become attached to teeth and grow as part of the plaque that then forms on the teeth is clearly also of great significance. In the gingival crevices and across a section of plaque, nutrient concentration gradients are no doubt set up. This is a further way in which the different microbial species effect each other's growth and may explain why facultative anaerobes are common in the oral cavity. Such organisms would be at an advantage in being able to grow in conditions which may change from aerobiosis to anaerobiosis during the periods between one tooth cleaning session and another.

Microbial interactions in the intestine have also been the subject of many investigations because of their importance in the etiology of disease. Most of the interrelationships that have been discussed on previous pages have been observed in the intestine and an understanding of some of these factors has led to improved therapeutic techniques.

In a series of papers Freter and his colleagues (see Ref. 168 for references) have demonstrated the importance of competition in controlling intestinal populations. This work explains why previous authors had demonstrated that starved mice, germ-free mice, or mice whose intestinal flora had been largely removed by antibiotic treatment were susceptible to fatal enteric cholera infections, whereas animals with a normal enteric flora were resistant to infection. The normal flora reduced the concentration of the growth-limiting carbon sources to such a low level that the *Vibrio cholerae* organisms could not become established in the intestine. It has often been proposed that the establishment or introduction of harmless species to the intestine will prevent the growth of enteric pathogens. Early work (of which there is a great deal) on this subject is inconclusive,[75] but controlled experiments show that the approach is feasible.[60] Another facet of what may be regarded as competition in the intestine has received wide popular attention in recent years. This is the rise in the frequency of infection rates by antibiotic-resistant microorganisms following treatment with broad spectrum antibiotics. Thus, a range of species of fungi, yeast, and bacteria, that before the antibiotic era were considered not to be highly pathogenic, has now been reported to cause disease. The reason for this phenomenon is readily appreciated if it is accepted that competition is an important factor controlling microbial growth in the intestine. When the growth of antibiotic-sensitive species is restrained, an ecological niche becomes available to species that were formerly excluded because of their inability to compete effectively for a position in that habitat. This leads to changes in the established intestinal population involving the replacement of antibiotic-sensitive strains either by antibiotic-resistant mutant progeny, or by antibiotic-resistant strains of different species. The transfer of drug resistance between different species of bacteria is widely reported and is a subject of profound significance that has been discussed in detail elsewhere.[240]

Amensalism, due to the formation of short chain fatty acids, has been reported to be the reason that *Salmonella typhimurium* cannot become established in the intestine of mice.[142] Bacteriocines may be isolated from the intestine. However, the role of bacteriocines either as growth inhibitors or as factors involved in the transfer of genetic material is subject to debate.

Commensal or mutualistic relationships probably explain why it has been observed that two weakly pathogenic species can cause death as a mixed infection, whereas an infection of either species on its own would be harmless.[8]

The intestinal tract contains substantial concentrations of protozoa many of which normally ingest cohabitant bacterial species. The influence of bacteria on the pathogenicity of *Entamoeba histolytica* was explained by Phillips and Wolfe.[177] When germ-free guinea pigs were inoculated with *E. histolytica*, no lesions developed and the amoeba could survive in the intestine. However, normal animals fed on the same diet developed acute disease on infection with amoebae. When the autoclaved caecal contents of normal guinea pigs were fed to germ-free animals, the authors concluded that the amoebae depended on bacteria in order to infect and become established in the host's tissues, but did not ingest bacteria as a source of nutrients.

Schaedler et al.[200] have provided data on the succession of bacterial species that takes place in the mouse intestine. The intestines were free of bacteria at birth but rapidly became infected with a variety of species. The population densities of some of the pioneer species fell after a period of time and other species then became dominant. The reasons for this type of succession are not well understood, but changes in the metabolic activity (e.g., enzyme excretion) or in the feeding habits of the maturing host species would be expected to produce such effects.

Aquatic Environments and the Soil

Stanier and Cohen-Bazire[211] have discussed what they describe as the complementarity of light absorption by some phototrophs. By this, these authors mean that algae absorb light of a wavelength in the middle of the photosynthetic spectrum, whereas light absorption by the photosynthetic bacteria is at wavelengths at the two ends of this spectrum. This means that competition for light of a given wavelength is decreased and the photosynthetic microorganisms utilize a major part of the spectral range where photochemical reactions are possible. Since photosynthetic bacteria are mostly obligate anaerobes and require suitable hydrogen donors, they are less well suited than algae to growth near the surface of ponds and lakes. Therefore, it is common to observe algae on the surface of standing pools of water with purple and green bacteria growing underneath. If the bacteria were to depend on light of the same wavelength as the algae, bacterial growth in such a situation would be restricted. However, bacteria utilize light of a wavelength that is transmitted by the algae and both types of organism can therefore coexist in what might be considered to be a neutral relationship. However, it may be that competition in the distant past led to the development of the various distinctive photosynthetic pigments.

Jannasch, Veldkamp, and their co-workers have studied in detail the competition for nutrients between marine bacteria. Jannasch[100] showed that with a given set of growth conditions, one species of bacteria became dominant during continuous culture experiments designed to simulate marine conditions. In another further paper, Jannasch[101] showed that the elimination of *Enterobacteriacae* from sea water was due to competition, and not to factors such as bacteriophage, microorganisms producing lytic enzymes, the heavy metal ions present in sea water, or inhibitory substances produced by other species, as had been suggested by other authors. However, more recently, Iturriaga and Garcia-Tello[97] have shown that amensalism can occur between members of the *Enterobacteriacae* and marine species. Therefore, it seems that both competition and amensalism may be significant in marine environments. Harder and Veldkamp[79] have shown that temperature is an additional factor which determines the outcome of competitive situations involving marine species. In an elegant series of experiments it was shown that obligate psychrophilic organisms outgrew facultative psychrophiles at low temperatures (less than 4°C),

whereas the reverse was true at higher temperatures. These results explain the fluctuations in the populations of marine bacteria that take place with seasonal temperature changes.[206] Brock[20] has suggested that some of the species found growing in hot springs do so not because they prefer high temperatures, but because they cannot compete effectively with other species at lower temperatures. In the high temperature environments competition is less severe because fewer species are available to fill a given ecological niche. The same kind of argument could be used when discussing other extreme environments. Indeed an extreme environment may be thought of as one in which only a limited number of different species are observed.

In aquatic environments and in the soil, concentration gradients exist for various environmental factors. The metabolic activity of microorganisms contributes to the establishment of these gradients and the gradients in turn affect the distribution of microorganisms in natural environments. Detailed studies of this topic have been made by Russian scientists. For example, Gorlenko[70] has shown that the redox potential in salt water lakes varied with depth. A layer of purple sulfur bacteria was observed only at the level where the redox potential was suitable for these organisms and where there was sufficient light for their growth. Similar types of gradients exist in lakes with regard to the concentration of hydrogen sulfide[111] and organic nutrients from sewage.[183] Since these gradients are partly caused by microbial activity, it is clear that delicately balanced microbial interrelationships affect their maintenance. Similar types of gradients are observed in rivers downstream of sewage outfalls,[95] and this phenomenon is similar in principle to a succession in which the time function is replaced by a space function.

Gorden et al.[71] have studied a heterotrophic succession in an aquatic ecosystem. The dominant primary producer, *Chlorella*, was dependent upon thiamine supplied by bacteria, which used algal excretions as a carbon and energy source. Successions are a characteristic feature of the decomposition of organic matter in the soil. A typical succession of the microorganisms that grow during a composting process has been described by Gray et al.[73] Acid-producing bacteria were the first to appear, but caused the temperature to rise because of their metabolic activity. At the higher temperatures thermophilic bacteria and fungi became dominant and these species had a form of commensal relationship with their predecessors. If the temperature rose above 65°C, only thermophilic spore-forming bacteria could grow. As the nutrients became depleted and the temperature declined, *Actinomycetes* became the dominant group. During these final stages of decomposition it has been postulated that amensalism is an important factor controlling the microbial population of compost heaps.[29] The work of Knoll,[107] who studied the death rates of some pathogenic species of bacteria in compost-refuse, supports this conclusion. Successions of microorganisms in soil have been observed during the decomposition of cellulose[225] and wood.[74] These successions were due to changes in the microclimate and physicochemical features of the environment brought about by the activity of the microorganisms. Another type of succession was observed following the flooding of soil.[218] In this system a series of electron acceptors was utilized in sequence. The sequence was oxygen, nitrate, manganic and ferric ions, carbon dioxide, and finally sulfate ions. Such a succession of reactions would be predicted from an understanding of the physiology of the bacterial species involved and this same series of microbial conversions also occurs in other habitats as anaerobic conditions become established.

Bdellovibrio species have been isolated from soil samples and Postgate[246] has suggested that the presence of these species is responsible for the absence of *Azotobacter* from many cultivated soils. The activity of predatory protozoa in the soil has been discussed by Nikoljuk[159] and by Mitchell[244] and bacteriophages have also been shown to be members of the soil flora. All of these factors, as well as amensalism,[242] commensalism,[30] and mutualism,[233] probably affect the growth of bacteria in the soil in the ways discussed in previous sections.

Carbon, nitrogen, and phosphorus are resources for which microorganisms compete in the soil and in aquatic environments. In this connection, it is interesting to note that several species of organisms have mechanisms which could increase their competitive ability under phosphorus- or ammonia-limited growth conditions. Thus, under phosphorus-limited growth conditions, *Bacillus subtilis* organisms could decrease their phosphorus content by synthesizing teichuronic acids instead of teichoic acids as part of the cell wall complex of polymers.[220] However, under magnesium-limited growth conditions, cell walls were produced with high phosphate content which bound magnesium strongly.[135] This latter property would enable cells to compete effectively for limited supplies of essential cations. When nitrogen limits growth, many bacterial species synthesize enzymes which enable them to assimilate ammonia by a route with a low K_m for this substrate.[134,138] Organisms able to accomplish this change in their metabolism will be at a competitive advantage in environments where the nitrogen concentration fluctuates.[21] The fact that bacteria have the genetic capacity to make these marked changes in their metabolism in response to changes in the nature of the growth-limiting substrate is evidence that such a capacity confers a selective advantage on strains able to effect changes of this kind in natural environments.

DISCUSSION

Generations of microbiologists have studied the growth of isolated bacterial species growing under favorable conditions in glass vessels. However, the voluminous literature on this subject can only be of limited value in understanding the growth of populations in natural environments where the much harsher conditions rarely allow unrestrained growth. Much of the literature describing naturally occurring microbial communities is taxonomically orientated and does not attempt to explain the interrelationships between the various species in a given community. Some years ago Alexander[3] said that there was a "great need in the field of microbial ecology for experimental information and suitable parameters to provide a basis to account for the occurrence and dominance of specific microbial groups in individual ecosystems, but, despite the large body of descriptive literature, few basic ecological criteria have been proposed to explain the adaptions of microorganisms to the physical environment, and the apparent adaptive value of distinctive physiological and genetic traits." Two years after Alexander's comments, Pipes[180] felt moved to say that ecological approaches to the study of activated sludge had failed to provide anything but a "few vague generalizations and much worthless speculation." He went on to suggest that the reason for the lack of understanding was due to the quality of the research in this area rather than to any defect in the ecological approach. In the last few years the situation has improved somewhat, partly because of the recent general awareness of the importance of matters associated with the human environment. This interest has meant that research funds have been provided for studies of microbial ecology and the literature on this topic has expanded markedly. Among this literature are papers which add significantly to previous knowledge of the ways in which one bacterial species influences the growth of another.

In natural environments, competitive pressures act in such a way as to select those species with physiological traits that are of value in the struggle for existence. The same kinds of physiological traits would be expected to affect the growth of mutants and a further understanding of the important factors influencing competition will lead to an understanding of evolutionary trends. In man-made environments one of the objections to the industrial use of continuous fermentations is that contamination and mutation can lead to the competitive displacement of the original commercial strain. Since it is not easy to avoid contamination and mutation in large plants, it would be advantageous to be able to select growth conditions that would enable the parent strain to be at a selective advantage over potential colonizers.

All microorganisms, and indeed all animal and plant species, can be considered to be growing in a state of mutual symbiosis. In other words, no species could grow alone on the earth indefinitely. This is because all species are involved in the cyclic transformations of essential elements. If this were not so, these essential elements would accumulate as particular compounds and ultimately further growth would be prevented because of the lack of suitable nutrients. As man alters the established rhythm of these geochemical transformations,

because of his increased demands on the natural resources of the earth, it becomes important to establish whether or not dynamic equilibria can be maintained by an alteration of the growth kinetics of the microorganisms involved.

Despite the widespread therapeutic usage of antibiotics, a clear understanding of the normal role of these compounds in natural environments is lacking. The recently noted side effects of the feeding of antibiotics to animals were largely unforeseen, but could have been predicted if the ecological significance of amensalism had been appreciated. It has been argued that antibiotics are not produced in natural environments and therefore the use of these compounds can be patented in those countries where companies making natural products cannot be given full patent protection. However, recent work seems to indicate that compounds such as penicillin can in fact be produced in the soil, and furthermore that antibiotic-resistant mutants appear in response to antibiotic production in the soil. Therefore, had a better understanding of microbial ecology been acquired before antibiotics were widely applied, then many of the results of their usage could have been predicted.

In conclusion, it is not difficult to argue the case for further work that will lead to an understanding of the ways in which bacterial species interact. What is now required is that the work done in this area be of the highest quality so that future speculations may be based on data of undoubted soundness. In many respects, natural ecosystems act as buffers that dampen down the effects of change in the environment. However, if the constraints offered to such a system are too great, then the natural community will be fundamentally altered. It is wiser to assess the probable outcome of man's ravages on natural ecosystems before any changes in the environment become irreversible.

REFERENCES

1. Adams, J. G., *Experientia*, 22, 717, 1966.
2. Alexander, M., *Introduction to Soil Microbiology*, John Wiley & Sons, New York, 1961.
3. Alexander, M., *Ann. Rev. Microbiol.*, 18, 217, 1964.
4. Alexander, M., *Microbial Ecology*, John Wiley & Sons, New York, 1971.
5. Amerine, M. A. and Kunkee, R. E., *Ann. Rev. Microbiol.*, 22, 323, 1968.
6. Anderson, E. S., in *Microbial Ecology*, 7th Symp. Soc. Gen. Microbiol., Williams, R. E. O. and Spicer, C. C., Eds., Cambridge University Press, London, 1957, 189.
7. Annear, D. I., *Aust. J. Exp. Biol. Med. Sci.*, 29, 93, 1951.
8. Arndt, W. F. and Ritts, R. E., *Proc. Soc. Exp. Biol. Med.*, 108, 166, 1961.
9. Bail, O., *Z. Immunforsch.*, 60, 1, 1929.
10. Ball, G. H., in *Research in Protozoology*, Vol. 3, Chen, T. T., Ed., Pergamon Press, New York, 1969, 565.
11. Barta, J., *Continuous Cultivation of Microorganisms*, Proceedings of Second International Symposium, Malek, I., Beron, K., and Hospodka, J., Eds., Czech. Acad. Sci., Prague, 1962, 325.
12. Bates, J. L. and Liu, P. V., *J. Bacteriol.*, 86, 585, 1963.
13. Benda, I. and Schmitt, A., *Weinberg Keller*, 13, 239, 1966.
14. Bergter, F. and Noack, D., *Stud. Biophys.*, 1, 257, 1966.
15. Burroughs, L. F. and Carr, J. G., *Ann. Rep. Agr. Hort. Res. Stat.*, Long Ashton, 1956, 162.
16. Braithwaite, C. W. D. and Dickey, R. S., *Phytopathology*, 60, 1047, 1970.
17. Braun, W., *Bacterial Genetics*, W. B. Saunders, Philadelphia, 1953.
18. Brian, P. W., *Microbial Ecology*, 7th Symp. Soc. Gen. Microbiol., Williams, R. E. O. and Spicer, C. C., Eds., Cambridge University Press, London, 1957, 168.
19. Brock, T. D., *Principles of Microbial Ecology*, Prentice Hall, Englewood Cliffs, N. J., 1966.
20. Brock, T. D., *Science*, 18, 1012, 1967.
21. Brown, C. M., Taylor Robinson, D., and Meers, J. L., *Adv. Microbial Physiol.*, in preparation.
22. Brunner, W., Oberzill, W., and Menzel, J., Continuous Cultivation of Microorganisms, *Proceedings of the 4th International Symposium on Continuous Culture*, Malek, I., Beren, K., Fencl, Z., Ricica, J., and Smrckova, H., Eds., Acedemica, Prague, 1968, 323.
23. Bryant, M. P. and Robinson, I. M., *Appl. Microbiol.*, 9, 96, 1961.
24. Bryant, M. P., Wolin, E. A., Wolin, M. J., and Wolfe, R. S., *Archiv. Mikrobiol.*, 69, 20, 1967.
25. Bungay, H. R. and Krieg, N. R., *Chem. Eng. Prog. Symp. Ser.*, 62, 68, 1966.
26. Bungay, H. R., *Chem. Eng. Prog. Symp. Ser.*, 64, 10, 1968.
27. Bungay, H. R. and Bungay, M. L., *Adv. Appl. Microbiol.*, 10, 269, 1968.
28. Burkholder, P. R., in *Symposium on Marine Microbiology*, Oppenheimer, C. H., Ed., Charles C Thomas, Springfield, Ill., 1963, 133.
29. Burman, N. P., *Town Waste Put to Use*, Wix, P., Ed., Cleaver-Hume Press Ltd., London, 1960, 113.
30. Burton, M. O. and Lochhead, A. G., *Can. J. Bot.*, 29, 352, 1951.
31. Canale, R. P., *Biotechnol. Bioeng.*, 12, 353, 1970.
32. Cantarelli, C., *Appl. Biochim.*, 2, 167, 1955.
33. Carr, J. G., Skerrett, E. J., and Tuchnott, O. G., *J. Appl. Bacteriol.*, 82, 136, 1969.
34. Carr, J. G., *Microorganisms: Function, Form and Development*, Hawker, L. E. and Linton, A. H., Eds., Edward Arnold Ltd., London, 1970.
35. Castelli, T., *Vini d'Italia*, 9, 245, 1967.
36. Chao, C. and Reilly, P. J., *Biotechnol. Bioeng.*, 14, 75, 1972.
37. Chet, I., Fogel, S., and Mitchell, R., *J. Bacteriol.*, 106, 863, 1971.
38. Chain, S. K. and Mateles, R. I., *Appl. Microbiol.*, 16, 1337, 1968.
39. Contois, D. E., *J. Gen. Microbiol.*, 21, 40, 1959.
40. Contois, D. E. and Yango, L. D., *Abstr. Papers, Am. Chem. Soc.*, 148th Meeting, 1964, 17Q.
41. Cooney, C. L. and Mateles, R. I., *Proc. Int. Congress for Microbiol.*, Mexico, 1971, 441.
42. Cornforth, J. W. and James, A. T., *Biochem. J.*, 63, 124, 1956.
43. Cornick, D. E. R. and Bowen, W. H., *Br. Dent. J.*, 130, 231, 1971.
44. Crombie, A. C., *J. Anim. Ecol.*, 16, 44, 1947.
45. Curds, C. R. and Cockburn, A., *J. Gen. Microbiol.*, 54, 343, 1968.
46. Curds, C. R., *Proc. Symposium on Methods of Study of Soil Ecology*, Philipson, J., Ed., Paris, UNESCO, 1970, 127.
47. Curds, C. R. and Cockburn, A., *J. Gen. Microbiol.*, 66, 95, 1971.
48. Curds, C. R., *Water Res.*, 5, 793, 1971.
49. Darwin, C., *The Origin of Species by Means of Natural Selection or the Preservation of Favoured Races in the Struggle for Life*, Murray, London, 1859.
50. Davis, J. G., Ashton, T. R., and McCaskill, M., *Dairy Ind.*, 35, 569, 1971.
51. Demain, A. L., Burg, R. W., Miller, I. M., and Hendlin, D., *Bacteriol. Proc.*, 34, 1964.

52. De Stoppelaar, J. D. and Gibbons, R. J., *Int. Assoc. Dent. Res.*, Abstracts of 43rd vol., Gen. Meeting, Toronto, 1965, 64.
53. Dias, F. F., Dondero, N. C., and Finstein, M. S., *Appl. Microbiol.*, 16, 1191, 1968.
54. Dias, W. J. and Bhat, J. V., *Appl. Microbiol.*, 12, 412, 1964.
55. Dworkin, M. and Foster, J. W., *J. Bacteriol.*, 15, 592, 1958.
56. Eadie, J. M., *J. Gen. Microbiol.*, 29, 579, 1962.
57. Ekstedt, R., *Ann. N.Y. Acad. Sci.*, 65, 119, 1956.
58. Elliker, P. R., *Practical Dairy Bacteriology*, McGraw-Hill, New York, 1949.
59. Evans, C. A., Smith, W. M., Johnson, E. A., and Griblet, E. R., *J. Invest. Dermatol.*, 15, 305, 1950.
60. Freter, R., *J. Infect. Dis.*, 110, 38, 1962.
61. Furukawa, T., Nakahara, T., and Yamada, K., *Agric. Biol. Chem.*, 34, 1833, 1970.
62. Gale, E. F., *Biochem. J.*, 34, 415, 1940.
63. Gandini, A. and Tarditi, A., *Ind. Agrar.* (Florence), 4, 411, 1966.
64. Gaudy, A. F., Obayashi, A., and Gaudy, E. T., *Appl. Microbiol.*, 22, 1041, 1971.
65. Gause, G. F., *The Struggle for Existence*, Williams & Wilkins, Baltimore, 1934.
66. Gause, G. F., *La Theorie Mathematique de la Lutte pour la Vie*, Hermann et Cie, Paris, 1935.
67. Gibbons, R. J. and Macdonald, J. B., *J. Bacteriol.*, 80, 164, 1960.
68. Gilliland, R. B., *J. Inst. Brew.*, 64, 304, 1966.
69. Girard, G., *Bull. Soc. Pathol. Exot.*, 51, 18, 1959.
70. Gorlenko, V. M., *Mikrobiologiya*, 37, 26, 1968.
71. Gorden, R. W., Beyers, R. J., Odum, E. P., and Eagon, R. G., *Ecology*, 50, 86, 1969.
72. Goswell, R. W., *Proc. Biochem.*, 2, 5, 1967.
73. Gray, K. R., Sherman, K., and Biddlestone, A. J., *Proc. Biochem.*, 6, 32, 1971.
74. Greaves, H., *Wood Sci. Technol.*, 5, 6, 1971.
75. Haenel, H., *J. Appl. Bacteriol.*, 24, 242, 1961.
76. Haldane, J. B. S., *The Causes of Evolution*, Longmans, London, 1932.
77. Halmann, M., Benedict, M., and Mager, J., *J. Gen. Microbiol.*, 49, 451, 1967.
78. Hamon, Y. and Léron, Y., *Ann. Inst. Pasteur*, 104, 55, 1963.
79. Harder, W. and Veldkamp, H., *Antonie van Leeuwenhoek*, 37, 51, 1971.
80. Hardy, P. H. and Munro, C. O., *J. Bacteriol.*, 91, 27, 1966.
81. Hegemann, F., *Z. Hyg. Infekt.*, 131, 355, 1950.
82. Hentges, D. J., *J. Bacteriol.*, 93, 2029, 1967.
83. Herbert, D., Elsworth, R., and Telling, R. C., *J. Gen. Microbiol.*, 14, 601, 1956.
84. Herbert, D., *Soc. Chem. Ind.* (London), 12, 21, 1961.
85. Hobson, P. N., in *Microbial Growth*, 19th Symp. of Soc. for Gen. Microbiol., Meadow, P. and Pirt, J. S., Eds., Cambridge University Press, London, 1969, 43.
86. Hobson, P. N., *Prog. Ind. Microbiol*, 9, 41, 1971.
87. Hoffman, H., *Adv. Appl. Microbiol.*, 8, 195, 1966.
88. Hollowell, C. A. and Wolin, M. J., *Appl. Microbiol.*, 13, 918, 1965.
89. Holman, W. L. and Meekson, D. M., *J. Infect. Dis.*, 39, 145, 1926.
90. Horne, M. T., *Science*, 168, 992, 1970.
91. Horne, M. T., *Science*, 172, 405, 1971.
92. Humenile, F. J. and Hanna, G. P., *Biotechnol. Bioeng.*, 12, 541, 1970.
93. Hungate, R. E., *The Rumen and Its Microbes*, Academic Press, London, 1966.
94. Hunter, G. J. E., *J. Hyg.* (Camb.), 45, 307, 1947.
95. Hynes, H. B. N., *The Biology of Polluted Waters*, Liverpool University Press, Liverpool, 1960.
96. Ichikawa, Y. and Kitamoto, Y., *J. Agric. Chem. Soc. Japan*, 41, 171, 1967.
97. Iturriaga, R. and Garcia-Tello, P., *Acta Microbiol.* Pol. B, 2, 243, 1970.
98. Jannasch, H. W., *Arch. Microbiol.*, 23, 55, 1960.
99. Jannasch, H. W., *J. Gen. Microbiol.*, 23, 55, 1960.
100. Jannasch, H. W., *Arch. Microbiol.*, 59, 165, 1967.
101. Jannasch, H. W., *Appl. Microbiol.*, 16, 1616, 1968.
102. Jarvis, B. D. W., *Appl. Microbiol.*, 16, 714, 1968.
103. Jenkins, D., in *Biochemistry of Industrial Microorganisms*, Rainbow, C. and Rose, A. H., Eds., Academic Press, London, 1963, 508.
104. Kadavy, J. L. and Dack, G. M., *Food Res.*, 16, 328, 1951.
105. Kielhöfer, E. and Wurdig, G., *Dtsch. Wein-Stg.*, 99, 1022, 1963.
106. Kirsch, E. J. and Sykes, R. M., *Prog. Ind. Microbiol.*, 9, 155, 1971.
107. Knoll, K. H., (1959), Results reported by Gray et al., 1971.
108. Kunkee, R. E., *Adv. Appl. Microbiol.*, 9, 235, 1967.
109. Kraus, F. W., Nickerson, J. F., Perry, W. I., and Walker, A. P., *J. Bacteriol.*, 73, 727, 1957.
110. Kurihara, Y., Eadie, M. J., Hobson, P. N., and Mann, S. O., *J. Gen. Microbiol.*, 51, 267, 1968.

111. Kuznetsov, S. I., *Die Rolle der Microorganismen in Stoffkreislauf der Seen,* V E B Deutscher Verlag der Wissenschaften, Berlin, 1959.
112. Larsen, D. H. and Dimmick, R. L., *J. Bacteriol.,* 88, 1380, 1964.
113. Latham, M. J., Sharpe, M. E., and Sutton, J. D., *J. Appl. Bacteriol.,* 34, 425, 1971.
114. Lewin, R. A., *Can. J. Microbiol.,* 2, 665, 1956.
115. Lewis, P. M., *J. Appl. Bacteriol.,* 30, 405, 1967.
116. Lightbrown, J. W. and Jackson, F. L., *Biochem. J.,* 63, 130, 1956.
117. Lilienthal, B., *Aust. J. Exp. Biol. Med. Sci.,* 28, 279, 1950.
118. Lobitz, W. C. and Osterberg, A. E., *Arch. Derm. Syph.,* 57, 387, 1947.
119. Lotka, A. J., *Elements of Physical Biology,* Williams & Wilkins, Baltimore, 1925.
120. Luthi, H., *Am. J. Enol.,* 8, 176, 1955.
121. Magasanik, B., *Cold Spring Harbor Symp. Quant. Biol.,* 26, 249, 1961.
122. Mah, R., 1969, Experiments reported by Kirsch and Sykes, *Progr. Ind. Microbiol.,* 9, 155, 1971.
123. Mann, S. O., *J. Appl. Bacteriol.,* 33, 403, 1970.
124. Marples, M. J. and Bacon, D. F., *Trans. R. Soc. Trop. Med. Hyg.,* 50, 72, 1956.
125. Marples, M. J., *The Ecology of the Human Skin,* Charles C Thomas, Springfield, Ill.
126. Marples, R. R. and Kligman, A. M., *Arch. Dermatol.,* 103, 148, 1971.
127. Mateles, R. I., Chian, S. K., and Silver, R., in *Microbial Physiology and Continuous Culture,* Powell, E. O., Evans, C., Strange, R. E., and Tempest, D. W., Eds., H.M.S.O., London, 1967, 233.
128. Mateles, R. I. and Chian, S. K., *Environ. Sci. Technol.,* 3, 569, 1969.
129. Maxted, W. R., *J. Gen. Microbiol.,* 13, 484, 1955.
130. Maynard Smith, J., in *Microbial Growth,* 19th Symp. Soc. Gen. Microbiol., 1969, 1.
131. McHugh, W. D., Ed., *Dental Plaque,* Livingstone, Edinburgh, 1970.
132. McLaren, A. D., *Proc. Soil Sci. Soc. Am.,* 33, 551, 1969.
133. Meers, J. L. and Tempest, D. W., *J. Gen. Microbiol.,* 52, 309, 1968.
134. Meers, J. L. and Tempest, D. W., *Biochem. J.,* 119, 603, 1970.
135. Meers, J. L. and Tempest, D. W., *J. Gen. Microbiol.,* 63. 325, 1970.
136. Meers, J. L., unpublished observations, 1970.
137. Meers, J. L., *The Assimilation of Cations by Microorganisms,* Ph.D. Thesis, London University, 1971.
138. Meers, J. L., Tempest, D. W., and Brown, C. M., *J. Gen. Microbiol.,* 64, 187, 1971.
139. Meers, J. L., *J. Gen. Microbiol.,* 67, 359, 1971.
140. Meers, J. L. and Kjaergaad Pedersen, L., *J. Gen. Microbiol.,* in press.
141. Meynell, G. G., *Br. J. Exp. Pathol.,* 44, 209, 1963.
142. Mian, F. A., Fencl, Z., and Prokop, A., in *Continuous cultivation of Microorganisms,* Proceedings of the 4th International Symposium on Continuous Culture of Microorganisms, Malek, I., Beran, K., Fencl, Z., Ričica, J., and Smrčková H., Eds., Academia, Prague.
143. Mitchell, R. and Alexander, M., *Soil Sci.,* 93, 413, 1962.
144. Mitchell, R., *Nature* (London), 230, 257, 1971.
145. Millar, M. W., *The Pfizer Handbook of Microbial Metabolites,* McGraw-Hill, New York, 1961.
146. Milne, A., *Cold Spring Harbor Symp. Quant. Biol.,* 22, 253, 1957.
147. Milne, A., *Symp. Exp. Biol.,* 15, 40, 1961.
148. Mondari, R. M. and Havighorst, C. R., *Food Eng.,* 43(2), 59, 1971.
149. Monod, J., *Recherches sur la Croissance des Cultures Bacteriennes,* Hermann et Cie, Paris, 1942.
150. Monod, J., *Ann. Inst. Pasteur,* 79, 390, 1950.
151. Moore, W. B. and Rainbow, G. J., *J. Gen. Microbiol.,* 13, 190, 1955.
152. Moser, H., *Cold Spring Harbor Symp. Quant. Biol.,* 22, 121, 1957.
153. Mossel, D. D. A. and Ingram, M., *J. Appl. Bacteriol.,* 18, 232, 1956.
154. Mueller, J. C., *Can. J. Microbiol.,* 15, 1114, 1969.
155. Munson, R. J. and Bridges, B. A., *J. Gen. Microbiol.,* 37, 411, 1964.
156. Neronova, N. M., Ibragimova, S. I., and Lerusalimskii, N. D., *Mikrobiologiya,* 36, 404, 1967.
157. Nevin, J., *J. Bacteriol.,* 80, 783, 1960.
158. Nicholson, A. J. and Bailey, J., *Proc. Zool. Soc.* (London), Part 3, 551, 1935.
159. Nikoljuk, V. F., *Acta Protozoologica* (Warsaw), 7, 99, 1969.
160. Noack, D., *J. Theor. Biol.,* 1, 18, 1968a.
160a. Noack, D., in *Continuous Cultivation of Microorganisms,* Proceedings of the 4th Symposium on Continuous Culture, Prague, 1968b, 233.
161. Northrop, J. H., in *Continuous Cultivation of Microorganisms,* Proceedings of the 4th Symposium on Continuous Culture, Prague 1968, 243.
162. Novick, A., *Ann. Rev. Microbiol.,* 15, 492, 1955.
163. Nurmikko, V., *Experientia,* 12, 245, 1956.
164. Oberhofer, T. R. and Frazier, W. C., *J. Milk Food Technol.,* 24, 172.

165. Okuda, A. and Kobayashi, M., *Mikrobiologiya*, 32, 797, 1963.
166. Ousby, J. C., unpublished observations, 1970.
167. Owen, T. R., personal communication, 1972.
168. Ozawa, A. and Freter, R., *J. Infect. Dis.*, 114, 235, 1964.
169. Pachtmann, E. A., Vicher, E. E., and Brunner, M. J., *J. Invest. Dermatol.*, 22, 389, 1954.
170. Parker, R. B. and Snyder, M. L., *Proc. Soc. Exp. Biol. Med.*, 108, 749, 1961.
171. Parker, R. B., *Biotechnol. Bioeng.*, 8, 473, 1966.
172. Parker, R. B., *J. Dent. Res.*, 169, 804, 1970.
173. Paynter, M. J. B. and Bungay, H. R., in *Fermentation Advances*, Perlman, D., Ed., Academic Press, 1969, 323.
174. Paynter, M. J. B. and Bungay, H. R., *Biotechnol. Bioeng.*, 12, 341, 1970.
175. Paynter, M. J. B. and Bungay, H. R., *Biotechnol. Bioeng.*, 12, 347, 1970.
176. Paynter, M. J. B. and Bungay, H. R., *Science*, 172, 405, 1971.
177. Phillips, A. W. and Wolf, P. A., *Ann. N. Y. Acad. Sci.*, 78, 308, 1959.
178. Picci, G., Melàs-Joamnidis, Z., Carbis, A., and Vasilatos, G., *Ann. Fa. Agar. Univ. Pisa*, 20, 9, 1959.
179. Pipes, W. O. and Jones, P. H., *Biotechnol. Bioeng.*, 5, 287, 1963.
180. Pipes, W. O., *Adv. Appl. Microbiol.*, 8, 77, 1966.
181. Pike, E. B. and Curds, C. R., in *Microbial Aspects of Pollution*, Sykes, G. and Skinner, F. A., Eds., Academic Press, London 1972, 123.
182. Pohunck, M., *J. Hyg. Epidermiol. Microbiol. Immunol.* (Prague), 5, 267, 1961.
183. Pokorny, J., *Arch. Hydrobiol.*, 3 (Suppl. 38), 336, 1971.
184. Pollock, M. R., *J. Gen Microbiol.*, 2, 23, 1948.
186. Poole, D. F. G. and Newman, H. N., *Nature* (London), 234, 329, 1971.
187. Powell, E. O., *J. Gen. Microbiol.*, 18, 259, 1958.
188. Powell, E. O., *J. Appl. Chem. Biotechnol.*, 22, 71, 1972.
189. Rainbow, C., *Proc. Biochem.*, 6(4), 15, 1971.
190. Randall, W. C. and McClure, W., *J. Appl. Physiol.*, 2, 72, 1949.
191. Roberts, D. S., *J. Infect. Dis.*, 120, 720, 1969.
192. Roberts, D. S. and Egerton, J. R., *J. Comp. Pathol.*, 79, 217, 1969.
193. Rockl, H. and Muller, E., *Arch. Klin. Exp. Dermatol.*, 209, 13, 1959.
194. Rogers, H. T. and Perkins, H. R., *Cell Walls and Membranes*, E. & F. N. Spon Ltd., London, 1968.
195. Rothman, S., *Physiology and Biochemistry of the Skin*, University of Chicago Press, Chicago, 1954.
196. Rountree, P. M. and Barbour, R. G. H., *J. Pathol. Bacteriol.*, 63, 313, 1951.
197. Sanders, A. C., Pelczar, M. J., and Hoefling, A. F., *Antibiot. Chemother.*, 12, 10, 1962.
198. Schlegel, H. G. and Jannasch, H. W., *Annu. Rev. Microbiol.*, 21, 49, 1967.
199. Seidler, R. J. and Starr, M. P., *J. Bacteriol.*, 100, 769, 1969.
200. Shaedler, R. W., Dubos, R., and Costello, R., *J. Exp. Med.*, 122, 59, 1965.
201. Shaeffer, P., *Bacteriol. Rev.*, 33, 48, 1969.
202. Sheenan, B. T. and Johnson, M. J., *Appl. Microbiol.*, 21, 511, 1970.
203. Shehadeh, N. H. and Kligmann, A. M., *J. Invest. Dermatol.*, 40, 61, 1963.
204. Shindala, A., Bungay, H. R., Krieg, N. R., and Culbert, K., *J. Bacteriol.*, 89, 693, 1965.
205. Siddiqui, M. A. Q. and Peterson, G. E., *Antonie van Leeuwenhoek*, 31, 193, 1965.
206. Sieburth, J. McN., *J. Exp. Mar. Biol. Ecol.*, 1, 98, 1967.
207. Simons, R. D. G., in *Handbook of Tropical Dermatology and Medical Mycology*, Elsevier, Amsterdam, Vol. 1, 1952, 464.
208. Slyter, L. L. and Wolin, M. J., *Appl. Microbiol.*, 15, 1160, 1967.
209. Slyter, L. L. and Weaver, J. M., *Appl. Microbiol.*, 22, 930, 1971.
210. Smith, P. H., *Dev. Ind. Microbiol.*, 7, 156, 1966.
211. Stanier, R. Y. and Cohen-Bazire, G., in *Microbial Ecology*, 7th Symp. Soc. Gen Microbiol., Cambridge University Press, London, 1957, 56.
212. Stent, G. S., *Molecular Biology of Bacterial Viruses*, Freeman, San Francisco, 1963.
213. Stojanovic, B. J. and Alexander, M., *Soil Sci.*, 86, 208, 1958.
214. Stolp, H. and Starr, M. P., *Antonie van Leeuwenhoek*, 29, 217, 1963.
215. Strauss, J. S. and Kligman, A. M., *J. Bacteriol.*, 51, 671, 1960.
216. Stutzenberger, F. J. and Bennett, E. O., *Appl. Microbiol.*, 13, 570, 1965.
217. Styczynska, D., Niemczyk, H., and Mazon, M., *Acta Microbiol.*, 13, 570, 1965.
218. Takai, Y. and Kamura, T., *Folia Microbiol.*, 11, 304, 1966.
219. Tempest, D. W., Dicks, J. W., and Meers, J. L., *J. Gen Microbiol.*, 49, 139, 1967.
220. Tempest, D. W., Dicks, J. W., and Ellwood, D. C., *Biochem. J.*, 106, 237, 1968.
221. Thompson, R. and Shibuya, M., *J. Bacteriol.*, 51, 671, 1946.
222. Toerien, D. F., Siebert, M. L., and Hattingh, W. H. J., *Water Res.*, 1, 497, 1967.
223. Toerien, D. F. and Hattingh, W. H. J., *Water Res.*, 3, 385, 1969.
224. Tramer, J., *Nature* (London), 211, 204, 1966. 105,129,146,149,151,153

225. Tribe, H. T., in *Microbial Ecology*, 7th Symp. Soc. Gen. Microbiol.. Cambridge University Press. London. 1967. 287.
226. Truby, C. P. and Bennett, E. O., *J. Gen. Microbiol.*, 14, 769, 1966.
227. Van Gemerden, H. and Jannasch, H. W., *Arch. Microbiol.*, 79, 345, 1971.
228. Van Houte, J., Gibbons, R. J., and Banghaut, S. B., *Arch. Oral. Biol.*, 15, 1025, 1970.
229. Varon, M. and Shilo, M., *J. Bacteriol.*, 95, 744, 1968.
230. Vaughn, R. H., *Adv. Food Res.*, 6, 67, 1955.
231. Veldkamp, H., in *Methods in Microbiology*, Vol. 3A, Academic Press, New York, 1969, 305.
232. Veldkamp, R. H. and Jannasch, H. W., *J. Appl. Chem. Biotechnol.*, 22, 105, 1972.
233. Venkataraman, G. S. and Neelakantan, S., *Phykos*, 7, 242, 1968.
234. Volterra, V., *Mem. Acad. Lineii* (Roma), 2, 31, 1926.
235. Waksman, S. A., *Biol. Rev.*, 23, 452, 1948.
236. *Water Pollution Research*, Report of the Director of the Water Pollution Research Laboratories, Stevenage. H.M.S.O., London, 1968, 153.
     Wright, J. M., *Nature* (London), 177, 896, 1956.
     Yeoh, H. T., Bungay, H. R., and Krieg, N. R., *Can. J. Microbiol.*, 14, 491, 1968.
239. Young, G., Krasner, R. I., and Yudkofsky, P. L., *J. Bacteriol.*, 72, 525, 1956.
240. Anderson, E. S., *Annu. Rev. Microbiol.*, 22, 131, 1968.
241. Groscop, J. A. and Brent, M. M., *Can. J. Microbiol.*, 10, 579, 1964.
242. Hill, P., *J. Gen. Microbiol.*, 70, 243, 1972.
243. Knorr, M., *Schweiz. Z. Hydrol.*, 22, 493, 1960.
244. Mitchell, R., *Nature* (London), 230, 257.
245. Oates, R. P., Beers, T. S., and Quinn, L. Y., *Bacteriol. Proc.*, 1963, 44.
246. Postgate, J. R., *Antonie van Leeuwenhoek*, 33, 113, 1967.
247. Tsuchiya, H. M. and Fredrickson, A. G., verbal communication at Am. Soc. Microbiol., 1971.

INDEX

temperature, 75
water, mud, sewage treatment, 75
Oxygen, 38, 41, 71, 145
Oxygen replenishment, 72
Oxygenase, 41

P

Pantothenic acid, 22, 62, 160
para-aminobenzoic acid, 63
Parasitism, 85, 92, 93, 137, 152–155
Paraoxon, 39
Particle size distribution, 80, 123
Particulates, 77–79, 80
Partition coefficient, 20
Partitioning, 21
Pasturella tularensis, 156
PCNB, 15, 19
Pedicoccus sp., 167
Pelagic food chains, 21
Peloscope/pedoscope method, 96, 98
Penicillin, 16, 152, 176
Penicillium sp., 1, 14, 35, 39, 42, 46–48
  chrysogenum, 45
  decumbens, 47·
  frequentans, 52
  janthinellum, 47
  luteum, 47
  notatum, 45
  roqueforti, 42, 66
  spinulosum, 47
Pentachloroaniline, 15, 52
Pentachloroanisole, 52
Pentachloronitrobenzene, 50
Pentose sugars, 167
Pepsin, 121
Percolating filters, 164
Periodontitis, 172
Permeases, induction of, 64
Peroxidase, 47, 53
Pesticides, 60
  accumulation, microbial, 18–22
  biodegradation, 12–14
  conversion, 14–18
  nitrogen metabolism, effects on, 24–26
  persistence, 24
  photosynthesis, effects on, 24
  toxicity, 22
pH, 62, 71, 73–75, 78, 80, 109, 116, 118
  amino acids, effects on, 74
  ΔpH, 74
  effects, direct and indirect, 73
  interface, charged, 74
  surface pH, 74
  toxicity, 74
  transitory changes, 74
Phagocytes, 149
Phenol, 41
  chlorinated, 54
Phenoxyacetic acid, 42
  chlorinated, 39
Phenoxyalkanoates, 42, 47, 48

Phenoxybutyric acid, 48
Phenoxyvaleric acid, 48
Phenylacetic acid, 45
Phenylacylanilides, 52
Phenylalanine, 159
Phenylcarbamates, 49, 52
Phenylureas, 35, 36
  herbicides, 47
Phobophototaxis, 70
Phoma sp., 47
Phosphorus, 61, 165
Phosphorylation, oxidative, 25
Photic zones, 70
Photography, infrared, 95
Photokinesis, 70
Photosynthesis, 24, 68, 70
Phototropism, 70
Phyllosphere, 86
Phytophthora sp., 72
Phytoplankton, 21
Phytotoxicity, 37, 49
Pigments, 91
Pili, 98
Plant residue, 57
Plasmids, 84, 86, 102
  episomal, 91
  transfer, 80
Plasmoptysis, 82, 91
Plastics, 60
Podzolic soils, 98
Poliovirus Type III, 66
Polymerization reactions, 53, 54
Polymorphic potential, 98
Polyphenol oxidase, 47
Polywater, 65
Population factor, 146
Populations, kinetics of, 101
  competitive, 109
  heterogeneous, 108
  primary, 71, 109
Pore space, 58
Porphyrins, 60
Potassium, 61
Predation, 92, 137, 148–152, 165, 172
Pressure
  atmospheric, 68
  osmotic, 68, 107, 109
Prey depletion, 85
Procaryotes, 83
Producers, primary, 70
Propanil, 16, 50, 53
Protective mechanisms, 92
Proteins, denaturation, 63
Proteus vulgaris, 14, 160
Protocatechuic acid, 47
Protocooperation, 137, 159
Protonation, 75
Protoplasts, 93
Prototrophs, 140
Protozoa, 72, 86, 91, 165, 172, 173
Pseudomonas sp., 7, 9, 10, 49, 155
  P. fluorescens, 4, 17

Substituent effect, 6
Substitution, isomorphous, 116
Substrate
  depletion, 90, 109
  movement, 13
  sorbed, 78
  utilization studies, 123
Succession, 173
Sulfate reducers, 76
Sulfonomides, 5
Sulfur dioxides, 168
Sulfur oxidizers, 2
Surface interaction studies, 116–122
Surface tension, 58
Survival of microbes, 66
Sweating, 169
Symbionts, 164
Symbiosis, 85, 87, 88, 93, 159, 161, 176
Synergism, 87, 137, 159, 161
Synthetic mechanisms, 52, 53
Syntrophism, 80, 87, 159
Syringaldehyde, 45, 113

T

2,4,5-T, 5, 6
Talaromyces wortmanii, 36, 52
Tannin, 167
TBA, 9, 10
TDE, 3
Temperature effects, 66–70, 93
  freezing of microbes, 60
  survival of spores, 67
  survival of vegetative cells, 67
Terregens factor, 63
Tetracladium setigerum, 5
3,3′,4,4′-Tetrachloroazobenzene, 53
Thermal death point, 67
Thermoactinomyces sp., 14
Thiamine, 62, 160
Thin-layer chromatography, 39, 41
Thiobacilli, 76
Thiocystis sp., 157, 158
Titrable acidity, 166
Tooth eruption, 172
Topophototaxis, 70
Torula sp., 46
  T. utilis, 161
Toxicity, 1, 37, 39
Toxins, 64, 152, 160
Toxophene, 2, 3, 24
Transaminases, 101
Transduction, 83, 93
Transformations, 83, 100
  xenobiotic, 54
Treponema pertenue, 169
Triazine herbicides, 47
  s-triazine herbicides, 37
Trichloroacetic acid, 7
Trichoderma sp., 14, 48
  T. lignorum, 39
  T. virgatum, 52
  T. viride, 18, 38, 47, 50, 52

Triethylene melamine, 155
Tripartite system (soil), 57
Trophic levels, 165
Trophic transfer, 21
Troposphere, 70
Trypticase, 4
TSB, 4
Turbidostat, 146
"Turn-over rate", 61
Tyrosinase, 47
Tyrosine, 45, 47

U

Ureas, 38, 41

V

Valeric acid, 164
Valine, 22
Vanillic acid, 39
Vanillin, 45, 113
Varicosporium eludeae, 5
Vectors, 108
Vegetative growth, 60
Veratric acid, 39
Vermiculite, 77, 78, 104
Vertical distribution, 72
Vibrio cholera, 173
Vinegar, 167
Virus-sensitive cells, 155
Vitamins, 171
Volatile fatty acids (VFA), 171, 172
Volatile organics, 60
Volatilization, 24

W

Warburg studies, 46
Water
  activity (a), 64, 67, 96
  free, 64
  gravitational, 72, 86
  metabolic functions, 64
  retention in soil, 66
  structure, 65
  surface tension, 66
  thermodynamic properties, 64, 66, 67
  turgor, maintenance, of, 64
Water bridges, 58
Water pockets, 65
Wine fermentation, 168
Wort, 168

X

Xenobiotic substances, 34, 35, 41, 48, 54
X-ray diffraction, 103, 121

Y

Yaws, 169
  Yeasts, alcohol forming, 166
Yeast extract, 62

9 780367 657482